FACTORY UNDER THE ELMS:
A HISTORY OF HARRISVILLE,
NEW HAMPSHIRE, 1774–1969

Published for The Merrimack Valley Textile Museum by
The M.I.T. Press, Cambridge, Massachusetts, and London, England

FACTORY UNDER THE ELMS:
A HISTORY OF HARRISVILLE,
NEW HAMPSHIRE, 1774–1969

JOHN BORDEN ARMSTRONG

I Harrisville, New Hampshire, ca. 1850. From an unknown painting
Courtesy of Mrs. Cecil P. Grimes, Pennacook, N.H.

F
44
H43
A8

To Helen

CONTENTS

Town Affairs

Physical Appearance; Changes in Transportation and Communication; Political Leadership; The Town Budget; Police and Fire Protection; Politics; National Events, The First World War; Conclusion

LIST OF ILLUSTRATIONS

FOREWORD

Miss Sarah Loring Bailey's *Historical Sketches of Andover (comprising the present towns of North Andover and Andover), Massachusetts* (Boston: Houghton Mifflin and Company, 1880) is one of the best of nineteenth-century New England town histories. It is a conscientious and serious work of more than six hundred pages, with excellent heliotype illustrations. Yet a glance at its ten chapters will quickly show how the approach of the local historian has changed in the past ninety years. After considering memorials of the early settlers, Miss Bailey deals with witchcraft and Andover's part in the early Indian wars, the French and Indian War, and the American Revolution. These subjects fill the first two thirds of the book. Then follow chapters on churches and ministers, public schools and libraries, academies, the Andover Theological Seminary, and, finally, thirty pages on mills and manufactures.

In the preface the author announced that her original intention was "simply to make a collection of sketches of the romance and poetry of Old Andover history," but that "subsequently, by request of friends, the task was undertaken of arranging these in chronological order in the form of a continuous history." Nevertheless "romance and poetry" predominated in her choice of subjects, for although mills and manufactures were included, out of consideration for the feelings of the mill owners who subscribed toward the publication of the book, the author noted on her final page that "to cover the entire history of the town in all its departments of enterprise, there should be added an account of the agricultural and general industries, and trade; stores, banks, post-offices, stage-routes, bridges, railways; of the professional men, physicians, lawyers, authors, editors, artists; and the town history and action under the Federal Constitution, including the War of 1812, the Civil War, and other important crises in the history of the nation and of the commonwealth, as well as biographical sketches of all the eminent citizens, and the natives of the Andovers who have become eminent in other communities." She thus clearly had a grasp of what needed to be done, although she actually produced only the beginning.

The preface further suggested that "public spirited citizens and the towns themselves (North Andover and Andover) could hardly make a better appropriation of money than to found and endow an Old Andover Historical and Genealogical Society, which would possess means and influence to prosecute the work of historical research and collection." Although the suggestion was not immediately followed, the North Andover Historical Society was established in 1913 by Samuel Dale Stevens (1859–1922), a local wool manufacturer, who bought for it a small late eighteenth-century cottage near the North Parish Church as a meeting place and a repository for its possessions. Two decades later the cottage barn was replaced by a brick museum building as a memorial to the founder and his wife. The Andover Historical Society in 1929 came into possession of the Amos Blanchard House, a pleasant dwelling of 1819 at 97 Main Street, Andover. These two separate organizations accumulated a variety of artifacts, which were placed in their houses on the theory that they would prove of interest to visitors and would somehow turn school children into upright citizens. Like most such local organizations, they inspired backward-looking to the (supposedly) "good old days" that had passed, and proved more active in promoting nostalgia than in prosecuting "the work of historical research and collection."

"The romance and poetry of Old Andover history" were the chief goals sought until 1960 when, through the energetic efforts of Mrs. Horatio Rogers, president of the North Andover Historical Society and a daughter of its founder, the Merrimack Valley Textile Museum was incorporated. Although the nineteenth- and twentieth-century history of North Andover had been intimately allied with the woolen mills of M. T. Stevens and Company, the new institution was not bound by such local considerations. Its purposes went far beyond the town, and beyond artifacts relating to the handicraft period, for it undertook to assemble material that would illustrate the wool industry in America after the impact of the Industrial Revolution and to develop a library and archives that would become a center of scholarly research in the history of the wool industry.

Construction began in September 1960 upon a story-and-a-half, red-brick building that blended pleasantly into the village landscape of North Andover Common and that provided 18,000 square feet of floor space, divided almost evenly between exhibit, study collection, and administrative areas. The initial staff consisted of a Hagley Fellow and a Winterthur Fellow from the University of Delaware, with a librarian experienced in the care of rare books and manuscripts from the Chapin Library, Williams College. The Museum was formally opened in September 1964.

Although the main-floor galleries are devoted to an exhibition
of objects that, through contemporary design techniques,
endeavors to convey to general visitors something of the development of textile technology, the serious activity of the institution
is carried on in its library, where books and manuscript business
records of the textile industry are assembled. While there is no
intention of enlarging the space for public exhibitions, it proved
necessary within a very few years of opening to add a sizable
new book stack to accommodate the fast-growing collection of
manuscripts and business records. This is inevitable and appropriate for an institution primarily concerned with research and
publication.

The Merrimack Valley Textile Museum warmly welcomes
scholars from other institutions who share its interests. Thus
Professor John Borden Armstrong of Boston University has been
for several years a frequent visitor to the museum and a good
friend and ally of its staff. The museum welcomes the opportunity
to join with the M.I.T. Press in the publication of Professor
Armstrong's study of the textile town of Harrisville, New
Hampshire. His ten chapters differ markedly in their emphasis
from Miss Bailey's ten, for his is a study of a community that
depended essentially upon textile manufacturing, with little
"romance and poetry" to distract the historian.

In 1799, just twenty-five years after the first settlers came to this
highland region, the attempt was made to use waterpower for
woolen manufacturing. Consequently, after an introductory
chapter on the eighteenth-century background, Professor
Armstrong moves directly into the mills of Harrisville, 1799–
1861. Although this is his main theme, he fully treats the village
and its people, their social, economic, and political life, the coming
of the railway, and almost everything that Miss Bailey omitted in
dealing with Andover. The scene is a modest one, but it is of
extraordinary interest, for Harrisville offers a crystallization of a
small New England textile town. The pattern is infinitely clearer
than in larger places that lie within the orbit of great cities.
Professor Armstrong has, however, had to start from scratch,
for the town had bred no Miss Bailey in its past, nor had any
preliminary work been done on its history. There was little in
the way of letters or personal narratives, nor was there even any
substantial body of old residents volunteering oral misinforma-
tion. The author has mined his facts chiefly from mill records,
public documents, and newspapers. From these he has created
a convincing and readable study of a small New England town
that, while far from typical, deserves careful consideration as an

example of the process of survival. As he points out in his preface, "it never became an industrial slum, a ghost town, or the fief of some outside industrial overlord." Although Harrisville looks today much as it did a century ago, it has, as he points out, "adapted to change without destroying its past." This is a lesson to ponder.

Walter Muir Whitehill

Boston Athenaeum
18 August 1969

PREFACE

This book describes the evolution of a small mill town in New England. The writer has sought to eschew the strictly antiquarian, to place the local developments in the larger context of the American experience, to portray the details of economic, political, social, and cultural history in a balanced and integrated fashion, and in the process to convey something of the quality of life in this community. The scholarly need for this type of town history seems self evident. The current popularity of urban history is understandable, but historians, among others, would do well to heed the words of Page Smith who, in his recent, distinguished study of the role of the town in American history, has pointed out that "if we except the family and the church, the basic form of social organization experienced by the vast majority of Americans up to the early decades of the twentieth century was the small town."

Harrisville's history includes in miniature many developments important in the history of the nation. Here one finds, successively, a frontier community of self-sufficient farmers, the utilization of waterpower to run small mills, the erection of a factory village, the rise and fall of a paternalistic authority, the color and conflict brought by mass immigration, the railroad mania, clashes between farmers and mill population, a separatist movement, the transformation of an economy punctuated by booms and busts, and the slow growth of a sense of community.

Central, however, to the town's experience, as to the nation's, has been the process of industrialization. The original settlement antedated the Revolution, but it was early in the nineteenth century that a controllable and reliable fall of water gave rise to an industrial village along the common boundary of the agricultural towns of Dublin and Nelson. The decline of these towns was accompanied by the growth of this awkward and demanding stepchild soon to be known as Harrisville. By 1850, two rival concerns were engaged there in woolen manufacturing. The next two decades saw the mill village's most vigorous growth. The parting of the ways for Harrisville and Dublin and Nelson came over the question of the railroad; in 1870, Harrisville was incorporated as a separate town. In the late nineteenth and early twentieth centuries, business panics, a fire, and the disruption of agriculture dealt heavy blows to the town's growth. Nonetheless, a century after the town's incorporation the woolen manufacture survives, and along with the virtual disappearance of agriculture there has been the compensating development of a summer tourist industry.

This process of economic growth and change acted upon, and was reflected in, every aspect of the community's life. The Harrises' patriarchal control was challenged by the Colony family in the 1850s. There were many points of friction between the mill village and the surrounding rural area; the immigrant population was troublesome, elections were turbulent, and the moral tone of the community sagged. Much of this pattern of life persisted after Harrisville became a separate town, although there were efforts at reform. By some indices life in Harrisville reached "low ebb" about the turn of the century. Since then a sharp population decline has leveled off, certain quickening influences, such as improved education, transportation, and communication, have improved the quality of life, and the hard times of the twenties and thirties served to bring about a new sense of community. In the last quarter century Harrisville has continued to experience change, but it has been change within a familiar framework.

Although it contains much of the American, and more of the New England, experience, the history of this mill town has in some respects been atypical, and therein may lie the town's real significance. It never became an industrial slum, a ghost town, or the fief of some outside industrial overlord. With an economy resting upon the numerous summer residents and upon the up-to-date woolen mill owned and operated by the same family that founded it in 1850, the town today is neat, harmonious, and business-like. The village itself looks very much as it did a century ago. In the faded brick of its buildings one is aware of that indefinable sense of continuity; the town has adapted to change without destroying its past. Through a process of time, adversity, and what might be called permissive paternalism, Harrisville has made terms with the industrial revolution that favor both the dignity of the individual and the strength of the community.

John Borden Armstrong

Hingham, Massachusetts
June 1969

ACKNOWLEDGMENTS

To acknowledge the contributions of all those who have helped in one way or another with the composition of this book is not practicable. Nonetheless, the interest, help, and encouragement of so many people has indeed made the writing a pleasure, and I must at least mention the following individuals and institutions for having contributed significantly to the enterprise. Many others, not unimportant but simply too numerous, are cited in the footnotes for the letters they wrote and interviews they granted me and in the credit lines for the valuable illustrative material they provided.

I am indebted to many at Boston University. Professor W. S. Tryon recognized the importance of the town, encouraged me to write its history, and served as first reader par excellence of the doctoral dissertation that was the original version of the book. Professor Tryon was all that a graduate student could ask for in a mentor. Professional mastery, stylistic elegance, dignity, joie de vivre, unfailing sense of humor, broad humanity, and personal kindliness all rolled into one man have made it a great joy and privilege to work for and with him.

Professor Robert E. Moody served as my second reader with a concern and competence that one would be fortunate to find in a first reader. His assistance, encouragement, and candid advice give him a special place in my affections as in my esteem.

Other professors and colleagues to whom I am indebted for encouragement and counsel on special problems include Robert V. Bruce, Sidney A. Burrell, Everett Burtt, Saul Englebourg, and Herbert Moller. The Boston University Graduate School generously provided for typing the manuscript.

I have used extensively the resources of several libraries. Scholars have, or should have, a special place in their hearts for librarians, and I am particularly grateful to those at the Boston Athenaeum, the Keene Public Library, the New Hampshire Historical Society, and the New Hampshire State Library. At the last-named institution I should especially like to mention the kindness of Miss Stella J. Schecter, Head of the Reference and Loan Division.

In the towns of Dublin, Harrisville, and Nelson people were generous with time and effort. The town authorities, particularly the town clerks, were helpful in giving me access to the town records. Many people gave me interviews that provided a valuable part of the twentieth-century record. The office staff and other employees of the Cheshire Mills were very helpful and very patient with my questions about details of the woolen industry.

Mr. and Mrs. John J. Colony, Jr., of Harrisville, made a contribution to the history to which it is not easy to do justice. Mr. Colony gave me as complete freedom to the company records and to the mill itself as a scholar could desire. Nothing that I wanted to include in the history was refused me. Mr. and Mrs. Colony both read the entire manuscript and made invaluable suggestions. To have made such friends as the Colony family has been one of the unexpected rewards of writing the history of Harrisville.

The Merrimack Valley Textile Museum's former and present curators, Mr. Ian M. G. Quimby and Mr. James C. Hippen, read the manuscript and gave me valuable counsel concerning the woolen manufacture. The Museum and its director, Thomas W. Leavitt, have courageously undertaken to publish the history and have in the lengthy process shown me every kindness and consideration.

Finally, the women in my life. My cousin, Miss Helen P. Burns, has given me aid, advice, and encouragement over the years to warrant the dedication of many volumes. My wife, Bonnie Barton Armstrong, has suffered and supported my endeavors as only a good wife can. Most particularly, she ably acquitted the onerous task of typing the many appendixes.

Such merit as the book contains is due to the help of many people. Its shortcomings remain the exclusive possession of the writer.

J.B.A.

FACTORY UNDER THE ELMS:
A HISTORY OF HARRISVILLE,
NEW HAMPSHIRE, 1774–1969

1 THE EIGHTEENTH-CENTURY BACKGROUND

In the southwest corner of New Hampshire there is a hilly, rocky, heavily forested region known to geographers as the "Monadnock Highlands."[1] It is anchored to the south by solitary Mount Monadnock. The valleys of the Merrimack and Connecticut Rivers form its eastern and western limits. To the north rise the White Mountains. The rough surface of the land provides many bowls that hold ponds or lakes.[2] The soil is in general thinner and poorer than in the river valleys, but there is considerable variation even within a small area. The climate is an important feature of the region: summers are cool and relatively short; winters begin sooner, last longer, and are more severe than in the lowlands.[3]

Several miles north of Mount Monadnock, spread along the banks of two ponds and the tumbling stream connecting them, lies Harrisville. This small mill village grew up without regard for political lines and divisions. For nearly a century before it finally set up on its own, the settlement lay astride the boundary between the towns of Nelson to the north and Dublin to the south. The decisive influence in its origin and history has been its geography, and particularly its remarkable supply of waterpower.

The source of this waterpower consists of about ten square miles of drainage area, which includes a chain of three large ponds plus three smaller tributary ponds and intervening meadow land, all available for use as storage reservoirs. The first in the chain is Spoonwood Pond, which lies in the town of Nelson at an elevation of 1,385 feet above sea level and covers 170 acres with a drawable depth of about 12 feet of water. The second and largest lies in the towns of Nelson and Hancock and is known as Long Pond or Lake Nubanusit. Just to the south of Spoonwood, Long Pond lies at an elevation of 1,365 feet and covers about 800 acres with an available storage depth of about 12 feet. The water from these two ponds drains across a mile and a half of meadowland and empties into Harrisville Pond,[4] which lies at an altitude of 1,320 feet and covers about 120 acres.[5] The water from these three ponds then runs off into Goose Brook,

1 Margery D. Howarth, *New Hampshire, A Study of Its Cities and Towns in Relation to Their Physical Background*, p. 8.

2 Cf. Katherine F. Billings, *The Geology of the Monadnock Quadrangle, New Hampshire*.

3 Howarth, *op. cit.*, p. 9.

4 This pond has been known successively as Brackshin Pond, Twitchell's Pond, Upper Pond, Harrisville Pond, and, most recently and most regrettably, Lake Nubaunsit.

5 John Humphrey, *Water Power of Nubanisit Lake and River*, pp. 1–2. (The name of the lake is generally spelled "Nubanusit," and this spelling will be used in this account.)

turbulently descends a steep ravine in a southerly direction, and soon empties into what is known as North Pond. From here, the water runs eastward into a branch of the Contoocook River and eventually into the Merrimack. In the nine miles before joining the Contoocook, the water falls a total of 600 feet.[6] The best place to harness the power created by this fall is where Goose Brook descends the steep ravine, and here it was that the mill village of Harrisville was built.

The remote and rugged nature of this highland region, together with the danger of Indian raids during the French and Indian War, served to delay permanent settlement of Dublin and Nelson until the 1760s.[7] Even then, especially for Nelson, growth was slow until after the Revolution.[8] For this early period, it is difficult to estimate the population of the area later to be the town of Harrisville. In the southeast part of Nelson, which was to make up more than a quarter of Harrisville, there were living in 1774 fourteen settlers, most of them with families.[9] In the northern part of Dublin, which was later to form nearly two thirds of Harrisville, settlement was still more sparse. Perhaps not more than seven families were living in this area in 1774.[10] Most of those who settled within the limits of Harrisville before the Revolution came from Massachusetts towns like Dunstable, Sherborn, and Sudbury. A few came from nearby towns in New Hampshire, such as Mason and Temple. Two came from England, and the names of the others—Adams, Bemis, Harris, Twitchell—suggest their English background.[11]

The first settler on the site of the village of Harrisville came there at the very end of the colonial period, in 1774. He was Abel Twitchell, from Sherborn, Massachusetts.[12] His father, Joseph Twitchell, Esq., was the leading citizen of Sherborn. He was also agent for some of the proprietors or holders of land in Dublin and sold land to those who wished to settle there, a number that eventually included two of his daughters and five of his sons. The eldest of these, Samuel, claimed to be the third permanent settler in Dublin and owned a sawmill in the southern part of town. Another son, Eleazer, lived for a time in Nelson. There he owned a large tract of land which surrounded Brackshin Pond, as Harrisville Pond was then called. Abel himself had moved to Dublin, returned to Sherborn, and finally returned again to Dublin to settle on Goose Brook.[13] The rough, rocky lot with its steep gorge and turbulent brook was not land to attract a farmer, but undoubtedly it was these features that attracted Twitchell. He soon had built a grist- and sawmill at the mouth of Brackshin Pond and also acquired his brother

6 Hamilton Child, *Gazetteer of Cheshire County, New Hampshire, 1763–1885. Part First*, p. 175.

7 Leonard, 1855, pp. 130–131.

8 Jeremy Belknap, MS Description of Packersfield, N.H. [ca. 1789].

9 *Masonian Papers*, VII, 36, in *N.H. State Papers*, XXVIII, 44.

10 Cf. Appendix 1.

11 *Ibid.*

12 Leonard, 1855, p. 273.

13 *Ibid.*, pp. 402–405; Humphrey, *op. cit.*; Simon Goodell Griffin *et al.*, *Celebration of the One Hundred and Fiftieth Anniversary of the Town of Nelson, New Hampshire, 1767–1917*, p. 17.

Eleazer's land in Nelson, which gave him control of the entire waterpower of the pond and the stream below it.[14]

Abel Twitchell also built a square, two-and-half-story, frame house. It still stands, perched on the rocky slopes above the stream. Despite changing architectural style and the use of different materials, this first house seems to have been used as a model for those later built around it. (Plate II) After Abel Twitchell, the next person to settle here was Jason Harris, of Framingham, Massachusetts. "At an early date," perhaps in 1778, he built a blacksmith and trip-hammer shop a short distance downstream from Abel Twitchell, but he did not stay long.[15] "Twitchell's Mills" was a logical name for this tiny settlement, and thus it was known for half a century.[16] A residence or two and two small mills drawing a meager custom from a sparsely settled countryside, this must have been the extent of the settlement until the end of the eighteenth century. Until then, also, the history of Harrisville is very much the history of Dublin and Nelson.

The early political and economic life of these two towns was quite similar, and they can be studied to advantage side by side. Both were part of the Masonian Grants, that is, the New Hampshire lands purchased from the heir of the original claimant, Captain John Mason, by a group of proprietors who thereafter made grants for settlement. Dublin, then known as "Monadnock Number Three," was chartered in 1749, and Nelson, or "Monadnock Number Six," was chartered in 1752.[17] The proprietors laid down very specific conditions in these original grants. Certain lots were reserved for the support of religion and education. Others were reserved for the proprietors. A meetinghouse was to be built within a specified time. The grants also contained detailed timetables for settlement: so many acres in each share[18] to be cleared, fenced, and prepared for mowing or tillage within so many years, dwelling houses to be built and occupied by such a time, and so on.[19] Considering the speculative nature of these grants and the difficulties of settling the region, many of these terms were probably impractical. At any rate, they were not fulfilled; when the time expired, the original grants were renewed for the current holders.[20]

The grantees, or shareholders, of Dublin and Nelson were apt to be men of some wealth living in the central or eastern part of the state. They were not the sort likely to move to the raw frontier, and, so far as is known, none of them ever settled on the lands they owned there.[21] Probably very few of them ever saw their lands. Instead, they sought to sell their shares, or lots,

14 Leonard, 1855, p. 273; Humphrey, op. cit., p. 6.

15 Leonard, 1855, pp. 144, 273. The shops of both men have long since disappeared.

16 Philip Carrigan, Map of New Hampshire (1816); Child op. cit., pp. 175–176.

17 Nelson was incorporated in 1774 with the name of "Packersfield," and in 1814 received its present name. (Griffin, op. cit., p. 5, n. 1.) To avoid confusion, if not anachronisms, it will hereafter be called Nelson.

18 In Dublin a share was equal to three lots, and in Nelson equal to two lots.

19 N.H. State Papers, XXVII, 171–174, in Leonard, 1855, pp. 5–6, 124–129; N.H. State Papers, XXVIII, 4–6, in Griffin, op. cit., pp. 5–6.

20 Masonian Papers, VII, 3, 38, in N.H. State Papers, XXVIII, 8, 47.

21 Leonard, 1855, pp. 5, 130. For names of Grantees of Dublin and Nelson see N.H. State Papers, XXVII, 175–176; XXVIII, 4, 7.

either to men who would settle there or to other specu-
lators.

Leonard, in *The History of Dublin, New Hampshire*, writes that
"After the close of the French War, there was a numerous
emigration from Massachusetts into New Hampshire. The proprietors of the unsold lands in the southern townships offered
strong inducements to young men to purchase farms, and remove
thither."[22]

It was this emigration that provided Dublin and Nelson with
their first permanent inhabitants. The earliest settlers in Dublin
were some Scotch-Irish who came from neighboring Peterborough during the French and Indian War. Perhaps the region
was too rugged even for the Scotch-Irish. Whatever the reason,
they all removed, leaving behind little but the name of the town
to indicate their presence.[23] By 1762, English settlers were coming
up from Sherborn, Massachusetts, to work on the roads, clear
land, and build homes. Several of them became residents within
the next two years, and thereafter the settlement proceeded
steadily if slowly.[24] Nelson lagged behind by several years.
Breed Batchelder, a man of exceptional ability from Brookfield,
Massachusetts, was the town's first resident. He came to Nelson
in 1766, and others came shortly afterward from Dunstable,
Hubbardston, and Marblehead, Massachusetts.[25]

Essential for orderly settlement was the early surveying and
laying out of the new towns. Dublin was surveyed in 1750 by the
proprietors' agent Joseph Blanchard.[26] The town was in the shape
of a rectangle, or more exactly, a parallelogram, seven miles
east to west and five miles north to south. This area was laid out
in ten ranges running east and west, and numbered one to ten
from south to north. Each range was divided into twenty-two
lots, numbered from east to west, making in all two hundred
and twenty lots of rather more than one hundred acres each.[27]
Seven lots lying on Monadnock mountain were excluded, leaving
213, or 3 apiece for each of the 71 shares. The survey completed,
the lots were distributed the same year by a sort of lottery.
The grantees drew lots for their turn to draw the three numbers
that should represent their respective holdings; then, in turn,
they again drew to determine where their lands should be. This
determined, the grantees could go or send to Dublin to find their
lots and see whether they lay under water, were boulder-strewn
hillsides, or whatever. Certainly not all were pleased with what
they found. It is not known what John Usher thought of his
Lot 13, Range 10, the site of the village of Harrisville, but he
soon disposed of it to Matthew Thornton, Esq., a large landowner

22 Leonard, 1855, p. 133.
23 *Ibid.*, p. 13.
24 *Ibid.*, pp. 8, 132–134.
25 Griffin, *op. cit.*, pp. 7–9, 12.
26 *N.H. State Papers*, XXVII, 177.
27 *Ibid.*; Humphrey, *op. cit.*, pp. 3–4.

in Dublin. He in turn sold the lot to Abel Twitchell in 1778, several years after Twitchell first settled there.[28]

Nelson, adjoining Dublin to the north, was a rectangle eight miles east to west and five miles north to south. Breed Batchelder, a trained surveyor, made the survey in 1768. The plan was more complicated than Dublin's, but this was due to a more piecemeal settlement and to plans drawn before the survey was made. In 1761, the grantees had divided the town into quarters by running intersecting center lines.[29] Batchelder laid out each quarter in gridiron fashion. He first divided a quarter into ranges running north and south, and numbered from west to east, and then divided these ranges into lots, numbered from south to north. Each lot was 160 by 104 rods, or 104 acres. This system resulted in equal shares, but there were strips of land left over in each quarter, and there was no coincidence of the lines in the different quarters.[30]

The rectangular land survey of New England was undoubtedly a more orderly system than some others used in the colonies, but, as can be seen, it had its complications. Nonetheless, the system worked well enough in a pioneer settlement where uncleared land was at a discount and where the immediate tasks were not to define boundaries but to clear and plant and build. Probably more troublesome than boundaries was the matter of titles. Charles A. Bemis, a local historian and resident of a century later, wrote of Nelson that "there was much difficulty about titles in the early days, and several families came to settle but got discouraged and left. Some had to pay twice for their land or lose it. . . ."[31] Incidents of this sort must have been unusual, but they suggest one more potential hazard for the early settler.

The prices paid by the settlers for their lands are mentioned only occasionally in the historical accounts. In 1763, Breed Batchelder purchased one third of the southwest quarter of Nelson, 2,135 acres, for £60 sterling.[32] In 1770, the widow of Agent Joseph Blanchard sold one hundred acres of land in Nelson for £10 sterling.[33] C. A. Bemis, citing no evidence, writes that land was very cheap, sometimes as low as thirty cents an acre but rising to as high as four dollars an acre in war times.[34] Sometime before 1774, James Bancroft is reported to have bought 416 acres—four lots—in southeast Nelson and paid for them with one pair of steers.[35] And in 1778, Abel Twitchell, for the entire of Lot 13, Range 10, in Dublin, paid Matthew Thornton just £15 sterling.[36]

Facing the new settlers were the arduous tasks of opening

28 Leonard, 1855, p. 6; Humphrey, op. cit., pp. 3–4. Thornton was also a signer of the Declaration of Independence.

29 Griffin, op. cit., p. 6.

30 Ibid., pp. 6, 8. For Breed Batchelder's Plan of 1768, see N.H. State Papers, XXVIII, 8.

31 Bemis MSS, Box 6, XLVIII, 235.

32 N.H. Records, May 1765, in Griffin, op. cit., p. 6.

33 N.H. State Papers, XXVIII, 9.

34 Bemis MSS, Box 6, XLVIII, 235.

35 Griffin, op. cit., p. 15.

36 Cheshire County Records, Book 23, pp. 123–125, in Humphrey, op. cit., p. 5.

roads, clearing fields, and building homes. Sometimes they began this work on a "non-resident" basis, as did the early Dublin settlers who came up from Sherborn during the slack seasons,[37] or Breed Batchelder who boarded with a friend in Keene while building his house in Nelson.[38] Others simply plunged in, living in a rough camp during the good weather. However they went about the job of making new homes in the wilderness, it was an undertaking unlikely to end in success without stamina, intelligence, doggedness, and good fortune—all in ample proportions.

The land to which Batchelder and his neighbors came must have been all but covered with unbroken forest, a mixed growth of oak, sugar maple, ash, birch, walnut, pine, and hemlock.[39] Griffin writes that Aaron Beal, in the southeast quarter of Nelson, pastured his cows on the site of the later village of Chesham, perhaps two miles distant, "that being the nearest place where grass grew. . . ."[40] While the forest hindered travel and farming, it provided wood for housing, fuel, and household articles. It also provided a cash crop in the form of potash, or "salts," when the settler burning over his land took the trouble to gather the ashes and boil them.[41] In short, the forest was a large part of the challenge facing the first settlers.

The early roads through the forest were as primitive as they were essential. There must have been a road running from Keene eastward through Dublin to Peterborough as early as 1759.[42] In like manner, the first road into Nelson also climbed northeasterly from Keene, passed the house of Breed Batchelder, and continued to the center of town.[43] Perhaps as early as 1768, and certainly by 1773, Dublin and Nelson were connected by a road. At that time a road ran from Dublin center through the northwest part of town to Nelson center.[44] In 1774, when Abel Twitchell settled at the mouth of Brackshin Pond, the town of Nelson planned a road that would connect him with the village. The committee that laid it out described the last stretch of this road in the usual fashion:

Beginning at a Large hemlock at the Brook Running into the Northwesterly side of Brackshin pond so called, and then runs Northerly to a spruce, then to a Large Black Birch, then to a small Beach . . . to a Large Red oak Tree southerly of James Bancroft's old Barn Near the Brook, it being the Last mentioned Bound of the Road Laid out from the Meeting House. . . .[45]

One writer, whose grandfather helped to build the first roads in Dublin, described them as being "little more than openings cut

37 Leonard, 1855, p. 133.

38 Griffin, op. cit., p. 7.

39 Howarth, op. cit., p. 5; Griffin, op. cit., p. 13; Leonard, 1855, p. 119.

40 Griffin, op. cit., p. 15.

41 Ibid., p. 14.

42 Leonard, 1855, pp. 8, 131.

43 Griffin, op. cit., pp. 16–17.

44 Leonard, 1855, pp. 8–9; Griffin, op. cit., p. 17.

45 Griffin, op. cit., p. 17.

through the dense, continuous woods, with some slight demonstration towards a partial removal of the rocks, logs and stumps, and leveling of the grosser inequalities of the surface."[46] Work on these roads was never really finished. It required the continual effort of those the roads served to keep them in passable condition. Mending the roads was also, in this land of little money, a common way of discharging one's highway taxes.[47]

Ingenuity and hard work were required to keep the roads open during the deep snows of winter time. Leonard describes how it was done in Dublin. Beginning at Twitchell's Mills

The oxen and young cattle were turned unyoked into the road, and one person went before them to commence a track, and he was followed by the cattle. When the man on the lead became tired, another took his place. At each settlement, the fresh cattle were put forward; and, by being thus driven in Indian file, a good horse-path was made.[48]

Winter and summer, for many years after Dublin and Nelson were first settled, most of the traveling was done on horseback or on foot.[49] Wheeled vehicles or sleighs were few. Heavy supplies were brought in on a horse-drawn rig similar to the *travois* used by the Indians.[50] Such roads meant hard traveling for a man on horseback, and even the experienced traveler was lucky not to lose his way. Probably it was a matter of mutual cause and effect that while the roads were so poor travel on them would be light, and so it seems to have been throughout the New England countryside.[51]

During those early years, especially through the long winters, the diet of the settlers was unappetizing and dull. Leonard cites one menu consisting of bean porridge for breakfast and supper, and a piece of pumpkin with a thin slice of pork baked upon it for dinner.[52] Griffin writes of those in Nelson that "they often lived almost wholly upon boiled rye. . . ." Game was "tolerably plenty," but not so the supply of lead and powder with which to hunt. Other commodities mentioned are salt pork and rum. This last, together with a hearty appetite, must have helped considerably with such a diet.[53]

The rudeness of the frontier homes is impressive. Leonard describes them in Dublin, citing as his source one who certainly saw this type of dwelling.

Neither bricks, nails, nor boards were accessible; and for dwelling places, the pioneers in the settlement built with logs what would now be called a pen, in dimensions about fifteen feet by twelve, having two doors, one on the south side and the other on the east. For a roof, they

46 Leonard, 1855, p. 8.

47 In 1772, the proprietors voted that the grantees, or those holding land under them, might work out their highway taxes. This was to be allowed at the rate of three shillings per day per man in summer and two shillings six pence in winter. Griffin, *op. cit.*, p. 11.

48 Leonard, 1855, p. 283.

49 Griffin, *op. cit.*, p. 17; Leonard, 1855, p. 283.

50 Griffin, *op. cit.*, p. 18.

51 Charles Francis Adams, *History of Braintree, Massachusetts*, p. 84.

52 Leonard, 1855, p. 283.

53 Griffin, *op. cit.*, pp. 13, 18.

took spruce-bark, and tied it to poles by means of withes or twisted twigs. . . . For a chimney they laid stone up to the mantel-tree, and then split laths, built them up cob-house fashion, and plastered inside and outside with clay mortar.[54]

During their first winter in town some families did not have even the dubious comfort of such a house but lived in rude shanties, or huts, without chimneys and providing little protection from the elements.[55] Such privation could hardly have lasted more than one season; without better shelter the family would have removed. More substantial housing had a high priority for the pioneer family. It is to be noted that even as early as 1774, of the settlers living in the southeast quarter of Nelson nearly as many had a "board house," as had a "pole house," or log house.[56] There is a substantial number of these improved dwellings still standing to testify eloquently of the skill of their builders.

The difficulties and hardships facing those who attempted to settle this highland region of southern New Hampshire seem to have been without number. Remoteness saved these towns from Indian attacks and enemy depredations during the Revolution[57] but had its own disadvantages. Also, various accounts mention the hazard of fire. Belknap, in 1789, wrote of Nelson that "several of the inhabitants have had their habitations burnt . . . a Disaster to which new Countries are peculiarly Subject."[58] A brief newspaper item of about the same time shows how this was likely to happen: "Last Friday night the house of Samuel Griffin Esq. of Packersfield was consumed by fire by reason of some hot ashes put in a tub near the side of the houses. . . ."[59] Abel Twitchell's mill also fell victim to fire, but was soon rebuilt.[60] The stories of early hardships tell of the alarms raised and the destruction done to livestock and crops by wild animals.[61] Bears and wolves were troublesome in the region to the end of the century.[62] However, it is probable that the long, severe winters and the meagerness of arable land discouraged more settlers than did anything else. All things considered, it is little wonder that towns like Dublin and Nelson grew slowly, remained sparsely populated, and were soon being forsaken by the young, the restless, the hopeful, and, sometimes, by the simply weary.[63]

Under the proprietors, political organization had proceeded apace. After proper petition to the royal governor, Dublin was incorporated in 1771, and Nelson in 1774.[64] In both places during these years before the Revolution, the main items of business before the town meetings were to open roads, build a meeting-house, and settle a minister.[65] The meetinghouse and the minister's

54 Leonard, 1855, pp. 279–280; Griffin, *op. cit.*, p. 18.

55 Leonard, 1855, p. 281.

56 *Masonian Papers*, VII, 36, in *N.H. State Papers*, XXVIII, 44.

57 Leonard, 1855, p. 281.

58 Belknap, *op. cit.*, p. 2.

59 *Newhampshire Recorder*, Sept. 16, 1790, in Bemis MSS, Box 2, XLVII, h.

60 Leonard, 1855, p. 273.

61 Griffin, *op. cit.*, p. 13; Leonard, 1855, pp. 28–29, 280–281.

62 Leonard, 1855, p. 28.

63 At least one out of four of the eighteenth-century settlers within the area of Harrisville eventually removed. See Appendix 1.

64 Griffin, *op. cit.*, p. 12; Leonard, 1855, pp. 139–140.

65 Griffin, *op. cit.*, pp. 10–11; Leonard, 1855, p. 146.

salary were paid for by laying a tax on each share or lot, and the first minister also received his one share of land, as provided in the charters. Dublin had built its meetinghouse, rough-boarded and lacking pews, and hired its minister by 1772. Nelson lagged behind only slightly.[66] It may be remarked that neither town kept its first minister for long. In these years also, as evidenced by the regular appearance of the same names in the town records, the natural leaders in these settlements were emerging.[67] If their political and religious institutions were still inchoate, the progress made by the settlers of Dublin and Nelson in their first decade was nonetheless impressive.

The role these towns played in the Revolution was minor but creditable. They both had formed revolutionary committees before hostilities commenced.[68] Nelson received word of the clash at Lexington and Concord on the day after the battle, and at sunrise on the twenty-first a company of twenty-seven men left town for the scene of action.[69] When the " Association Test," the loyalty oath of the Revolution, was submitted to them, the residents of Dublin and Nelson signed it, with one conspicuous exception.[70] Breed Batchelder, Nelson's first resident, leading citizen, and principal landowner (and a Major in the Provincial Militia to boot), remained stubbornly loyal to the Crown. This was to cost him dearly. After being imprisoned, and hiding in a cave for three months, he fled the colony, served in the British army, and died in the exile of Nova Scotia. His lands, amounting to nearly one quarter of the entire town, were confiscated.[71] In their loyalty to the American cause, Dublin and Nelson had a somewhat better record than various other towns in the same part of New Hampshire.[72] In furnishing men and means for the Revolutionary cause, both did well for towns remote from the scene of any action. Leonard gives the names of twenty-six men who went to war from Dublin, and Griffin lists fifty-five men then living in Nelson who served enlistments.[73]

The original settlement of Dublin and Nelson had come with the expansion of the colonies that followed the French and Indian War. Now, in the two or three decades after the Revolution, these towns were to experience a new and much larger growth. The census figures show a dramatic rise in population. At the time of the New Hampshire census in 1775, Dublin had a population of 305 and Nelson 186. By the time of the first federal census in 1790, Dublin had boomed to 905 and Nelson to 721.[74]

Confronted with this new migration, Dublin took steps to avoid becoming freighted with paupers and to keep down the

66 Griffin, op. cit., pp. 10–12, 16; Leonard, 1855, pp. 11, 30, 154ff.

67 Cf. Dublin Town Records, 1771–1806, passim; Town Records of Monadnock Number Six and Packersfield, 1751–1801, passim.

68 Griffin, op. cit., p. 18; Leonard, 1855, pp. 18, 148.

69 Griffin, op. cit., p. 19.

70 N.H. State Papers, VIII, ix, 228, 263. In Dublin there were fifty-seven signers and no nonsigners; in Nelson thirty-eight signers and one nonsigner. These were all the males over twenty-one years of age in the two towns.

71 Griffin, op. cit., pp. 21–25.

72 Ibid., p. 21, note.

73 Leonard, 1855, pp. 20–23, 148–151; Griffin, op. cit., pp. 31–161, passim.

74 Jeremy Belknap, The History of New Hampshire, III, 238–239.

taxes. Taking advantage of a law that enabled towns to remove their liability for the support of paupers by warning from town all persons moving in for a settlement, the town for some years issued warnings-out-of-town indiscriminately to all newcomers, regardless of wealth or station. There must have been some droll scenes when town officials served these warnings upon substantial newcomers. Other towns also followed this practice, but apparently Nelson did not. It is a matter of conjecture whether this was due to the town's greater solicitude for its poor, or the lesser likelihood of attracting such paupers.[75]

Jeremy Belknap, the early great historian of New Hampshire, wrote in 1789 a description of Nelson that was apparently based on personal observation. It was succinct and of practical outlook. He described the location and terrain and remarked on the number of ponds there. He noted the "clear and healthy" air and appraised the possibilities for cultivation. He told of the town's rapid growth since 1780, and he pointed out that "the inhabitants have been somewhat remarkable for their care to provide Schooling for their children, which has been commonly by a Female or Woman's school in the summer, and a Master's school in the Winter."[76]

Concerning this matter of education, Nelson also had incorporated in 1797 a social library,[77] which would seem to corroborate Belknap's statement that the town showed a concern for education. Dublin, beginning with a small appropriation for schooling in 1773, had provided for public education and, a few years before Nelson, also formed a social library.[78] These developments were consistent with the general state of education in New Hampshire. During most of the eighteenth century there was great apathy about education, and the laws were very imperfectly enforced, but interest increased and conditions improved in the last decade of the century.[79]

From this evidence, the town records, the accounts like Belknap's, the homes that were built and the lands that were cleared, there emerges a picture of these hill towns as industrious, growing, and hopeful agricultural communities. Another generation would see the opening up of the West and, for towns such as Dublin and Nelson, the beginning of a decline that would bring great changes.

In that "second wave" of settlers that came at the close of the Revolution was one Erastus Harris, father of the Jason Harris who had come to Dublin in 1778. Supposedly Erastus was the descendant of the Thomas Harris who came to America from Kent County, England, in 1630, with Roger Williams. Thomas

75 Leonard, 1855, pp. 26, 144–146; Daniel B. Cutter, *History of Jaffrey, N.H., 1749–1880*, pp. 167–168.

76 Belknap, MS Description of Packersfield, N.H., pp. 1–4.

77 *Laws of New Hampshire, 1792–1801*, VI, 459–460.

78 Leonard, 1855, pp. 246–261.

79 George Garvey Bush, *History of Education in New Hampshire*, pp. 9ff.

had followed Williams to Rhode Island, but some of the family later returned to Massachusetts. Erastus was born in Wrentham in 1731.[80] Before he was twenty-four, he had married and moved to nearby Medway. Here he worked as a carpenter with occasional flings at soldiering. He marched with several expeditions in the French and Indian War, including the one to Crown Point. No longer a young man when the Revolution began, he nonetheless answered the Lexington alarm and served eleven days as a sergeant in the militia. It is probable but not certain that he also served other enlistments during the War.[81] Then, "about the time of the close of the Revolution," he moved once more, following his son Jason, to settle on a farm in the southeast quarter of Nelson. His home was a short distance away from Twitchell's Mill. Here in Nelson, working at farming and carpentering, Erastus Harris spent the last twenty-five years of his life.[82] Little else is known about this progenitor of the Harris family that was to raise a new town.

Bethuel, his son and Jason's brother, was born in Medway in 1769 and followed his father to Nelson in 1786.[83] Why, as a boy of twelve, he had not come with his father is not clear. Perhaps Erastus came alone and the rest of the family later. Whatever the explanation, 1786 was a year of economic distress and social disorder in Massachusetts that provided good reasons for leaving then. Bethuel worked at carpentering with his father for several years before beginning on his own.[84]

For one who, as his son later put it, came to Nelson "destitute of pecuniary ability," Bethuel did well for himself. Early in the 1790s, he bought a 280-acre farm that lay athwart the Nelson–Dublin town line.[85] In 1794, he married a daughter of neighbor Twitchell. (Plates III, IV) Deborah was then not eighteen years old but was certainly of hardy stock. (She bore her husband ten children, survived him by several years, and died at the age of seventy-nine.)[86] They lived on his farm at Nelson, while Bethuel combined farming and carpentering with an increasing role in town affairs. In 1800, he attempted to sell his farm, of which the advertisement in the *New Hampshire Sentinel* gives a description:

A farm in the southeasterly part of PACKERSFIELD containing about 160 acres of Land, with two small houses, a Barn, and other out houses, in good repair. Said Farm is well situated for carrying it on to advantage. There is a corn mill within 60 rods of the boundaries, and a blacksmith and clothier within a very short distance. A handsome credit, with good security, will be given for a large part of it. For further particulars enquire of the subscriber living on the premises

Packersfield, Nov. 8, 1800   Bethuel Harris[87]

80 Albert Hutchinson, *A Genealogy and Ancestral Line of Bethuel Harris of Harrisville, New Hampshire, and His Descendants*, p. 10.

81 Griffin, *op. cit.*, p. 91.

82 Child, *op. cit.*, p. 178; Hutchinson, *op. cit.*, pp. 14–15.

83 D. Hamilton Hurd (ed.), *History of Cheshire and Sullivan Counties, New Hampshire*, p. 213; Hutchinson, *op. cit.*, pp. 15–16.

84 Hurd, *op. cit.*, p. 213; Hutchinson, *op. cit.*, p. 15.

85 Hurd, *op. cit.*, p. 213.

86 Leonard, 1855, p. 405; Leonard, 1920, p. 787.

87 *NHS*, Nov. 29, 1800.

There are several possible explanations for this advertisement. Bethuel may have wanted to move to his father's farm. His health may have prompted this move, for he was troubled with sciatica.[88] Or he may already have had his eye on other enterprises.

At the end of the eighteenth century, then, the essentials were present for a new and different village in this highland region. Abel Twitchell's mill stood at the head of a roaring stream. To the north and south lay two agricultural towns that by great effort were still growing. And near the stream lived a man and his family with the ambition and the intelligence to build an industry on this waterpower.

88 Hurd, *op. cit.*, p. 213.

2 THE MILLS OF
HARRISVILLE, 1799–1861

Beginnings of a Woolen Industry

Abel Twitchell and Jason Harris had made the first attempts to turn the waterpower of Goose Brook to the elementary needs of a new settlement. As the eighteenth century drew to a close, Jonas Clark, from Townsend, Massachusetts, tried to apply this waterpower to a more specialized enterprise—the manufacture of cloth. Little is known about Jonas Clark; he seems to have been plagued with hard luck[1] and was probably a better mechanic than businessman. Whatever his story, he married Mary Twitchell, another daughter of Abel, and in 1799 built a small establishment on the banks of Goose Brook, between the two already there, for the fulling and finishing of cloth.[2]

Ordinarily, the major processes in the manufacture of woolen cloth are scouring, picking, carding, spinning, warping, weaving, fulling, washing, dyeing (also possible at earlier stages, as stock or yarn dyeing), napping, and shearing. Curiously, in the removal of the industry from home to factory, which came with the development of specialized skills and power-driven machinery, some of the last steps of the manufacture were the first to leave the home. The earliest fulling mills in America dated from the middle of the seventeenth century, and they were quite numerous by the time Jonas Clark began his business.[3]

The object of the fulling operation is to shrink and strengthen the loosely woven cloth as it comes from the loom and to remove the grease that was added after scouring to reduce the static electricity. Before the development of the rotary fulling mill in the 1830s, there were two basic types of water-driven fulling mills, one with a vertical and one with a horizontal action. In the former, two or more large wooden hammers were alternately raised and dropped by cams attached to the shaft of a waterwheel. The hammers struck the cloth that was soaking in a trough of water to which had been added a fulling solution of soap and other agents. In the latter type, the hammers were pushed against the cloth. The vertical mill produced a heavier fulling action than the horizontal, although the time that the cloth was

1 *NHS*, April 20, 1822.

2 Leonard, 1855, pp. 322–323.

3 Arthur H. Cole, *The American Wool Manufacture*, I, 11; Letter from James C. Hippen, Curator, Merrimack Valley Textile Museum, Aug. 2, 1968.

in either mill also helped to determine the extent of the fulling action.

There were also optional finishing operations done in these fulling establishments. One was dyeing. Another was napping, or the brushing of the fulled cloth by teasels, a variety of thistle, which were held in a wooden frame. Shearing was the highly skilled process of cutting off the raised nap on the cloth with a large pair of shears. It was because the fulling and finishing processess required power-driven machinery, special tools, or skills beyond the possibility of the household manufacture that they early became a specialized business.[4]

Jonas Clark not only used the waters of Goose Brook to full cloth, but also in his shop "Mrs. Clark spun linen thread by waterpower, a single thread at a time." This thread may have been used in linsey-woolsey, "a linen warp material with thick wool filling, which was usually well-fulled, and strong as iron."[5] The brief history of his enterprise can be traced in several advertisements in the Keene *New-Hampshire Sentinel*. These notices were similar to those of other entrepreneurs of the period who were seeking either trade, help, or payments. Early in 1800, he announced,

The Subscriber informs the Public, that he colours Cotton, Linen, and Tow Yarn DEEP BLUE, on the shortest notice. The other branches of his business carried on as usual at his Clothier's Works in DUBLIN.[6]

At intervals, Clark advertised for a journeyman clothier and for apprentices—"one or two smart active lads from about 14 to 16 years of age."[7] The clothier's trade was one that required a long apprenticeship, and clothiers were probably difficult to hire and retain in such a remote spot as Twitchell's Mills.[8]

The third notice, tinged with bitterness, indicated the issue of Clark's enterprise:

The Subscriber, unwilling to give offence to even his most negligent Customers, by threatening to sue them, would inform them, that by waiting much longer than they contracted for, or the nature of his business would allow of, his own property has been attached—they may make their own conclusions.[9]

Unfortunately, there is no evidence that this appeal to the New England conscience was successful. On the contrary, after another season there appeared in the *Sentinel* a more strongly worded notice.[10] But it was not strong enough. In 1804, Jonas Clark

4 Cole, *op. cit.*, I, 12–13; Monte A. Calvert, "The Technology of Woolen Cloth Finishing. . . .," p. 19; Letters from James C. Hippen, Curator, Merrimack Valley Textile Museum, Aug. 2, 1968, March 13, 1969; Merrimack Valley Textile Museum, *Wool Technology and the Industrial Revolution: An Exhibit.*

5 Leonard, 1855, p. 273; Calvert, *op. cit.*, p. 24, citing William T. Davis (ed.), *The New England States*, I, 194.

6 NHS, June 7, 1800.

7 *Ibid.*, Aug. 9, 1800; Aug. 8, 1801.

8 Albert Annett and Alice Lehtinen, *History of Jaffrey, New Hampshire*, pp. 373–374.

9 NHS, Sept. 25, 1802.

10 *Ibid.*, March 1, 1804.

sold out and took his family to live in Canada.[11] His venture, though abortive and short-lived, was an important first step in the growth of this manufacturing town.

Even while Jonas Clark was trying to collect his bad debts, and before he sold out, another business was started nearby that was to be much more significant and much more successful. Enterprising Bethuel Harris joined with his father-in-law, Abel Twitchell, to set up a carding machine in Twitchell's grist mill.[12]

Carding is the third step in the woolen manufacture. It follows the tub-washing or scouring of the raw wool in an alkali solution to remove the grease and impurities, and picking, the process of pulling apart the tangled wool fibres and simultaneously removing dust and dirt particles. The object of carding is the straightening and intermixing of the wool fibres. In the household manufacture it was done by combing a small quantity of wool between two hand cards, which resembled hairbrushes with bent-wire bristles, until the individual fibres had been laid somewhat parallel and fibres of various lengths mixed in a fairly uniform manner. It was slow and arduous work, and the "carding" produced was often of indifferent quality.[13]

The first power-driven carding machines appeared in England in the second half of the eighteenth century. The first ones to appear in America were crude and inefficient, but in the 1790s improved models were developed. By the first years of the nineteenth century their use was spreading rapidly throughout New England. The early carding machines consisted of a wooden cylinder or cylinders which were held in a frame and covered with "card clothing," namely, leather in which were set fine bent-wire bristles. The wool was carried around on the clothed surface of the revolving main cylinder while being worked by the combing action of smaller, similarly clothed cylinders, which were separated from the main cylinder by small, carefully regulated spaces. The carding machine not only resulted in a great saving of time but produced a much finer and more uniform "carding." Its introduction caused the speedy disappearance of the process from the home manufacture.[14]

The carding machine of Harris and Twitchell must have been the same one advertised in the *Sentinel* in 1804 as having been erected in Dublin by Kneeland and Company of Jaffrey, New Hampshire.[15] Kneeland was a clothier already running a carding mill in nearby Jaffrey.[16] Neither Twitchell nor Harris had any experience in the trade, but as Cole states, "the operation of carding by machine was not a difficult one to learn passably well; and the machine itself neither was expensive nor required

11 Leonard, 1855, p. 273.

12 *Ibid.*

13 Cole, *op. cit.*, I, 7–8; Merrimack Valley Textile Museum, *Wool Technology and the Industrial Revolution: An Exhibit.*

14 Cole, *op. cit.*, I, 87–89; Merrimack Valley Textile Museum, *Wool Technology and the Industrial Revolution: An Exhibit*; Letters from James C. Hippen, Curator, Merrimack Valley Textile Museum, Aug. 2, 1968, March 13, 1969.

15 *NHS*, June 2, 1804.

16 Annett, *op. cit.*, pp. 374–375.

an exceptional amount of power."[17] One local historian has called this machine in Twitchell's mill the second one erected in the country; another, the second one in the state.[18] The first claim must be erroneous; the second is possible but unlikely.[19] But even without these claims the machine was an important innovation. The historian of the woolen industry in America has called the introduction of the carding machine the "initial point, the first proper introduction of the woolen manufacture."[20] Certainly it was for the future town of Harrisville.

In this early period of American woolen manufacturing, it was the normal course for factories to grow gradually from carding and fulling mills to full-fledged factories, during which period the mills supplemented and cooperated with the earlier household manufacture.[21] Thus, in Harrisville it was a logical next step for the carding mill to combine with the fulling mill. Bethuel Harris and Abel Twitchell soon bought the fulling mill from Jonas Clark's successor and combined the operations in one building, which was probably where Clark's shop stood.[22]

This business must have been run by Abel Twitchell and his son; at least the town of Dublin occasionally taxed them during these years for "mills" at the modest rate of seventy-five cents or a dollar.[23] Bethuel Harris paid no such tax, and it is likely that he either retained only a minimal interest in the business or else had gotten out of it altogether. He was still doing farming and carpentering and may have been waiting to see how the venture would go before committing himself. Competition was keen and returns uncertain. Two notices run by Abel Twitchell, Jr., in the *Sentinel* are the principal record of the company. In 1805, he called on his customers to pay their debts, "all kinds of country produce taken in payment."[24] The next year Twitchell advertised that since "others in his line of business have lately reduced the price of carding wool from 8 to 6 cents [per pound], he is determined to offer the same terms. . . ."[25] For the next seven years there is silence, and by the end of that time another party occupied, if he did not own, the clothier's shop.[26]

Finally Bethuel Harris made his move, decisive for the town as well as himself. A notice in the *Sentinel*, June 19, 1813, announced that "B. Harris & Co." had purchased the "Clothier's Works in Dublin," and would carry on the dressing of cloth and carding business.[27] Harris's return to business was at least partly due to his health. Sciatica forced him to give up the strenuous occupations of farming and carpentering.[28] Also, he must by now have been convinced of the possibilities of the woolen industry. Harris was a shrewd businessman, and the

17 Cole, *op. cit.*, I, 97.

18 Leonard, 1855, p. 273; Bemis MSS, Box 6, XLVIII 230.

19 Cf. Cole, *op. cit.*, I, 87–93; V. S. Clark, *History of Manufactures in the United States*, I, 561.

20 Cole, *op. cit.*, I, 96–97.

21 *Ibid.*, 222–225.

22 Leonard, 1855, p. 273.

23 Dublin Town Records, Town Meetings and Tax Invoices, 1807–1827.

24 *NHS*, Dec. 14, 1805.

25 *Ibid.*, July 26, 1806.

26 *Ibid.*, June 19, 1813.

27 *Ibid.*

28 D. H. Hurd (ed.), *History of Cheshire and Sullivan Counties, New Hampshire*, p. 213.

prospects were promising. The interruption of trade during the Napoleonic Wars and the War of 1812 gave a great boost to the domestic woolen industry, and profits were high.[29] This combination of push and pull moved Bethuel Harris to commit himself to a new career, in which he was to become a pioneer in the manufacture of woolen goods in this country.[30]

Despite the scantiness of the evidence, it is possible to trace the growth of this small business built up by a purposeful Yankee. The first step was to meet the competition of nearby carders. The interesting combination of business sagacity and moral indignation, so characteristic of the Harris family, shows plainly in a notice in the *Sentinel*:

Carding at Four Cents

The owners of the neighboring carding machines, viz. those of Peterborough and Swanzey, have been in the habit of underworking the subscribers the present season, not because they can afford to card for less than six cents a pound, but mainly to draw custom from the vicinity of other machines.

Therefore the undersigned determined to commence *carding* on Monday the 9th inst. common Sheeps Wool for money in hand, at *Four cents the pound*, for three months—trust at five cents, and to take pay in labor, or produce, or trust a year at six cents the pound, and find oil for oiling at two cents the pound.

Packerfield, Aug. 2, 1813   B. Harris & Co.[31]

Bethuel again had Twitchell, Jr., for a partner when he reentered the business in 1813, but this lasted less than three years. In 1816, the partnership was dissolved "by mutual consent," and Bethuel was now in complete control of the enterprise.[32] This appears to have been an amicable separation, for the Twitchells then signed an agreement to "maintain a gate in our dam for the said Harris to draw water at all times for the purpose of carding wooling [sic] cloth in all its branches, and for no other use."[33] The next year Harris enlarged the building and added new machinery, probably another carding machine.[34] The business continued to be carding wool in the summer, dressing cloth in the autumn and winter, and dyeing "deep blues."[35] There are no records to indicate how much business was done or its source. However, this was the period of greatest growth for Nelson and Dublin, and the business must have come from the farmers of these and other nearby towns. Bethuel Harris advertised occasionally for a "journeyman clothier" or an apprentice,[36]

29 Cole, *op. cit.*, I, 229.

30 John Humphrey, *Water Power of Nubanisit Lake and River*, p. 7.

31 *NHS*, Aug. 6, 1813. The oil was applied to the wool before carding to reduce the static electricity.

32 *Ibid.*, April 20, 1816.

33 Cheshire County Records, Book 84, p. 268, in CMR.

34 Hurd, *op. cit.*, p. 213.

35 Hamilton Child, *Gazetteer of Cheshire County, New Hampshire* p. 178; *NHS*, Aug. 23, 1817.

36 *NHS*, Aug. 10, 1816; Oct. 30, 1819; Aug. 3, 1822.

but to a considerable extent he met the problem of a labor supply by "putting his sons, as fast as old enough, at work in that business."[37] Of his six sons, at least the three eldest, Cyrus, Milan, and Almon, must have worked in Bethuel's carding and fulling mill. Although the woolen industry was no longer growing in these years as it had before 1815,[38] the concern of "B. Harris and Company" was growing faster. It is a testimony to Bethuel's business ability that by 1822 the company was ready for a considerable expansion, the erection of a new mill.

The Middle Mill

The story of this first complete woolen mill in Harrisville can be told briefly. It was small, had an independent existence for only thirty-five years, and left behind little record of itself. Its significance lies in its being the original factory, the one from which the later factories sprang.

Bethuel and his eldest son Cyrus built the new mill on the site of the old shop of Jonas Clark. It was brick and of modest size, but nothing else is known about the building. Because of its position between the factories built later, it was always known as the "Middle Mill."[39]

When it began operation in 1823, it was far from being a complete woolen mill; Bethuel and Cyrus Harris simply resumed dressing cloth there. By the next year Bethuel's second son, Milan, had put in some power looms and so supplied the cloth for finishing.[40] That is all that is known definitely about the mill's machinery. Bethuel must still have had his carding machine from the earlier shop. For spinning the cardings into yarn, the progressive Harrises may have had an early spinning jack, an adaptation of the English cotton mule, partly power-driven, partly hand-driven. More likely, either they had a hand-driven spinning jenny, a small machine that was "in substance a combination of a number of spinning wheels,"[41] or else they depended on the spinning wheels of the farmers' wives and daughters in the neighborhood.[42]

Milan Harris's power looms are worth remarking. These water-driven machines were probably not more than two in number, of narrow width and simple construction.[43] Power looms had been patented in the previous decade, and by the early 1820s their use was spreading. The dates given by Cole indicate that Milan Harris, if not a pioneer in adopting them, was at least quite progressive in bringing these machines to the Middle Mill in 1823.[44]

37 Hurd, *op. cit.*, p. 213.

38 Cole, *op. cit.*, I, 229; Clark, *op. cit.*, I, 564.

39 Leonard, 1920, p. 573; Child, *op. cit.*, p. 176; Humphrey, *op. cit.*, p. 8.

40 Leonard, 1855, p. 273.

41 Cole, *op. cit.*, I, 107–111.

42 Clark says that despite earlier experiments the general introduction of power-spinning lagged through the decade ending with 1830. Clark, *op. cit.*, I, 565; Merrimack Valley Textile Museum, *Wool Technology and the Industrial Revolution: An Exhibit.*

43 Not a great deal is known about the early power looms, and it is unfortunate that these first ones in Harrisville were not described in detail by contemporaries. For an account of the early looms and what Milan Harris's looms might have looked like, see Cole, *op. cit.*, I, 120–126.

44 Cole, *op. cit.*, I, 123–125. For a comparison, the Slater Mill at Webster, Massachusetts, began to weave by power in 1825.

Shortly after its erection, Bethuel and his two sons formed a partnership and together operated the mill down to 1832.[45] For the time, they did quality work. At Cheshire County Fairs during these years the Harrises repeatedly won premiums in the field of "Domestic Manufactures" for the "Best fulled Cloth," or the "best piece of Cassimere."[46] This last was the best of the cloths manufactured by the new American woolen industry—cloths that were of medium quality, free from any considerable style influence, and subject to a large domestic demand. Cassimere was less finished than broadcloth and of looser texture. As made then, it was heavy weight and resembled overcoating.[47] With something of a reputation in the area, the Harrises could easily have disposed of their limited output, either directly from the mill "for individual customers in Harrisville" or through stores in towns such as Keene.[48]

Cole laments the scarcity of information on the labor supply of the pre-1830 woolen mills,[49] and such is the case with the Middle Mill in this period. It was a small concern, for even twenty years later the mill employed only seventeen workers.[50] For the time, however, the Harris mill may have been as large as the average woolen mill in the country.[51] Yet even a small work force was evidently difficult for the early woolen mill owners to recruit.[52] The few foreign-born artisans available, because of their valuable skills, had an importance in the growth of the industry far out of proportion to their numbers, but the great majority of the operatives had to be drawn from the native population. It was difficult and expensive to hire, train, and retain a work force from among the native population, and the problem was aggravated by the fact that the early woolen factories required an unusually large proportion of adult males in comparison with the proportion of women and children employed.[53] For the operators of the Middle Mill, the most difficult of these problems was probably, after training them, to retain these workers in the face of the westward and urban migrations which already, before 1830, had put the populations of Dublin and Nelson on the decline.[54]

There is a similar lack of information regarding the workers' wages and hours. Cole estimates that in 1828 wages for men in the industry averaged about $20.00 to $25.00 a month and for women and children about half that amount.[55] For the area of Harrisville in this period, the only figures are those of C. A. Bemis. He has some notes that are too detailed to be hearsay or recollection, and are probably based on some local source. His estimate of wages in the woolen industry in the "early nineteenth

45 Leonard, 1855, p. 273; Child, op. cit., p. 176.

46 NHS, Oct. 13, 1826; Oct. 12, 1827; Oct. 16, 1829.

47 Cole, op. cit., I, 155, 197.

48 Ibid., 254–255; NHS, July 2, 1868.

49 Cole, op. cit., I, 233–234.

50 Census, 1850, Vol. XIV, Dublin, Schedule 5.

51 Cole states "the number of workers per establishment according to the earliest figures available, those of 1840, was only fifteen." Cole, op. cit., I, 275.

52 Ibid., 234.

53 Ibid., 234–235, 237–238.

54 See Appendix 4.

55 Cole, op. cit., I, 240.

century" is much lower than Cole's: for men, 33-1/3 to 50 cents a day, and for women, 50 cents a week.[56] Much later, in 1850, the Middle Mill paid an average wage of $26.00 a month to male help and $14.28 a month to female help.[57]

To earn their wages, the workers spent long hours in the mills. Cole quotes some "representative replies" to the questions asked woolen mill operators by the Congressional Committee on Manufactures in 1828:

We expect to get sixty-eight hours per week, the year round, and this is all we insist on;

In the summer time we work twelve hours, over and above the time for meals, and we fall but little short of that in the Winter;

In the Summer we begin to work at about sun-rise, and continue until about sun-set, allowing half an hour for breakfast, and from one hour to one and a half hours for dinner. In the Winter we begin as soon as possible after light, and work until about nine o'clock at night, allowing about half an hour for meals.[58]

There is no record of the hours worked at the Middle Mill, but judging from the statistics of later years, there is no reason to believe that such replies could not have been written by Bethuel Harris.

The partnership was dissolved in 1832 when Milan Harris withdrew and put into operation a second mill of his own. This was to be an important juncture for the village.[59] Within a few years the Milan Harris mill was to become the principal one in the village down to the Civil War.

The ownership and management of the Middle Mill from 1832 to 1857 is clouded with confusing changes. According to the Dublin tax records, "B. Harris and Son" continued to run the mill after Milan's departure, until 1841. Then for four years it was listed as "Cyrus Harris and Co." Finally, from 1846 to 1858, it was listed as "Harris and Hutchinson."[60] Bethuel may well have retired in 1842 (he was then 72) though retaining an interest in the business. The succeeding "Cyrus Harris and Co." could have included Bethuel's sixth son, Charles Cotesworth Pinckney, and a son-in-law, Abner Hutchinson.[61] When Cyrus dropped out, these two men ran the mill as "Harris and Hutchinson" until its failure.[62] Behind this confusion of changing partnerships must lie a story of a family of ambitious, hard-working, and strong-minded men, each seeking to get ahead in a young and highly competitive industry.

56 Bemis MSS, Box 6, XLVIII, 238.

57 Census, 1850, Vol. XIV, Dublin, Schedule 5.

58 *State Papers, Finance*, V, 821, 819, 817, in Cole, *op. cit.*, I, 243.

59 It is curious that the two dates Cole says were the important milestones for the woolen industry in the nineteenth century, 1830 and 1870, were also milestones for this manufacturing village. In 1830, Milan Harris gave his family's name to the village, and started his mill, and 1870 saw Harrisville become a separate town. As for the industry, by 1830 the small factory had become the typical form of organization in the woolen industry; 1870 is deemed a transition year because of changes in the quality of goods, the disappearance of the household manufacture, a dispersion of woolen mills throughout the country, and a charge in tariff policy. Cole, *op. cit.*, I, 56, 265.

60 Dublin Town Records, Vol. 4, Town Meetings and Tax Invoices, 1828–1846; Tax Invoices, 1847–1862.

61 Child, *op. cit.*, p. 180.

62 Leonard, 1920, p. 573. The revised edition of the *History of Dublin* is almost certainly wrong in stating that the partnership of Harris and Hutchinson was formed "about 1835." The 1855 edition of the *History* makes no such statement, and it is difficult to reconcile with the tax records. The only evidence in those are two brief entries for the years 1833 and 1835.

The Middle Mill did not expand in this later period[63] but, with one set of machinery,[64] continued a modest output of plain woolens,[65] cassimeres, and doeskins.[66] Bethuel Harris died in 1851, and two years later Milan, the administrator of his estate, announced the sale at public auction of "the Woolen Mill . . . belonging to the estate of Bethuel Harris . . . together with the fixtures and machinery, dye house and fixtures, Store house, land and water privilege. . . ."[67] Whatever the issue of this auction, the mill continued to operate until the Panic of 1857 put it out of business.[68] In the absence of any direct evidence, it seems reasonable to conclude that, unlike its neighbors which survived the Panic, the Middle Mill failed because of its small and more antiquated plant and a less vigorous management. Whatever the causes, the first mill passed from the scene and was quietly swallowed up by those it had spawned.[69]

The Mill That Milan Built

The mill that Milan Harris built still stands. The bell in the cupola is silent, the windows are boarded over, and except for occasional bales of wool the darkened interior is empty. (Plate V) A century ago it was the most important mill in the village, and the man whose name it bore was Harrisville's leading citizen.

Milan Harris is the only one of the founding family to emerge from the records as an individual. He had traits characteristic of many successful businessmen of his time. He was strong-minded, acute, upright, and hard-working.[70] A close associate said of him, "As a business man and financier he had few equals for his time. Strictly honest and just so he had the confidence of all with whom he dealt."[71] This may all be true, but there is also evidence that he could be touchy, self-righteous, querulous, and petty. Nor is it too much to suspect that although he named the village for the Harris family, he came to think of it as a monument to Milan Harris.

The early experience of Milan Harris also resembled that of many men who became prominent in the last century. He was the second of Bethuel Harris's ten children who were born on the farm in Nelson. He received only a common school education and then worked alternately at his father's farm and woolen mill. At twenty-one he tried his hand at school teaching in Dublin and Nelson but evidently got his fill of that in one year. It is reported that "in 1821 he began the manufacture of woolen goods at Saxton's River, Vermont."[72] Whatever it amounted to, that venture was a brief one, for in the next year or so he was

63 Dublin Town Records, Vol. 4, Town Meetings and Tax Invoices, 1828–1846; Tax Invoices, 1847–1862.

64 Census, 1850, Vol. XIV, Dublin, Schedule 5. A "set" was a rough unit for measuring or stating the capacity of a woolen mill. Cole says that "a 'one-set mill' was one containing one set of woolen carding machines, with the complement of spinning machinery and looms that would be kept busy on the wool which carding machines would normally turn out." Cole, *op. cit.*, I, 200, note 2.

65 *NHS*, Aug. 30, 1843.

66 Leonard, 1920, pp. 573–574. For the statistics of Harris and Hutchinson in 1850, see Appendix 3. By 1855, the capital stock had doubled in value, but the capacity and output of the mill were virtually the same as in 1850. Cf. Edwin A. Charlton, *New Hampshire as it is. . . .*, p. 318.

67 *NHS*, May 27, 1853.

68 Leonard, 1920, p. 574.

69 There is little evidence of what later happened to the plant of the Middle Mill, but most of it was probably taken over by Milan Harris. The tax records show Milan Harris's "factory buildings and machinery" and "stock in trade" nearly doubling in value between 1858 and 1859. Also, his payroll book in 1861 mentions a "Mill No. 2." See Dublin Town Records, Tax Invoices, 1847–1862. The Middle Mill

back home putting power looms in his father's new mill. A half dozen years there, and Milan was ready to move again. He bought the old saw- and gristmill built by Abel Twitchell and ran that for a year.[73] Then the village's first shop came down, and work was started on a large new woolen mill. It was a propitious time for such a start, for the ups and downs that the industry suffered in the 1820s were over, and 1830 marked the beginning of seven years of prosperity.[74] Whatever the proportions of luck and acumen, Milan's early career showed promise of his emulating the success of old Bethuel.

The new mill was known locally as the "Upper Mill,"[75] and like the Middle Mill its management was a succession of partnerships that changed its name. Even incorporation, when it came, did not greatly affect this pattern. Unlike the Middle Mill, however, the new mill was to be dominated for forty years by one individual, despite the fact that it was owned by from two to five partners for most of its existence. Deficiency of capital handicapped growth in the woolen industry,[76] and such a deficiency would explain some of the partners that Milan took into the business with him. For help in building the mill, Milan turned to Henry Melville, a wealthy storekeeper in Nelson. He probably had little to do with planning or running the mill, but he held a one-third interest in it until his untimely death in an explosion in 1838.[77] At the same time, Milan also took into partnership his own brother-in-law, Abner S. Hutchinson, who had come to Nelson from Wilton, in 1829.[78] It is possible that Hutchinson also put up some capital; it is certain that he helped Milan build the mill in 1832 and 1833.[79]

Once formed, the partnership underwent numerous changes. Abner Hutchinson soon left to work in the Middle Mill.[80] His place was taken by Bethuel's third son, Almon. A year younger than Milan, Almon had left home at the age of twenty-one, in 1821. He returned home in 1835 and became a partner at the Upper Mill.[81] The next year the company was incorporated as "Milan Harris and Company," under the names of Milan and Almon Harris and Henry Melville.[82] When Henry Melville died, the brothers apparently assumed his share. They continued the business together until 1847, when Almon removed to Pennacook to begin his own mill there.[83] Milan must have bought out his brother, and for the next twelve years he had no partners. Then, near the end of this period he took into partnership his eldest son, Milan Walter.[84] He had been working in his father's mill,[85] the business was rapidly expanding at this time,[86] and this was an opportune time for Milan to bring his son (and

was taken down in 1866 to give place to Milan Harris's New Mill. Humphrey, *op. cit.*, p. 8.

70 For the civic virtues of Milan Harris, see ch. 3, *passim*.

71 Bemis MSS, Box I, LV, Letter from J. K. Russell to C. A. Bemis, April 28, 1908.

72 Child, *op. cit.*, p. 178.

73 Hurd, *op. cit.*, p. 214.

74 Clark, *op. cit.*, I, 569.

75 Leonard, 1855, p. 273.

76 Cole, *op. cit.*, I, 229.

77 *NHS*, April 10, 1839; April 5, 1838.

78 Child, *op. cit.*, p. 180.

79 The early ownership of the mill is uncertain. Hurd, *op. cit.*, p. 214, says that Melville was a partner of of Milan Harris; Leonard, 1855, p. 273, and Child, *op. cit.*, p. 176, say that Abner Hutchinson was a partner. None of the sources mention both men, but it is likely that all three were partners in the company.

80 Leonard, 1920, p. 573.

81 *Ibid.*

82 [N.H.] Register of Deeds, Book 130, p. 388, in Michael G. Hall, "Nelson, New Hampshire, 1780–1870," p. 111.

83 Leonard, 1920, pp. 573, 788; Hurd, *op. cit.*, p. 214. During this partnership, the company went under the name of "M. & A. Harris."

84 Hurd, *op. cit.*, p. 214.

85 Leonard, 1920, p. 787.

86 Dublin Town Records, Tax Invoices, 1847–1862.

presumably his successor) into the management of the firm. Such is the hazy record of the ownership of the Upper Mill. Two particulars stand out. The changing partnerships were most likely due to the need for capital and to personal ambitions and personal conflicts. And throughout the whole period Milan Harris was the man who counted and the man who kept the mill running without interruption.[87]

The new mill was built of brick brought from a yard in Nelson several miles distant.[88] When it was finished, Milan had a building well placed, well constructed, and of good proportions. It stood at right angles to the Dublin–Nelson road and straddled the stream which there ran closely parallel to the road. The first two floors were well lighted by sizable small-paned windows, originally six on a side in addition to those on the ends. The third floor under the eaves was lighted by a low roof window, situated in the middle of the sloping roof and running nearly the length of the building. The mill originally had no tower at the front. The small cupola housing the bell stood on the ridge pole at the front end of the building.[89] The stairs now in the tower must have been cut through the floors. There were three doors on the front of the mill, one for each floor. Those on the second and third floors were used for raising and lowering bulky goods with the aid of a hoist. The builder added a touch of elegance by flanking the third floor door with lights and surmounting it with a rounded arch. There is reason to believe that as a factory building the Upper Mill would have rated high. In a comparison with the Lowell cotton mills of the late forties and fifties, the mill of Milan Harris must have been far superior in sanitation, safety, and comfort.[90]

Though there is no description of the Upper Mill's power system, some idea of it can be given from what is known. Brackshin Pond being so nearby, it was conveniently used as a millpond as well as for storage.[91] A short canal carried the water from the pond to Twitchell's milldam at the head of the ravine. (Plate VI) Milan may have used Twitchell's old waterwheel, but more likely he built a new and larger one. This would probably have been a wooden pitchback wheel, a type commonly used by American manufacturers until about 1840. This type turned inward toward the fall, the water striking just short of the highest point. The impact or kinetic energy of the stream was not utilized, the power being produced almost entirely by the simple weight of water in the buckets. Such a wheel utilized about three quarters of the power applied to it, and this was transmitted from the wheel to the machinery by a system of belts.[92]

87 Leonard, 1855, p. 273.

88 NHS, Aug. 30, 1843. Bemis describes the source of brick as "the Nathaniel Woods brickyard on the farm beyond the A. C. Tolman farm." Bemis MSS, Box 6, XLIII, 14.

89 See Plate I, Frontispiece. There are signs in the masonry to suggest that the original building was shorter than now and was lengthened by one bay when the tower was added.

90 Edith Abbott, *Women in Industry*, p. 126. The Upper Mill was probably not overcrowded, poorly ventilated, or poorly lighted, at least in daylight. The *Sentinel* had no mention of any fatal accidents in the mills of Harrisville in this period (though it recounted many accidents elsewhere); and only one accident, involving a loss of an arm, has been discovered, that in *one* of the Harrisville mills in 1838. (Bemis MSS, Box 3, LV, 8–9.) Neither do there appear to have been any serious fires.

91 Humphrey, *op. cit.*, p. 2.

92 Clark, *op. cit.*, I, 406.

According to a local authority on the water system in Harrisville, despite its great potential the stream at that time proved insufficient to afford a constant supply of water for the machinery Milan had put into his new mill. So, to make the supply more dependable, in 1836 he bought certain water rights from the owner of a sawmill located at the outlet of Long Pond, the one next above Brackshin Pond in the chain that supplied Goose Brook. This agreement gave Milan and others a right to draw as required for use of their mills and restricted the sawmill's use of water after the pond had been reduced to a certain height above the outlet channel.[93] Unfortunately, the agreement did not restrict the water's use early enough to provide for protracted seasons of little rainfall. These at times hampered operations at the mills, and in fact water shortages and rights were to be a recurrent difficulty and source of friction among the mill owners.

Like its structure, the operations of the Upper Mill can be better described than those of its predecessors. The manufacturing processes of a woolen mill require the close attention of its management,[94] and this Milan Harris knew. In 1843, the *Sentinel* said of the owners of the Harris mills, "The secret of their success in business is just no secret at all. They are their own designers, their own clerks, their own agents, and see personally to every minutiae of their concerns."[95]

By such close and able supervision Milan Harris and Company was able to weather the Panics of 1837 and 1857,[96] to buy wool at the best prices,[97] to expand plant and production at the right times, and to turn out quality goods. By the time the company was incorporated, Milan Harris had installed two sets of machinery which, with the one set of the Middle Mill, gave the village as large a concentration as existed in any town in the state except one at that time.[98] By the end of the 1850s, he had built a dye house just below his mill, erected two brick storehouses, put up a large brick boardinghouse for his workers, and acquired the plant of the Middle Mill.[99] (Plate VII) According to the tax records and the census schedules, most of this growth came in the 1850s. In that decade the invested capital increased more than fivefold, the machinery and labor force doubled, and production more than doubled.[100] According to Cole the 1850s was a decade of "small total expansion" and "relative backwardness" for the woolen industry, so the growth of Milan Harris and Company is that much more impressive.[101]

As the capacity of the mill increased, the quality of the woolens produced improved. There is no evidence that the company ever

93 Cheshire County Records, Book 130, p. 338, in Humphrey, *op. cit.*, p. 9.

94 Cole, *op. cit.*, I, 272.

95 *NHS*, Aug. 30, 1843.

96 Bemis MSS, Box I, LV, Letter from J. K. Russell to C. A. Bemis, April 28, 1908.

97 *NHS*, Aug. 30, 1843.

98 C. Benton and Samuel Barry, *A Statistical View. . . .*, p. 110. Gilsum had four sets of machinery.

99 *Map of Cheshire County, N.H.* (1858); also note 69.

100 Census, 1850, Vol. XIV, Dublin, Schedule 5; Census, 1860, Nelson, p. 6, Schedule 5. See Appendix 3. The total value of Milan Harris & Co. was appraised at $7,100 in 1851, and at $19,800 in 1861. See Dublin Town Records, Tax Invoices, 1847–1862.

101 Cole. *op. cit.*, I, 269.

produced either the cheapest grades of cloth, such as satinets, linseys, and negro cloths, or the finest goods, such as broadcloth. Instead, production was keyed to the growing national demand for medium grade woolens. As Cole states it, "the country was growing richer, the market broader, and the demand more insistent for semi-fine cloths."[102] For some years "common cassimeres" were still produced,[103] but new fabrics were produced that again showed Milan Harris's business acumen.

In 1840, the Middlesex Mills in Lowell made one of the notable American achievements in the woolen industry by producing fancy cassimere, a cloth of French origin, on a new power loom invented for the weaving of this cloth by William Crompton. The fancy cassimere's distinguishing characteristic was its wide range of possible patterns. In this country it became the principal competitor of broadcloth.[104] Within three years of this innovation Milan Harris had a Crompton loom and was also producing fancy cassimeres.[105] Nearly forty years later the man then running the Upper Mill wrote to the son of William Crompton,

I have made inquiries . . . about the loom. . . . Mr. [Fred] Harris said he remembered of your father coming up here with a horse and sleigh drove up all the way from your city Worcester, Mass. to set the loom up and started it he was then quite small boy and said it was the first Crompton loom started in New Hampshire. . . .[106]

Milan Harris was undoubtedly still the progressive manufacturer.

In the 1850s, the mill turned to production of black doeskin, a firm woolen cloth with a smooth, soft surface which was used for men's wear. It was by all accounts of high quality,[107] and those samples shown at an agricultural fair in 1854 the *Sentinel* called "splendid specimens of home manufacture" and "especially beautiful."[108] Prices are seldom mentioned in any accounts of the wool manufacture, but these Harris doeskins of 1852 were described as selling at one dollar per yard, a price that was said to outsell "nearly every other mill in the United States, and even the majority of the German doeskins."[109]

How Milan Harris & Company sold its goods is something of a mystery. At least as late as 1843 the company did not have an agent. Its woolens may have been sold through stores in Keene, but there is no mention of Milan Harris's woolens in the advertisements of these stores in the *Sentinel*, and with a good reputation there was no need to hide the manufacturer's name. Some woolens may have been sold at retail from the mill but hardly the bulk of its production. The answer is probably some

102 *Ibid.*, 323.

103 Nelson Town Records, XII, deed 1843, in Hall, *op. cit.*, p. 120; *NHS*, Oct. 23, 1839.

104 Cole, *op. cit.*, I, 305–308.

105 *NHS*, Aug. 30, 1843; Nelson Town Records, XII, deed 1843, in Hall, *op. cit.*, p. 120. The first Crompton looms were narrow, about forty inches across. Crompton's Company introduced an improved broad loom about 1857, but Milan Harris & Co. retained their narrow looms. In 1880, the mill bought broad looms and offered for sale "43 narrow Crompton looms." Cole, *op. cit.*, I, 313–314; CMR, C. & W. Copybook "C," p. 34, Letter from Craven & Willard to George Russell, Jan. 22, 1880.

106 CMR, C. & W. Copybook "D," p. 403, Letter from Michael Craven to George Crompton, March 8, 1881.

107 Leonard, 1855, p. 274; Charlton, *op. cit.*, p. 318.

108 *NHS*, Oct. 13, 1854.

109 Leonard, 1855, p. 274.

combination of these possibilities, with a steadily decreasing proportion of the woolens being sold directly from the mill.

It is interesting to note that, progressive as it was, the Upper Mill did not abruptly abandon the "custom work" of the earlier and more primitive industry. In the summer of 1837, Milan Harris and Company advertised that they would "Dress Cloth for customers the present season."[110] They were still providing the same service for the home manufacture that Jonas Clark had done forty years before. Hall also cites an example of the mill doing commission work, or working on shares:

John Hale . . . was a peddler who travelled about the country with a wagon load of goods of all description. He brought eight hundred pounds of wool to Milan and Almon Harris's mill, where it was spun and woven into three hundred yards of "fancy and common cassimeres." The mill kept part of the cloth as payment for the work done, and John Hale took the remainder.[111]

The continuing of such practices, Cole notices, was characteristic of concerns located in the country districts, and of depression years, and both of these circumstances applied here.[112]

Information on the workers, their wages and conditions of labor remains spotty, but there are some items of interest concerning the Upper Mill. The census schedules of 1850 and 1860 give the average number of hands employed by the company and their cost. The labor force at the Upper Mill grew from fifteen men and nine women to thirty men and thirty women during that decade of expansion. The "monthly cost" of the labor force in these years, figured for the average, was as follows:[113]

Year	Male workers	Female workers
1850	$27.73	$16.66
1860	$18.33	$16.66

The figure for the male worker in 1860 is unaccountable. It is out of line with the rate paid by the other mill in the village,[114] and there is no evidence that Milan Harris was hiring an exceptional number of either young or foreign-born male workers. Mistakes occurred in these census returns, and one can only conclude that a mistake was made in this instance.

Valuable as some of these statistics are, two letters written by a J. K. Russell, of Massilon, Ohio, in 1908 tell more about wages and working conditions at the Milan Harris mill, and they tell it in a refreshingly firsthand and convincing manner. Russell was

110 *NHS*, Aug. 24, 1837.

111 Nelson Town Records, XII, deed 1843, in Hall, *op. cit.*, p. 120.

112 Cole, *op. cit.*, I, 223.

113 Census, 1850, Vol. XIV, Dublin, Schedule 5; Census, 1860, Nelson, p. 6, Schedule 5. See Appendix 3.

114 See Appendix 3.

nearly eighty-five years old when he wrote these letters, and he was recalling events of more than sixty years before, but his memory was as clear as his fine, firm handwriting.

In the first letter Mr. Russell says that he knew Milan Harris from 1845 until the latter's death in 1884, that he was "an intimate acquaintance for almost forty years," and that he was for fifteen years "in his employ as a trusted servant."[115] The second letter is worth quoting at length:

On the 15th of September, 1846 I entered into an agreement with Mr. [Milan] Harris for a term of three years, to learn manufacturing. . . . My special trade in the mill was that of Weaver. My pay was for the first year $150, 2nd $300, 3rd $450. That included board. The third year I took the manufacturing of the flannel by the job, that is I took the wool at the Picker and delivered it ready for the fulling and finishing at 7cts. per yard. I furnished only the labor. I did considerably better than under my former agreement. This arrangement continued but one year on acct. of hard times and reducing his production. But I continued the management as though I was doing it by the job until 1858. This arrangement seemed to give good satisfaction to all parties. It relieved Mr. Harris of responsibility, except for the pay, and me from any appeal from the help, as they knew no other boss. They respected me more and I guess I treated them better. I know we got along finely. . . . I was born June 26, 1823 in Alstead, N.H. . . .

If you wish to know why I left Harrisville I will say it was not because I disliked the business for myself I did not choose it for my children and I wished to leave them in a better country, which I had had a taste of. Mr. Harris often expressed the wish that I would buy an interest in the factory. But we were most [sic] too much mixed up in family relationship My sister being my wife's step mother; there possibly might arise some jealousy, and I saw some things that did not trouble me as a hired man that might as a partner, and I prized their friendship more than money. . . .[116]

Housing for workers in such a small mill village was a problem that was met in various ways. In his later years in Harrisville, Mr. Russell lived in a large, attractive frame house that stood by itself on the edge of the village.[117] The other workers must have lived in the village or its vicinity. Some boarded with overseers of the mill or perhaps with Milan Harris himself. One "L. K. Hatch, M.D." had ten people besides his own family living with him in 1850, and some of these were mill workers.[118]

About 1852, Milan Harris built a large (really a double) brick house that served as a boardinghouse.[119] It overlooked the mill and the stream and was in turn overlooked by the home of Milan Harris and presumably by the watchful eye of its owner.[120]

115 Bemis MSS, Box 1, LV, Letter from J. K. Russell to C. A. Bemis, April 28, 1908. Russell was apparently writing in reply to a letter from Bemis. He promised Bemis a biographical sketch of Milan Harris, but, if he sent it, no such sketch has come to light.

116 Bemis MSS, Box 3, LV, Letter from J. K. Russell to C. A. Bemis, June 9, 1908.

117 *Map of Cheshire County, N.H.* (1858).

118 Census, 1850, Vol. XV, Nelson, Schedule 1; Vol. XIV, Dublin, Schedule 1, House # 735.

119 Dublin Town Records, Tax Invoices, 1847–1862; *Map of Cheshire County, N.H.* (1858).

120 *Map of Cheshire County, N.H.* (1858).

In 1860, twenty-eight people, nearly half the number of workers at the Upper Mill, lived in this boardinghouse. The census of that year gives their names, ages, occupations, and birthplaces. The workers were young—rural youths who had already had enough of New England farming.[121] Except for the housekeeper, only one was older than twenty-nine, the median age was twenty-one, and the youngest was a boy of fourteen.[122] The sex ratio was about even. The men and women were listed separately, just as apart as they must have been kept in the quarters. Some of the workers, perhaps a quarter, had names long familiar in the neighborhood. They were still native-born in overwhelming proportions—only two of them came from England and Ireland—and a majority of them were born in New Hampshire.[123] Thus, the operatives at the Upper Mill in 1860 must have been almost entirely native-born, for immigrants working there would most likely have lived in the boardinghouse. By contrast, half the operatives in the Lowell mills were Irish-born even in the early 1850s.[124] The immigrant generally took longer to reach the hinterland, but it may also have been that Milan Harris preferred to hire the native-born. He was a man with strong views on that sort of thing.

As the 1860s began, Milan Harris might well have congratulated himself on the woolen industry he had built up in Harrisville. However, despite his position there was now a new mill expanding along the banks of Goose Brook, one that would soon outstrip him and one that would run long after the Upper Mill fell dark and silent.

The Cheshire Mills

Asa Greenwood was one of the best stonemasons in New England. He had built fine granite buildings in Marlborough and stone bridges in Keene and Peterborough.[125] Then in 1846 he set to work on something different, a granite factory building for Cyrus Harris of Harrisville, the eldest son of old Bethuel Harris.[126] At night that summer ox teams straining with the large granite blocks could be heard coming up the eight-mile winding dirt road from the quarry in Marlborough to Harrisville.[127] To avoid the heat, a driver and team would leave Harrisville in late afternoon, pick up the blocks that had been made ready and, if nothing broke down, arrive back at the construction site sometime before work began the next morning.[128] Greenwood had a special interest in this job for he was also part owner of the mill he was building.[129] Thus, he wasted no time and built it well.

121 Cf James D. Squires, *The Granite State of the United States*, I, 289.

122 The absence of child workers here was not remarkable. The proportion of children in the woolen industry had never been large and declined further in the period after 1830. Cf. Cole, *op. cit.*, I, 371–372.

123 Census, 1860, Dublin, 65, Schedule 1.

124 Abbott, *op. cit.*, p. 139.

125 Leonard, 1920, p. 595.

126 *Ibid.*, p. 574.

127 Leonard, 1855, p. 274.

128 Interview with Arvo Luoma, August 1960. He had this account from a man whose father had done the hauling.

129 Cheshire County Records, Book 170, p. 357, in CMR.

Cyrus Harris, superintending the construction,[130] must have answered many question about the mill's appearance, for it was unlike that of the factories of Massachusetts and northern New England. The four-story building had fine large windows on all sides, roof windows running its length, a slate roof, and, instead of the usual wooden ones, solid granite eaves. At the front of the building was a square granite tower which housed the stairs and supported the cupola.[131] (Plate VIII) Architecturally, the mill resembled those found in Rhode Island and their English proto-types. In fact, it has been called "the purest form of old English industrial architecture north of Rhode Island."[132]

The granite mill was also an innovation in a village of brick and wood and underscored Cyrus Harris's new beginning in the woolen business with a new building, a new company, and new partners. The "Harrisville Manufacturing Company" was incorporated by act of the New Hampshire Legislature in 1847 under the names of Cyrus Harris, Robert Appleton, "and their associates. . . ."[133] Strangely, there had been an almost identical act chartering the same company in 1836.[134] Was this earlier company stillborn because of the depression that hit the country the next year? There can be no accounting with certainty for this anomaly, but the reason for Cyrus Harris incorporating in 1847 is clear. The company was capitalized at a maximum of $200,000, twice the capitalization of Milan Harris's mill in 1861.[135] To launch such a large enterprise called for more capital than Cyrus Harris could raise except by incorporating and securing outside investors. It even called for taking in the builder of the factory. The times were certainly more propitious for starting a new mill than they had been in 1836. The country was at war with Mexico, and the woolen industry was growing.[136] However, Cyrus Harris was not in luck. The dread scourge of consumption, which claimed so many lives in town, was on him. His health failed, and he died in April 1848.[137]

With Cyrus Harris dead, the new venture faltered. The mill stood completed but empty. The remaining members of the Company held their first meeting in November of 1849, at the home of Cyrus Harris's widow, and occasional meetings during that winter. They appointed committees to raise capital and to plan the "machining up" of the mill, but now there was no Harris in charge.[138] Furthermore, the prosperous times seem to have faded.[139] In March 1850, the mill, land, and water rights were sold to Faulkner and Colony, woolen manufacturers in Keene.[140] Why did not the remaining Harrises try to retain control of the mill? Perhaps they did try, or perhaps they were

130 Leonard, 1855, p. 274.

131 Florence C. Maxwell, "Most Paintable Mill Town," *The Christian Science Monitor*, Magazine Section, July 27, 1946. The building was 111 feet long and 40 feet wide. Leonard, 1855, p. 274.

132 *New Hampshire Sunday News*, Feb. 8, 1953, from Professor William Pierson of Williams College; *Becco Echo*, Vol. 4, no. 2, June 1953.

133 *Laws of New Hampshire, 1847*, Ch. 563.

134 *Ibid.*, 1836, Ch. LIV.

135 *New Hampshire Laws, 1861*, Ch. 2542.

136 Cole, *op. cit.*, I, 269; Clark, *op. cit.*, I, 571.

137 Albert Hutchinson, *A Genealogy and Ancestral Line of Bethuel Harris. . . .*, p. 17; Leonard, 1855, p. 274.

138 CMR, Minutes of Meetings of the Harrisville Manufacturing Co., Nov. 9, 1849; Dec. 13, 1849.

139 See above, p. 27.

140 Cheshire County Records, Book 170, p. 357, in CMR.

in no position to assume such a large undertaking. By 1850, the Harrises had built up their manufacturing village to a point where they could no longer exercise exclusive control.

The Colony family, destined to play an important part in the reduction of the Harrises' control of the village, had long lived in Keene. The American progenitor of the family, John Colony, had come to this country from Kilkenny, Ireland, about 1732. Here he served with Roger's Rangers in the French and Indian War, attained the rank of captain in that famous battalion, and was rewarded for his services by a grant of land in Saxton's River, Vermont. For some reason he traded this grant with another Ranger for one in Keene, and there the first John Colony settled. His grandson, Josiah, entered the woolen business in 1815, in partnership with one Charles S. Faulkner, and the company was soon both well known and well established.[141]

Under the aegis of Faulkner and Colony, a new company for the empty granite mill was organized and chartered in June 1850 under the name of the "Cheshire Mills." The company was also known locally as the "Lower Mills."[142] Initially the owners were old Josiah Colony, his partner Charles S. Faulkner, two of Josiah's sons, Timothy and Henry, and his son-in-law, Joseph W. Briggs, all of Keene.[143] In December of that year they met in Josiah Colony's "Counting Room" in Keene. At this first "annual meeting" they chose three directors, a president, agent, and treasurer. They also voted that the capital stock should consist of forty shares of $1,000 each, "to be paid by assessments as the Directors shall authorize."[144] The next year and a half was spent completing the plant and "machining up" the mill. When it was finished, Briggs, Faulkner, and Josiah Colony relinquished or sold their interests to two more of Josiah's sons, Alfred and John E.[145] The younger Colonys, each owning ten shares, took control, with Henry Colony serving as President.[146] Josiah Colony was as prolific a father as Bethuel Harris, and his progeny were at least as good businessmen. The Harrises have died off or scattered; the descendants of Josiah Colony still own the mills.

Milan Harris and the people in the village must have seen quickly that the new owners of the mill were in earnest, for the Colonys set about putting in four sets of machinery and twenty-four looms, doubling the capacity of the town's mills.[147] They built a dye house, wood shed, dry room, boiler room, brick storehouse, and a large brick boardinghouse. They contracted to have the Davis and Furber Machine Company in North Andover build and install the card machines and looms. They bought bobbins and spools from a manufacturer in Winchendon.

141 Interview with John J. Colony, Jr., Sept. 21, 1961.

142 Leonard, 1920, p. 574.

143 *Laws of New Hampshire, 1850*, Ch. 1057.

144 CMR, "Records of Cheshire Mills," Dec. 9, 1850. Josiah Colony, Henry Colony, and J. W. Briggs were chosen Directors. The officers were Josiah Colony, President, Timothy Colony, Agent, and J. W. Briggs, Treasurer and Clerk.

145 *NHS*, Aug. 6, 1884, "W. S. B."; Child, *op. cit.*, p. 176.

146 Cheshire County Records, Book 170, p. 357; Book 167, pp. 33, 35, 32, 31, 316, in CMR; *KES*, Aug. 1, 1950, p. 5; CMR, "Records of Cheshire Mills."

147 Leonard, 1855, p. 274.

If the mill architecture came from Rhode Island, the machinery and works came from the advanced technology of Massachusetts.[148]

For their power system the Colonys installed a water turbine. Turbines were originally water motors that revolved in a horizontal plane, the axis being vertical. They were based on the principle that "when water issues from a vessel, there is a reaction on the vessel tending to cause motion in a direction opposite to that of the jet."[149] Turbines had long been studied experimentally in this country and had been brought to greater perfection in France, especially by an inventor named Fourneyron. The first turbine used practically in New England was built by George Kilburn of Fall River, Massachusetts, in 1843.[150] Kilburn based his design on French models, but other models soon appeared, and the turbine was widely adopted to replace the pitchback waterwheel. Even in the early years of their development, turbines were quite as efficient as the pitchback type. In addition, they were more cheaply installed, occupied less space, turned with greater velocity, and were less impeded with backwater.[151]

Just eight years after Kilburn built his first turbine, the Colonys purchased their machine from his firm, now Kilburn and Lincoln. It was a forty-eight-inch Fourneyron turbine.[152] In this type "the water is directed downward along the central axis, or shaft, of the wheel and, turned outward by fixed, curved guide vanes, was discharged outward through the curved blades of the wheel."[153]

Unfortunately, "as this wheel was about three times the requisite capacity for driving the mill and a very wasteful wheel when used at much less than full gate, the demand for water far exceeded the ordinary supply. . . ." This not only interfered with the proper operation of the mill but caused "some serious unpleasantness" between the Colonys and Milan Harris respecting their right to draw water from his pond when the Milan Harris mill did not furnish sufficient water. Milan knew his rights and stood by them, locking down his water gate when necessary. He won his point, and the trouble was largely overcome by the Colonys modifying their wheel and then building a dam at Spoonwood Pond to increase their controllable reservoir capacity.[154] By the end of 1851, the "machining up" of the Lower Mill was finished, and soon after that the mill was in operation.

In its correspondence and cash books there is a wealth of information about the Cheshire Mills' organization, business arrangements, and plant during these early years. Although

148 CMR, Letter book "A," Letters Dec. 6–Dec. 15, 1851, *passim.*

149 "Hydraulics," *Encyclopaedia Britannica*, 11th ed., XIV, 97.

150 Clark, *op. cit.*, I, 407.

151 *Ibid.*

152 Humphrey, *op. cit.*, p. 12.

153 James Kip Finch, *The Story of Engineering*, p. 350.

154 Humphrey, *op. cit.*, pp. 12–13.

incorporated from the beginning, the company was a family concern. The four brothers, Timothy, Henry, John E., and Alfred, managed the mill[155] from both Harrisville and Keene where they kept an office. Henry and Alfred T. lived in Harrisville[156] and handled that end of the business.[157] John E. also lived in Harrisville.[158] Timothy lived in Keene and worked there part of the time. He had succeeded J. W. Briggs as treasurer, probably the most responsible position in the company.[159]

Working closely with the Colonys in the running of the mill was Faulkner, Kimball, and Company, commission agents. They were located in Boston, the wool center of the country. Following a growing practice in the industry, this commission house handled the entire output of the Cheshire Mills.[160] Far from being a parasitic middleman, Faulkner, Kimball and Company not only sold the mill's woolens but also rendered the manufacturer invaluable aid and advice on a variety of problems. They advanced funds to the Colonys for the purchase of wool.[161] They gave counsel on the new art of corporate management. Thus, concerning a note they were settling, they pointed out to the Colonys that "it should be signed by the Treasurer, or some other authorized person for the company—It is usual with corporations to appoint a Treasr. who is the only authorized person for signing the name of the corporation."[162] They went to great trouble to help the Colonys buy things they needed, even securing them a second-hand mill for grinding indigo.[163] They gave advice on market conditions and hiring. Sometimes their advice was taken and sometimes it was not, but the relationship between the mill and the commission house seems to have always been mutually cordial, understanding, and helpful.

Because the woolen industry was a highly seasonal business, the financial operations of the Cheshire Mills depended heavily on the commission agents for long-term credit. Short-term loans—perhaps up to three months—were secured from nearby banks in Keene, Bellows Falls, and Winchester.[164] However, wool purchases were a large expense requiring considerable time to meet, longer than the banks were sometimes willing to grant.[165] Alternately, Faulkner, Kimball, and Company would advance the Cheshire Mills an "acceptance," engaging to pay a draft for a certain amount on their firm in six months. This note was secured by Cheshire Mills goods consigned to their house but as yet unsold. When the Lower Mill was getting under way, the commission house even allowed the mill to draw beyond the value of their consignments, unprofitable as this was for them.[166] The Cheshire Mills would send such an acceptance to one of

155 *KES*, Aug. 1, 1950, p. 5.

156 Dublin Town Records, Tax Invoices, 1847–1862; *Map of Cheshire County, N.H.* (1858); *NHS*, Sept. 21, 1855.

157 CMR, Cash book "A."

158 Dublin Town Records, Tax Invoices, 1847–1862; Census, 1860, Dublin, p. 67, Schedule 1.

159 CMR, Copybook "A," p. 10.

160 Cole, *op. cit.*, I, 290; CMR, Copybook "A," p. 56. The only apparent exception to this arrangement was that the Colonys occasionally sold some of their flannel through stores in Keene in which they had an interest. See *NHS*, Sept. 10. 1852; Sept. 10, 1853.

161 CMR, Copybook "A," pp. 13, 16.

162 CMR, Letter book "A," Oct. 10, 1851.

163 *Ibid.*, Feb. 19, 1852.

164 *Ibid.*, Jan. 15, 1852; March 23, 1852.

165 *Ibid.*, March 23, 1852.

166 *Ibid.*, Dec. 11, 1851.

its agents purchasing wool, and he would get it discounted at some bank in whose vicinity he was buying wool.[167]

During the 1850s, the plant of the Cheshire Mills steadily expanded. When they had the opportunity, the Colony brothers bought pieces of land surrounding their mill.[168] They bought a machine shop and a handsome brick blacksmith shop that stood on the bank of the stream just below the mill.[169] In 1855, they helped to ensure an adequate water supply for the Harrisville mills by purchasing the mill property and entire water rights of the millowners on Long Pond.[170] In 1857, they bought from Abner Hutchinson his half interest in the land belonging to the Middle Mill.[171] In 1860, they even owned the local store.[172] Then, as the 1850s drew to a close, they began construction on a new mill.[173]

This structure was built of brick and stood at right angles to and adjoining the granite mill. Its three floors were well lighted with large windows. The loft did not have the old English features of the earlier mills in the town, but the building did have its touches of elegance. The windows were small-paned and had sills and lintels of granite. The cornice and eaves were elaborately stepped, and under the eaves at each end of the building were three fancy lancet windows. By 1860, the machinery was being installed in this building.[174] (Plate IX)

It was probably about this time that another small building was added. This was the picker house, which held machinery designed to loosen the matted locks of wool in preparation for carding.[175] Similar in design to the new brick mill, it stood on the north bank of the stream adjoining the rear corner of the granite mill—whose fine design and proportions were by now nearly as obscured by wings as Bulfinch's State House in Boston. As in the case of the Upper Mill, the expansion of the Cheshire Mills during this decade of stagnation in the woolen industry was testimony to its vitality.[176]

It was incumbent on the Colonys to protect their expanding mill with adequate fire insurance. With help from their commission house, the Colonys shopped around for the best rates[177] and finally divided the amount to be insured among several companies. Between 1851 and 1860, the Cheshire Mills had policies with a dozen different companies, ranging from Aetna to the "Liverpool and London Insurance Co.," and in the latter year had a coverage of $25,000 on the buildings and their contents.[178] Soon after the mill began to operate, a pump was installed that could be used in case of fire.[179] Because of construction, arrangements, and the pump, the Cheshire Mills were regarded as a good risk by the insurance companies.

167 CMR, Copybook "A," p. 60.

168 Cheshire County Records, Book 210, p. 538; Book 187, p. 259; Book 179, p. 360; Book 188, p. 421; Book 190, p. 621; Book 170, p. 212; in CMR.

169 Cheshire County Records, Book 188, p. 420; Book 198, p. 60; in CMR.

170 Humphrey, op. cit., p. 10.

171 Cheshire County Records, Book 187, p. 603, in CMR.

172 CMR, "April Letters, 1860," Homer S. Lathe to Mr. Colony, April 2, 1860.

173 Child, op. cit., p. 176. The building was 75′ long by 42′ wide.

174 CMR, "Letters, 1860," passim.

175 Cole, op. cit., I, 93; II, 287.

176 In 1852, the first year of operation, the Cheshire Mills were appraised at a total value of $20,000. In 1861, after the brick mill was completed, the total value was $35,000. See Dublin Town Records, Tax Invoices, 1847–1862.

177 CMR, Letter book "A," Letters of Jan. 15, Feb. 4, and Feb. 11, 1852.

178 Cash book "A," passim; Letters of March 5, March 17, in packet marked "March Letters, 1860"; Letters of Sept. 10, Dec. 5, 1860, in packet marked "Misc. letters 1860."

179 CMR, Letter book "A," H. Snow to J. W. Briggs, Feb. 11, 1852; J. C. Hartshorn to J. W. Briggs, Feb. 16, 1852.

The records of the Cheshire Mills surviving from the 1850s also cast considerable light on its business operations, that is, the purchasing of wool and the production and marketing of woolen fabrics.

Wool was purchased in several different ways and in a number of different places. In these years the hillsides of New Hampshire were covered with flocks of sheep,[180] and the Colonys frequently bought wool from the local farmers.[181] At least in the mill's early days, Faulkner, Kimball, and Company bought wool for the Colonys.[182] Soon the mill had agents who traveled about through New Hampshire and Vermont, buying wool from the farmers.[183] Sometimes the Colonys bought wool in the "western" markets, such as Utica, New York, or Plattsburg and Mansfield, Ohio.[184] There is no evidence that the mill was in these years buying wool in any more distant markets, or abroad.

The agents were supplied with either cash or acceptances to pay for the wool they bought. The agent received a commission and also a fee of one cent a pound for sacking and delivering the wool on board railroad cars for shipment to Keene.[185] Money was tight, and there was a concern to see that the money paid out went to the farmers, as a letter to an agent indicates:

James Lovell, Esq.

Dear Sir

Enclosed I send you by "Express" two thousand dollars *in cash* I want you to pay it out to the *wool growers personally* and use your best endeavors to give it "*a good circulation*" and *not let the banks* get it as I told the Bank folks that I would try and give it a good circulation. . . .

Yours Respy.

Timothy Colony Treas.[186]

Buying wool through a third party in these days of *caveat emptor* was a considerable act of faith. Timothy Colony could only set the maximum price to be paid for what he wanted and then hope for the best. (Curiously, as late as 1858 these prices were sometimes reckoned in shillings.)[187] His frequent letters to these agents are full of anxious, if not very helpful, warnings:

All I can say is this: if others pay as high as 40 cts. [per pound] for "*choice, fine, light, and clean*" wool that "*you must do the same*" "*use much caution*" and buy "*as low as possible*" if you go to Lebanon and Lyme buy some if possible. all I can say Gentlemen is this "*use judgment*" let me hear from you often[.][188]

180 See ch. 5.

181 CMR, Cash book "A," List of purchases beginning April 24, 1855; Letter book "A," Edward M. Casey to Josiah Colony, Feb. 24, 1852; Letter from Henry to Timothy Colony, June 23, 1860, in packet marked "June Letters, 1860."

182 CMR, Letter book "A," Nov. 6, 1851; Dec. 8, 1851; Dec. 10, 1851; Jan. Jan. 11, 1852.

183 CMR, Letter book "A," S. A. Lillie to Messrs. Colony & Sons, Oct. 29, 1851; Nov. 17, 1851; Copybook "A," p. 10, T. Colony to Messrs. Perkins and Standish, July 3, 1855.

184 CMR, Copybook "A," pp. 32, 38, 42: letters of Aug. 4, Sept. 9, and Nov. 18, 1856.

185 *Ibid.*, pp. 53–54: Letter to P. T. Washburn, April 27, 1858.

186 *Ibid.*, p. 17, Letter of July 19, 1855. Timothy Colony emphasized his points with quotation marks and underlining. At the risk of some confusion, his letters are here reproduced exactly as written.

187 *Ibid.*, p. 52, Letter to J. H. Robinson, April 7, 1858.

188 *Ibid.*, p. 10, T. Colony to Messrs. Perkins and Standish, July 3, 1855.

Despite these injunctions, the wool shipments received at the mill were sometimes less than as advertised. When this happened, Timothy lost no time in getting after the agent, suiting the reproof to the offense. When Perkins and Standish sent some bags of wool that were a few pounds light, he simply asked them sardonically, "are your scales correct, or does the wool shrink coming down here."[189] When an outfit in Ohio stuffed some wool pulled from dead sheep into the fleeces they sent to Harrisville, Timothy wrote, "I suppose that you are aware that wool put up in that manner is considered in a court of Law 'Fraud' ...."[190] The deceptions were sometimes more ludicrous than costly. Thus he wrote to an agent who had failed to look into some bags of wool bought from a wool grower that he had found in one bag "the worst lot of stuff we ever had in the shape of wool. It was *coarse covered with manure* ... etc., etc. In another sack there was nearly a peck of '*old dried Apple or Peaches*,' '*Such is the fact*'. ..."[191] It was the principle of the thing. Big or small, such deceptions were not to be tolerated. In these different markets and different ways, the Cheshire Mills were soon buying annually some 200,000 pounds of wool—exclusive of the dried apple.[192]

Central to its success or failure was the woolen cloth produced by the Cheshire Mills. With the experience of Faulkner and Colony behind it and with the constant and diligent counsel of its commission house, the new company chose its fabrics carefully. During the 1850s, it produced flannel at the considerable rate of a thousand yards a day.[193] Cole points out that flannel was an ideal cloth for American manufacture in this period. It used the medium grades of domestic wool which were readily available. It was simple to manufacture. It did not suffer greatly from the effects of changing style and fashion. And, it was a fabric of wide and steady consumption. "Flannel shirts, underwear, and petticoats absorbed huge weights of the material."[194]

Faulkner, Kimball, and Company, as much as the Colonys, wanted flannels produced that would sell. Especially in the early days of the mill they sent to Harrisville a steady stream of advice on production. Thus they might write,

We also enclose patterns of five mixt Flannels which we think you would do well to make—we fear the stock of Flannels will be large next season & if so you will do better to try and make such styles as are not generally made—all the shades of color we now send are good—we would advise your making 100 bales of each of these colors—*say 50 bales of* each grade— It will not do to have but one of these colors about your mill at the same time, as you would be liable to get your goods

189 *Ibid.*, p. 6, T. Colony to Messrs. Perkins and Standish, June 23, 1855.

190 *Ibid.*, p. 45, T. Colony to B. & A. Judy, Plattsburg, Clark Co., Ohio, Nov. 18, 1856.

191 *Ibid.*, p. 35, T. Colony to Messrs. T. S. Hickey & Co., Aug. 20, 1856.

192 Census, 1860, Nelson, p. 6, Schedule 5.

193 Leonard, 1855, p. 274.

194 Cole, *op. cit.*, I, 316–317.

striped by mixing the different kinds of yarn—We should like to have some of each of the above goods in market in May next, as our own fall sales commence as early as June 1 & unless we have the goods for our first sales we cannot do as well with them later in the season–. . . .[195]

In its endeavor to produce flannels that would sell, the Cheshire Mills produced plain flannel of two qualities and twilled flannel, which had a diagonally ribbed appearance.[196] It turned out bales of "Blue mixt" and "Black mixt" plain flannels,[197] as well as more lively fabrics such as "twill scarlets."[198]

The commission agents were constructive in their advice but conservative in what they promised: "Your plain goods look well for the quality—they are *well finished* and they will sell at their market value when there is any demand for flannels." They advised the mill that its twilled flannels would not sell for more than thirty-two cents per yard, even though they might appear to be worth more.[199] Their caution was founded in experience, for the woolen manufacture was a highly competitive industry. When the agents found that a rival, the Bay State Company at Lawrence, Massachusetts, was turning out great quantities of twilled flannels, they promptly warned the Colonys of the depressing effect this would have on prices.[200] Timothy Colony may have grumbled that it was too bad the commission house had not gotten its intelligence sooner, but he would not have grumbled long. Faulkner, Kimball, and Company was too great a help to the new mill.

Much has been written of the mill workers of New England, of the highly lauded native-born workers in the early textile mills, of their displacement during the forties and fifties by low-paid immigrant workers, of the attending decline in the conditions of factory and mill town, and of the changing relationship between mill worker and mill owner as the latter abjured the role of patron for the less responsible one of mere employer, and as the expansion of the mills caused a divorce between management and ownership. The Cheshire Mills' records of the fifties—especially a collection of letters written by applicants for jobs in the new mill—reveal something of the background and circumstances of the mill operatives in Harrisville and provide a basis for cautious comparisons.

Recruitment of a skilled labor force in such a remote region was no easy task, and the Colonys went about it in a variety of ways. They must have hired both skilled and unskilled workers from among the local populace. They advertised for weavers in the *Sentinel*.[201] They had Faulkner, Kimball, and Company

195 CMR, Letter book "A," F. K. & Co. to J. W. Briggs, Jan. 17, 1852.

196 *Ibid.*, F. K. & Co. to Josiah Colony, Sept. 15, 1851.

197 *Ibid.*, F. K. & Co. to J. W. Briggs, Dec. 24, 1851.

198 CMR, Copybook "A," p. 21, C. M. to F. K. & Co., Aug. 2, 1855.

199 CMR. Letter book "A," F. K. & Co. to C. M., March 24, 1852.

200 *Ibid.*, F. K. & Co. to J. W. Briggs, April 9, 1852.

201 *NHS*, Jan. 15, 1852.

interview a candidate for a responsible position.[202] They asked some of the hands they had already hired to scout about in their home towns for other workers.[203] They even received an unsolicited offer of spinners from a fire-stricken company in Lowell.[204]

The score or so of workers who wrote to the Cheshire Mills about jobs have by their letters brought to light valuable evidence about themselves, evidence that will allow construction of a rough sort of profile. Most of them wrote from within New Hampshire; only a few were living in towns or cities in Vermont, Massachusetts, or Rhode Island. They had names that would not have been thought alien or unusual in Dublin and Nelson seventy-five years earlier. Nearly all those writing about jobs had already worked in woolen mills and were now seeking skilled work or supervisory positions in the Lower Mill. Thus they may have been somewhat better educated than the average of woolen mill workers. At any rate, with few exceptions they wrote letters that did credit to their schooling.[205]

Most arresting in these letters is a certain tone or quality that is more easily recognized than defined. It is a spirit of independence and sense of self-sufficiency, combined with a total absence of the traditional hat-in-hand approach to an employer. A man in Vermont wrote,

Sir, Should you think best to engage me, please answer by return mail as I have a School in view, in this place. . . . can have $15 dollars per month, but prefer spinning. Harrison says you will not expect to pay less than $12 dollars, & board them. Shall not take the school until I hear from you. . . .[206]

So much for the relative attractions of school teaching and spinning in the 1850s.

A man in the unlikely town of Rock Bottom, Massachusetts, offered some bait:

Private   I have been given to understand that you are commencing manufacturing at your New Mill in Harrisville–

If you will want a Colourman & Finisher at the above place– I shall be glad to serve you either by the *day* or pr *piece*   I have something new in colouring & finishing line that I wish to have a conversation with you about–. . . .[207]

And the women were no less independent than the men:

202 CMR, Letter book "A," F. K. & Co. to J. W. Briggs, Jan. 1, 1852.

203 *Ibid.*, Letters from Stephen Kimball, Gonic, N.H., Sept. 29, 1851; Joseph Garnett, Rock Bottom, Mass., Jan. 23, 1852; Abby J. Butler, Gilsum, N.H., Jan. 27, 1852.

204 *Ibid.*, C. P. Julfer & Co. [?] to Mr. Colony, Dec. 18, 1851.

205 *Ibid., passim.*

206 *Ibid.*, O. K. Kendall, Enosburgh Falls, [Vt.], Nov. 4, 1851.

207 *Ibid.*, Joseph Garnett, Rock Bottom, Mass., Nov. 10, 1851.

Sir   I think I cannot go to work for you at present, as it is but four days since I left the mill, and I would like to stay at home a while before commencing again– I have spoken with several girls but have been unable to find any that wish to go into a mill now.[208]

In the sometimes delicate negotiations over wages, the applicants showed themselves to be good Yankees. A woman in New Alstead declined to work for what the Colonys offered, but she did it tactfully and politely:

I received a letter from you saying you would like to have me go to Harrisville to work for you. I should like to go if I could afford to work for the price you pay but I think I could not. I have not worked for less than three dollars a week for some years and I think I cannot come down to two just yet, if you would pay what I can make in other plaices I would like to go and work for you a while.[209]

A man in Fisherville (Pennacook) disdained to sing his own praises, and expressed his contempt for prevailing wage rates:

you say you want a good Spinner   my qualifications in that respect i always leave to the Judgement of my employers   i give you full liberty to enquire if you Please from . . . my late employer who gave me one $ Per day and has often expresst himself satisfied Both in quantity and quality and i must say it is very low as low as a man can maintain a family Honestly. . . .[210]

And a spinner writing from Peterboro showed both his self-assurance and his bargaining skill:

I am ready to come to spinning as soon as you wish   you wished to know my price and that is what you can get other spinners for that can do the same work and do it as well   I would like to come and work a few days and let you see how my work satisfies and if it gives good satisfaction I will work for you as reasonable as eney other hand that you can employ that will give as good satisfaction. . . .[211]

These workers may not have rivaled the operatives at Lowell in the early days, "when printed regulations were necessary to prevent the bringing of books into the mills, when young girls pasted their spinning frames with verses to train their memories to work with their hands, when mathematical problems were pinned up in the 'dressing room.' "[212] But the letters do indicate that at least as late as the 1850s the workers who came to the Harrisville mills were literate, independent, largely old-stock Americans, all with a sense of their own dignity.[213]

208 *Ibid.*, Abby J. Butler, Gilsum, [N.H.], Jan. 21, 1852.
209 *Ibid.*, Deborah P. Kidder, New Alstead, [N.H.], Jan. 20, 1852.
210 *Ibid.*, C. T. Hurst, Fisherville, [N.H.], Dec. 11, 1851.
211 *Ibid.*, W. [?] H. Orcutt, [or Orcott] Peterborough [N.H.], Dec. 8, 1851.
212 Abbott, *op. cit.*, p. 137.
213 Michael Hall, in his thesis on Nelson, says that the Cheshire Mills "imported in the 1850s dozens of immigrant workers from mill towns all over the country." The present writer could find no evidence of this in the mill's records or in the census schedules of 1860.

Since the Cheshire Mills kept no regular payroll book during the 1850s, any study of the workers and their wages must be sketchy and remain studded with questions. There is more evidence than in the case of the Harris mills, but there can be no systematic study of workers and wages before the 1860s. This fact itself is suggestive of the expansion that came with the Civil War. During the 1850s, the Colonys kept their payroll records in their cash book along with various other expenses. A sample from the cash book for 1852 illustrates the casualness of the payroll:[214]

Feb. 22	Paid Irishman for work	1.88
March 1	Wm. Orcutt cash [spinner]	17.00
„ 5	S. & E. W. Mason for 26 cords wood	31.42
March 10	2 Irish Girls	12.66
April	Wm. Orcott (cash)	5.00
„ 22	Fanny Hawkins	8.08

The same cash book shows long payrolls about four times a year—in January, April, July or August, and November. (Milan Harris paid his workers quarterly until after the Civil War.) There is no indication of job or days worked. The book simply lists the names of the workers, numbering about thirty, and the amount paid, which varied from a matter of cents to over a hundred dollars. The entries suggest that the workers could get modest advances from the company during the long months between pay days.[215]

Some idea of the wage rates at the newly opened Lower Mill can be derived from these letters written by job applicants:[216]

Spinner	$1.00 per day	Asked
Spinner	$12.00 per month minimum + board	Offered
Spinner or Finisher	7/6 per day	Asked
Finisher	$1.50 per day	Asked
Asst. Finisher or Scourer	5/- per day	Asked
Man	$.58–.66 per day	Asked
Woman	$3.00 per week	Asked
	$2.00 per week	Offered

More important to the worker at the Cheshire Mills than the wage rate was the amount he actually earned. Following are the earnings of several employees between March 1852 and February 1853:[217]

214 CMR, Cash book "B."
215 Ibid.
216 CMR, Letter book "A," passim.
217 CMR, Cash books "A" and "B," passim.

Joseph Garnett, Finisher	$354.71
Fanny Hawkins, Weaver(?)	80.32
Stephen Kimball, Overseer(?)	541.96
William Orcott, Spinner	153.99

In the manufacturing schedules of the Census of 1860, there are figures on the average number of employees and their average wage. The Cheshire Mills reported employing an average of forty male and fifteen female hands. The average "cost of labor" per month came to $25.00 for male workers and $18.00 for female workers.[218] Unfortunately, in all these figures on wages and wage rates there is one large unanswered question. Did they include board? It is not much more than an educated guess to say that in most cases they did not. But without a definite answer to the question, conclusions about the wages paid at the Lower Mill must bear that considerable reservation.

Concerning hiring practices, the evidence suggests that, although occasional families might be hired, most of the operatives taken on in the first days of the Lower Mill were single men and women.[219] This may have been due to the reluctance of persons with families to move to such a remote area. And in the early 1850s there may simply not have been accommodations for many new families in the village. In these first days the children of some of the operatives also worked in the mill. In April 1852, the Cheshire Mills paid "Cyrus Osborn for children $23.38."[220] There were probably never more than a few children in the mill, and, barring fraudulent statements in the 1860 census, there were none working there in that year. Indeed, there were only three mill workers listed in the village who were sixteen years or younger—and those three lived in Milan Harris's boardinghouse.[221]

Housing their workers in the tiny village was an immediate concern for the Colonys. Even before the mill was fully operating, they erected on the high ground across the road from the mill a long three-story brick boardinghouse.[222] (Plate X) Today it stands in the quiet shade of tall and graceful elm trees and still houses a few of the mill workers. Outside, its appearance is almost academic. Inside, the empty silence of the rooms and halls, the long bare tables, and the deeply worn staircases give the impression of a barracks whose company has long since marched away.

A century ago its attraction would have been less its peaceful atmosphere than its dinner table. This was important, for, as one writer has said, "workers could be induced to leave home,

218 Census, 1860, Nelson, p. 6, Schedule 5.

219 CMR, Letter book "A," Sept. 29, 1851.

220 CMR, Cash book "B," April 22, 1852.

221 Census, 1860, Dublin, pp. 65–68; Nelson pp. 61–64.

222 Humphrey, *op. cit.*, pp. 11–12.

friends, and family only if there were an A-1 boarding house on the mill premises. . . . Small town mills back country were forced to spread a good table to attract expert workmen."[223] Anyone familiar with boardinghouse cuisine may suspect that the Cheshire Mills' food was not the principal attraction for workers coming to Harrisville, but it may have been good enough to keep the mill hands from getting homesick. The boardinghouse's account with the local store in 1859 showed purchases of ham, mackerel, tripe, cheese, rice, eggs, lemons, ginger, molasses, tea, and coffee,[224] all of which allows at least the possibility of decent meals.

As for its other arrangements, the boardinghouse in these years housed between fifteen and thirty-five workers and was run by a married couple with a few young girls to assist with the chores. In 1854, the mill paid one George M. Pratt a bonus of fifty dollars to run the boarding house for a year, and some such contract was probably customary.[225] Between 1855 and 1860 the male boarders paid twenty-five cents a day and the female hands about twenty-one cents a day for living there, with allowances for meals missed.[226] There were probably complaints about the rates.

The occupants of the boardinghouse are listed in the 1860 census, providing some interesting facts and comparisons with the boardinghouse of Milan Harris and those in Lowell. About a third of the workers at the Cheshire Mills lived in the house. Again, their youthfulness is striking. Of the nineteen workers (again excluding the housekeeper, his wife and child, ages thirty-three, thirty, nine) the two oldest workers were in their early thirties, and the median age was twenty-four. The male workers were much in the majority as they had been during the fifties, but five girls, weavers, lived there, and were probably a feminizing influence out of proportion to their number. Most of the boarders had names unfamiliar in the area, only a bare majority were born in New Hampshire, and a third of them were foreign-born. All this was measurably at variance with the residents in Milan Harris's boardinghouse.[227] The foreign-born were just about evenly divided between English, Irish, and French Canadians. But in their jobs it is significant that the English and the Irish, too, were the skilled workers—spinners. The French Canadians were listed as "laborers," as might be expected of the most recent immigrant group.[228] One more fact is worth noting. Living in this boardinghouse in 1860 were two of the Colony family. John E. Colony was one of the original officers of the company, and Josiah T. Colony was the nineteen-year-old son of Timothy Colony.[229] Their living there may be

223 Florence C. Maxwell, *op. cit.*

224 CMR, Eben Jones's Ledger, Harrisville, 1858, p. 136.

225 CMR, Cash book "A," Oct. 18, 1854.

226 CMR, [Boarding-house Ledger], 1855–1860.

227 See above, p. 28.

228 It should be emphasized that these generalizations concerning the foreign-born are made from a small statistical base, there being only seven foreign-born workers in the Cheshire Mills boardinghouse in 1860.

229 Census, 1860, Dublin (Harrisville P. O.), p. 67, Schedule 1. See genealogy of Colony family, Appendix 2.

taken as evidence not so much of the Colonys' egalitarianism as of the decent accommodations in the boardinghouse. The overcrowded, unhealthy boardinghouses of Lowell in the 1850s had no counterpart in Harrisville.[230]

Building on the village's half century of experience in the woolen manufacture, the Cheshire Mills, like Milan Harris and Company, achieved an impressive expansion and development during the slack times of the 1850s. Now, as the war clouds boiled up on the horizon of the 1860s, both mills were ready to capitalize on the accompanying boom.

230 Cf. Abbott, *op. cit.*, pp. 128–129.

3 THE VILLAGE AND
ITS PEOPLE, 1800–1870

The Village

The village of Harrisville is attractive to the point of being picturesque. The mills and homes of mellowed red brick, the ponds joined by the tumbling stream, the church with its fine lines reflected in the waters of the canal, the side roads lined with the neat houses of the mill workers, the surrounding circle of low green hills all combine to made a scene that is rare indeed in mill towns. Equally rare, except for things related to modern communication—cars, paved roads, telephone and power lines—one sees today a factory village largely unchanged in appearance by the passage of a century.

The form and growth of the village was determined by the woolen mills. As Cole writes, "Woolen mills, being attracted to water-power sites, were usually located some distance from the larger towns, and in the little communities that subsequently grew up, the factory was always the center of the village."[1] The first house on the site of Harrisville had been built by Abel Twitchell about 1774, but that had not been the real beginning of the village, for it stood alone for nearly half a century. Only after the woolen industry had been well begun were other houses built. Following the steady growth of the mills, the greatest part of the village grew up in the half century between 1820 and 1870.

It was appropriate that the next house after that of Abel Twitchell should have been built by Bethuel Harris. Increasingly involved along with his sons in his carding and fulling mill, he finally decided to leave the farm in southeast Nelson and move nearer to his work. In 1819, he built a large, square, unpretentious Federal-style house on the north side of the canal, close by Twitchell's milldam.[2] (Plate XI) Either by chance or design, Bethuel had his house built squarely on the line dividing Dublin and Nelson.[3] He continued to consider himself a resident of Nelson, where he was active in town affairs. Presumably, if his growing woolen business on the Dublin side of the line were unfairly treated by that town, he could declare a change in his domicile and so better protect his interests. When Bethuel

1 Arthur H. Cole, *The American Wool Manufacture*, I, 241.

2 D. H. Hurd (ed.), *History of Cheshire and Sullivan Counties, New Hampshire*, p. 213.

3 Leonard, 1920, p. 787.

Harris, his wife Deborah, and their ten children moved into their new home sometime between 1819 and 1821, the village of Harrisville was fairly started.

Keeping pace with the modest expansion of the mills, the village grew slowly during the next thirty years. Tracing that growth involves some conjecture. Bethuel Harris must have already decided to erect his mill when he built his new home, for the Middle Mill went up the next year. As Bethuel's sons grew to maturity, went into business, and married, they too built homes along Goose Brook. Cyrus and Milan, Almon and Charles, Lovell and Calmer, all of them had homes in the village, and at least one of Bethuel's sons-in-law, Abner Hutchinson, also built a home there.[4] At one time in this early period it is said that three fourths of the inhabitants of the village bore either the name of Harris or Twitchell.[5] Well might Milan Harris, in 1830, have felt entitled to name the village for his family![6]

The homes built by the sons of Bethuel Harris are among the most handsome in Harrisville. Lovell's was a rambling frame house on the Dublin side of the line. All of the others were on the Nelson side and made of the same soft red brick from Nelson that was used in Bethuel's house and in the mills.[7] They were all similar, and their style has been described as a "simple, provincial reflection of the solid Bulfinch tradition of early nineteenth century Boston."[8]

The most attractive of these houses may have been built in the 1820s by Cyrus Harris. Its materials, design, and location combine in an eloquent testimonial to its builder. It stands two-and-a-half stories high and has two chimneys at each gabled end. Its broad front of rusty brick is set off by a white-paneled door, flanked by side lights, and nine windows lightly veiled by their white sashes. It stands on the high ground above Bethuel's house and faces south. From its front windows one overlooks much of the village and, from the rear windows, the upper pond and the woods beyond. The plain but handsome rooms, with their wide pine floorboards and shallow brick fireplaces, are made bright and cheerful through the day as the ample windows catch the advancing sunlight. (Plate XII)

The other Harris homes are scarcely less attractive. Alongside of this house just across the road is the house built in 1833 by Milan Harris. When he was first married, in 1823, he had lived in one of the few dwellings then available, the blacksmith shop built by his uncle, Jason Harris.[9] He built his own home at the time he was building his Upper Mill, and the one directly overlooks the other.[10] It is similar in design to its neighbor but in

4 Leonard, 1920, pp. 787–789; *NHS*, Aug. 30, 1843.

5 *NHS*, June 24, 1880.

6 Hamilton Child, *Gazetteer of Cheshire County, New Hampshire*, pp. 175–176.

7 Leonard, 1920, p. 787; Bemis MSS, Box 6, XLIII, 14.

8 Ada Louise Huxtable, "Progressive Architecture in America: New England Mill Village, Harrisville, New Hampshire," *Progressive Architecture*, Vol. 38 (July 1957), pp. 139–140.

9 *NHS*, June 7, 1823.

10 *NHS*, June 24, 1880. The present owner, Mr. John Clark, says that he has seen a deed giving the date of the house as 1818. This is sharply at odds with all other evidence. Interview, Nov. 11, 1962.

total effect is not its equal. (Plate XIII) The third son, Almon, built his home on the westward side of his father's house, just back from the canal. It is much like those built by his elder brothers, and, if it lacks their commanding situation, it has the canal—its clear waters fringed with the blue-flowering pickerel weed—to reflect and enhance its beauty. (Plate XIV)

Of the other early homes those of Charles C. P. Harris and Abner Hutchinson were farther west, facing the village with their backs to the upper pond. Although they were smaller and only a story-and-a-half high, their design resembles the other Harris homes. In the mid-nineteenth century, with white picket fences around their dooryards, they formed a neat enclosing line for the westward end of the village. Here and there were a few other houses belonging to early families like the Twitchells, the Farwells, and the Yardleys. These were frame houses but otherwise similar to their neighbors and like them well placed to suit the natural features of the landscape.[11] (Plate XV)

As the village slowly grew, business and public buildings appeared. A woodenware shop was started in 1838.[12] A fire company was chartered and organized in the next year.[13] Jason Harris's blacksmith shop was replaced by a larger one that also made machinery.[14] A store was supplying the wants of the villagers before 1842.[15] A chapel was built in 1840, and after the church was built in 1842, the smaller building was used as a schoolhouse.[16] Charles C. P. Harris built a bridge, 120 feet long, across Goose Brook. Before that, a fallen pine tree that lay across the narrow stream was the only direct means of going from his house to the lower part of the village.[17]

Thus, by the late thirties and early forties, although still tiny, Harrisville was assuming the dimensions of a real village rather than a pair of woolen mills with outbuildings. The proof of this appears in the *Sentinel* during these years. The first time that "Harrisville" was mentioned was in 1837, seven years after Milan Harris had changed its name.[18] Two years later, in 1839, the newspaper carried its first real comment on the place. That the piece was taken from the *Nashua Telegraph*, and probably by that paper from the *Boston Times*, simply illustrates the roundabout ways, full of opportunities for error, in which small-town papers got their news items. This brief account describes the village as being "almost exclusively manufacturing. . . . Two factories and some fifteen or twenty dwelling houses, mostly of brick."[19] Even these modest totals must have been an exaggeration; the observer either counted the barns, or the numbers grew in the copying. More helpful in determining the size of the village is an

11 *Map of Cheshire County, N.H.* (1858); Plate I, Frontispiece.

12 Hurd, *op. cit.*, p. 211.

13 *Laws of New Hampshire, 1839*, c. XII.

14 Census, 1850, Vol. XIV, Dublin, Schedule 5.

15 *NHS*, Jan. 26, 1842.

16 Hurd, *op. cit.*, p. 217.

17 *NHS*, June 24, 1880.

18 *NHS*, March 9, April 14, 1837.

19 *NHS*, Oct. 23, 1839.

account written four years later. The editor of the *Sentinel* occasionally found time to ride through the nearby countryside and gather some news for himself. In August of 1843, perhaps seeking relief from the summer heat in Keene, the sturdy old editor and founder of the paper, John Prentiss, made such a trip through Harrisville and later wrote a description of it for his readers:

> This new village is about 4 miles South [of Nelson], built around the outlet of a chain of ponds . . . extending . . . for seven or eight miles. The road to it is pretty level, extending through one of the valleys of this mountain town to Peterborough. . . . We were surprised and delighted, coming in view of the last of this chain of ponds . . . to see a little city, . . . well built of substantial two-story brick houses, a very handsome brick church and schoolhouse, two woolen factories, a large store, machine shop, and other buildings necessary to woolen factories, with some 10 or 12 dwelling houses, clustered together; and more so to learn that nearly all these buildings have been erected by the Messrs. Harris, father and sons. There are but three or four wooden buildings in the place, though the bricks are burnt some three or four miles distant. The Messrs. Harris must have expended something like $75,000 in this village. . . .[20]

From 1850 forward, quantitative evidence is available that allows a more precise following of the growth of both the village and the area that was to become the new town. The census schedules for Dublin and Nelson list the houses and occupants just as the census taker came to them on his rounds, and a careful collating with detailed maps of the period allows a close estimate of the size of the village.

At the beginning of the period there appears to have been in the village about 25 dwellings and a population of about 120.[21] (Plate I) During the decade that followed, the growth of the village still kept pace with that of the woolen mills. Other enterprises vital to a growing village appeared, such as a new schoolhouse,[22] a livery stable,[23] a freight line to Keene,[24] and a "public house."[25] The population mounted steadily, but only a few new houses were built. These were built by the Colonys and Milan W. Harris, which is to say, by the owners and supervisory personnel of the mills. The mill workers increased from 40 in 1850 to 115 in 1860,[26] but for them there were few houses available. Instead, the two principal mills each put up a large boardinghouse in the early fifties, and some of the families in the village also took in boarders. It must have been the exceptional mill hand who had a house for his family. The village almost

20 *NHS*, Aug. 30, 1843.

21 Census, 1850, Vol. XIV, Dublin, Vol. XV, Nelson; *Map of Dublin, N.H.* (1853 and 1906); *Map of Cheshire County, N.H.* (1858).

22 *NHS*, Jan. 29, 1858.

23 *Map of Cheshire County, N.H.* (1858).

24 *NHS*, June 24, 1880.

25 Edwin A. Charlton, *New Hampshire as it is. . . .*, p. 317. Charlton estimated the population of the village to be 350 in 1855. This figure is widely at variance with the number suggested by an examination of the census schedules.

26 Census, 1850, Vol. XIV, Dublin, Schedule 5; Census, 1860, Nelson, Schedule 5.

seems to have been gathering itself up for a new and greater expansion.

The frame house built about 1852 by the eldest son of Milan Harris is architecturally interesting.[27] It stands on a high bank above the main road, next door to the Cheshire Mills' boarding-house. The gabled end faces the road, the front door and hall are to one side, and two nearly full-length windows serve to light the front parlor. They are needed, for the house faces north and the second story overhangs the first so that, supported by pillars, it provides a front porch but darkens the interior. The gable and eaves are decorated with "gingerbread," which helps somewhat to bring down the disproportionately high upper story. Thus did the Gothic arrive in Harrisville! A mild form of it to be sure and hardly at odds with the prevailing style, but it is a matter of conjecture what Milan might have said about his son's new house. (Plate XVI)

The village underwent a considerable expansion in the decade of the sixties. Both the population and the number of houses just about doubled.[28] The two companies each erected a new brick mill, other industry appeared, and a hotel was opened.[29] Milan Harris put up an enormous white frame boardinghouse behind his mill.[30] But boardinghouses alone would no longer suffice for the mill hands, and the war-prosperous mill owners began to erect tenements.

In 1864, the Cheshire Mills built five identical frame tenements, which soon won the apt name of Peanut Row.[31] Indeed, the houses were small; each one was a story and a half, set on a quarter-acre lot,[32] and valued at six hundred dollars.[33] But they were both well built and attractively located, as can be seen today. Their design is simple and standard. The gabled end of the house faces the road. The front of the house has the door to one side with two windows beside it. A single window under the eaves lights the second floor. These tenements were built behind the fine Cyrus Harris house, then occupied by Henry Colony, on a road that curves along the shore of the upper pond. There is the space of a house between them, with the pond before and the woods behind them. (Plate XVII) About this same time another row of four houses was built on the Dublin side of the village on the road that led up to the new schoolhouse. Two of these were brick, and two were frame. Otherwise they were the same and identical with those on Peanut Row. They were probably built by the Colonys also, for the Cheshire Mills owned them in the 1870s.[34] (Plate XVIII)

In the late sixties, Milan Harris also put up behind his mill a

27 Dublin Town Records, Tax Invoices, 1847–1862.

28 Census, 1870, Vol. XV, Nelson, Vol. XVI, Dublin.

29 NHS, Oct. 13, 1870.

30 Census, 1870, Vol. XV, Nelson, Schedule 1, pp. 10–12; C. H. Rockwood, Atlas of Cheshire County, N.H.

31 CMR, General Journal, 1858–1866.

32 Nelson Town Records, Tax Invoices, 1867–1889 (1870).

33 Nelson Town Records, Vol. VIII, Tax Invoices, 1843–1866.

34 Rockwood, op. cit.

row of four frame tenements. They were larger than the Cheshire Mills' tenements, being a full two stories and valued at seven hundred dollars each.[35] Of older design these houses had the gable at the side, and the front door and hall were in the center, flanked by two windows on each side. They stood on lots of only one-eighth acre, but there was ample room between the houses. If their occupants missed the lovely view enjoyed by those living on Peanut Row, they also missed the cold winds that in winter drive across the frozen pond. (Plate XIX)

By 1870, both mills were renting other houses to their employees[36] and were sufficiently involved in the matter of tenements to allow some appraisal of the practice to be made. First, although the use of the word is proper, the usual connotations of the word "tenement" do not apply in these cases. These houses were well built, healthful, and attractive. As for rents the Cheshire Mills in the 1860s commonly charged five dollars a month. This was not exorbitant. In the month of September 1867, the company's spinners earned an average of thirty-three dollars and their male weavers thirty-six dollars.[37] Although these tenements may have been crowded, and they may have been inadequate in number, the mills nonetheless provided their workers with decent housing at a fair rent.

Surrounding the village and destined to be included in the new town in 1870, there was an area of small farms and smaller industry. The farming done was usually of a mixed variety and included the raising of grains, fruits and vegetables, cows and sheep. Despite the rough, rock-strewn terrain and the short growing season, these farms could produce good crops.[38] The small industry depended heavily on two features of the region, forests and waterpower. There were a number of small mills along the streams, which turned out lumber, clothespins, washboards, and the like. In a word the outlying parts of Harrisville were much like other towns in that part of New Hampshire.

In addition to the cluster of houses along Goose Brook, there was one other village that was to be included in the new town of Harrisville. This was Pottersville, later Chesham, and it lay in the northwest corner of the town of Dublin. The earliest settlement there dated from the 1760s.[39] In the last years of the eighteenth century a pottery business was begun there, which for about fifty years was a thriving little industry.[40] The potteries, the farming interests, and perhaps the Baptist Church established there gave rise to a sprawling village, which was until the 1860s larger than the village of Harrisville.[41] By then, however, it was on the decline, while Harrisville was still growing. There were

35 Nelson Town Records, Tax Invoices, 1867–1889.

36 *Ibid.*

37 CMR, Payroll.

38 Child, *op. cit.*, p. 175.

39 J. E. Conant & Co. (comp.), *Phelps Catalogue,* p. 17.

40 Child, *op. cit.*, p. 177; Hurd, *op. cit.*, p. 218.

41 *Map of Cheshire County, N.H.* (1858).

no special ties between the two villages, and Pottersville was probably included within the boundaries of the new town because of the route of the railroad and to simplify the town lines.

For such an exclusively industrial village as Harrisville, growing up in the remote and rural area of the Monadnock highlands, communication with the "outside world" was a matter of vital and increasing concern. Local roads were always an important and lively item in town affairs. "The location of every road was decided in town meeting after a petition had been presented to the selectmen."[42] Since the larger part of the cost of constructing and repairing roads was borne by the towns and paid for by an annual tax on the polls and estates,[43] every new road proposed was the object of close public scrutiny and debate. Because Harrisville was a sort of stepchild, although a paying one, to both Dublin and Nelson, the roads that its citizens wanted came in for extra scrutiny and debate in the town meetings. Beginning about 1835, the Dublin town records contain numerous references to new roads desired by the residents of the growing mill village.[44]

In 1850, for example, a connecting road, all of five eighths of a mile, was proposed which would benefit Harrisville residents. According to the record, it seems that Dublin first voted to grant it, then to discontinue it, then voted not to discontinue it, and then again reversed the decision, all in a space of two years.[45] In 1854, Dublin passed a resolution concerning another proposed road:

> 1st that the town of Dublin is willing to make all reasonable alterations in the road between Harrisville and Marlboro, and fulfill faithfully all obligations on it by the laws of the state.
>
> 2nd that an Agent be chosen to oppose the laying out of the road petitioned for by Josiah Colony and others in this town and Marlboro, and perform all other acts he may judge expedient to induce the Road Commissioners to decline laying out the same.[46]

Dublin's agent apparently did not convince the Commissioners, for the town continued to oppose the road, sometimes in comical fashion, during the next few years.[47] These disputes are interesting, for they illustrate how important the matter of roads was in these small towns and what close attention they received. More significantly, they also show that even by 1850 the separateness of Harrisville was a fact and that there were sharp conflicts between the mill village and its parent towns.

The main roads, which led to Keene, Peterborough, and Concord, being of more common concern were less a matter for

42 Bemis MSS, XIII, 6, in Michael G. Hall, "Nelson, New Hampshire, 1780–1870," p. 5.

43 Maurice H. Robinson, *A History of Taxation in New Hampshire*, pp. 209–212.

44 Dublin Town Records, Vol. IV, Town Meetings and Tax Invoices, 1828–1846, *passim*.

45 Dublin Town Records, Vol. V, Town Meetings, 1847–1882, pp. 64–70, 99.

46 *Ibid.*, p. 130.

47 *Ibid.*, pp. 160ff.

dispute. By the early years of the nineteenth century these roads were good enough for regular travel by horse–drawn vehicles.[48] The main road from Keene to Concord passed through Nelson village, and stagecoach routes ran through Dublin, Harrisville, and Nelson.[49]

By 1817, Dublin and Nelson had become important enough to have their own post offices,[50] and Harrisville had its own post office by 1848. As might be expected, this too was a Harris enterprise. After Cyrus and Milan Harris, and two others, had held the job of postmaster briefly, Charles C. P. Harris took the position and held it for the rest of his long life.[51] His office was the front room of his little brick house.

Soon after the Civil War the village's isolation was further reduced by the arrival of the telegraph. The *Peterboro Transcript* recorded the event:

> The poles are now being set for a line of telegraph from this place through Dublin, Harrisville and Marlboro to Keene. We understand this has been brought about mainly through the instrumentality and enterprise of the Messrs. Harrises and Colonys at Harrisville, and will be of great advantage to that community.—Harrisville is a thriving little place and its people are wide awake, and notwithstanding the natural obstacles which surround them, will some day secure the advantages of railroad as well as telegraphic communication with the rest of the world. The spirit is at work.[52]

When this telegraph line was completed, the residents of Harrisville were able to send messages of ten words or less to Boston for $.35, to New York for $.55, and to distant Chicago for $2.10.[53]

Long before this, railroad fever had hit New Hampshire, as it had other parts of the East. As early as 1835, the *Sentinel* was discussing a railroad that would run from Brattleborough, Vermont, through Keene, to Nashua.[54] Although actually it was not until 1878 that the railroad finally limped into Harrisville, as in the case of the telegraph, its champions had been the owners of the woolen mills. Indeed, it is perfectly clear that whether it was roads, postal service, telegraph, or railroads, Harrisville was progressive, and it was the Harrises and the Colonys who took the lead in seeking and obtaining better communications for their manufacturing village and for the region.

Harrisville in 1870 was a village of about sixty houses, plus mills, stores, and shops, and a population of over four hundred people.[55] It was a community not only large enough and vital enough to make itself into a new town, but it was a village of neat and harmonious design. The mills, their outbuildings, the

48 Bemis MSS, Box 6, XLVIII, 237.

49 Hall, *op. cit.*, p. 19; Leonard, 1855, pp. 245–246.

50 *NHS*, June 5, Nov. 15, 1817.

51 *NHS*, June 24, 1880; Leonard, 1920, p. 501.

52 *Peterboro Transcript*, in *NHS*, May 21, 1868.

53 *NHS*, Oct. 21, 1869.

54 *NHS*, Aug. 6, 1835.

55 Census, 1870, Vol. XV, Nelson, Vol. XVI, Dublin; Rockwood, *op. cit.*

old Harris homes, the shops, the church, the boardinghouses and tenements, all were fitted by an irregular but purposeful plan into the natural fall of the land on the steep banks of the gorge lining Goose Brook. Harrisville is a remarkable example of the planning of an industrial community, which also achieved a handsome, and unforced, homogeneity of appearance. It is the more remarkable for being largely the work, not of a single mind trained in such planning, but of two generations of the two leading, and rarely cooperative, families of practical-minded Yankee businessmen.[56]

The People

A study of the population that lived in the area of Harrisville before 1870 is the normal study of growth and change. Like most farming communities in northern New England, Dublin and Nelson had a heyday that was brief and a decline that was irreversible.[57] In both towns the population reached a peak and then began to decline even within the lifetimes of the first settlers. While this decline proceeded, Harrisville was growing apace with its mills and emerging as a community quite different from its neighbors.

Nelson, which had really begun to grow only after the Revolution, reached its maximum population of 1,076 people in the census of 1810 and has been declining ever since then.[58] Charles Bemis, the local historian, wrote in his notes that "Nelson was in its best estate in the early years of the 19th century. . . ."[59] Writing as an old man in the desolation of 1900, he nostalgically magnified this "golden age," but he was essentially correct. The years of the Embargo, followed by the War of 1812, were the turning point. Jefferson's Embargo worked a real hardship on local farmers, for they depended heavily on selling their surplus produce mainly for export to meet their financial obligations. Aroused, Nelson's town meeting in 1809 chose a committee to draft a protest to Congress. Bethuel Harris, his interest still in farming, was one of the committee.[60] Both the interruption of trade and the War itself fostered domestic manufactures elsewhere, particularly in the cities, and the peace settlement was followed by a westward migration on a vast new scale. All of these developments served to reduce the population in rural New England towns. Thereafter, except for the decade of the sixties when it held its own, the population of Nelson slowly but steadily dwindled.

The year 1820 saw more New Hampshire towns reach their

56 Huxtable, *op. cit.*, pp. 139–140.

57 Cf. Harold F. Wilson, *The Hill Country of Northern New England*, p. 54.

58 See Appendix 4.

59 Bemis MSS, Box 6, XLVIII, 237.

60 Nelson Town Records, Town Meetings and Tax Invoices, 1792–1890, pp. 499, 501.

maximum population than any other decennial year.[61] Dublin, with a population of 1,260, was one of these towns.[62] As it had a larger base and certain other advantages, it declined later than Nelson, fell in a more irregular fashion, and maintained a numerical superiority, but its decline was no less sure. Dublin's long-time minister, Levi W. Leonard, kept a careful record of the town's vital statistics. In 1840, he sent to the *Sentinel* a summary of the town's population by ages, noted the declining numbers since 1820, and pointed to the reason: "The great falling off between the ages of 20 and 30, and also of 30 and 40, indicates a large emigration, and such is the fact, as attested by those who have observed the annual departure of our young men and young women for the West, for manufacturing villages, and for cities on the seacoast."[63] With this erosion of population, wealth, and vitality, and an increasingly noncompetitive agriculture, both Dublin and Nelson underwent great change. The adjustments that these towns were having to undergo in the nineteenth century help to explain some of the difficulties, like new roads, they were having with the anomalous little mill town growing up between them.

The demographic history of Harrisville in this period is the opposite of its parent towns. It drew its first breath at the time they were beginning to decline. While they were being deserted by their young people, it was steadily growing. The population of the village totaled 120 in 1850, 210 in 1860, and 406 in 1870.[64] As for the countryside surrounding the village, including Pottersville that was to be included in the new town of Harrisville, its population was declining much the same as the rest of Dublin and Nelson.[65] The census of 1870 was taken just before the division of Dublin and Nelson to make the new town, but the area that was to be Harrisville had a population of just over seven hundred people, and it was still growing.[66] The separation was a hard blow to both Dublin and Nelson, but the Harrises' red brick village had outgrown their control and with divergent views and interests was by 1870 strong enough to go its own way.

Concerning matters of health, disease, and death in the area of Harrisville during this half century, the evidence is fragmentary but interesting. Thanks to the painstaking efforts of the Reverend Mr. Leonard, there are statistical records of the population of Dublin between 1820 and 1852. These include the great majority of the residents of the area of Harrisville in the period, and had such records been kept there, they must have been similar.

Although Leonard's statistics are not detailed enough to permit

61 Wilson, *op. cit.*, p. 52.

62 See Appendix 4.

63 *NHS*, March 18, 1840; cf. Wilson, *op. cit.*, pp. 19–26.

64 Census, 1850, Vol. XIV, Dublin, Vol. XV, Nelson; Census, 1860, Dublin, Nelson; Census, 1870, Vol. XV, Nelson, Vol. XVI, Dublin; *Map of Cheshire County, N.H.* (1858); Rockwood, *op. cit.*

65 *Ibid.*

66 Census, 1870, Vol. XV, Nelson, Vol. XVI, Dublin.

many comparisons with mortality rate studies, they do show the town's crude death rate, that is, the average annual death rate per thousand population without adjustments for age or sex. During this thirty-two-year period, Dublin had a crude death rate of 13.9. Without explaining his source, Leonard says "the United States census" put the crude death rate of the New England states at 15.5 and the Middle States at 13.9.[67] In Sweden, the country that kept the best records, and which was probably typical of western Europe, the crude death rate during these years was 23.25.[68] If Dublin could not be billed as a health resort, it was probably no less healthy a place to live than much of eastern United States or the birthplaces of its immigrants.[69]

In these years when infant mortality rates were high and life expectancy was less than forty years, the Harris family fared rather better than the average. Bethuel and Deborah Harris had a family of six sons and four daughters. Either they came of hardier stock or else they lived better than most of their neighbors, for none of their offspring died in childhood. One daughter died at twenty-six and Cyrus at fifty-one. Consumption caused his death and probably hers also. The other eight lived to ripe old ages, as had their parents, dying at ages that varied from seventy-five to eighty-six. When he was seventy-five, Milan Harris fell out of his hayloft and landed on his head. No one expected him to pull through, but tough old Milan recovered[70] and lived another ten years.

The same longevity did not apply to their families. All six of Bethuel's sons married, and all except Almon chose wives from Dublin or Nelson. Almon may have been well advised to court in Hancock, for all his brothers' wives predeceased them. Four of his brothers remarried, and one of them, Calmer, had a total of three wives. None of the Harris brothers had a family the size of their father's; three or four children were the average for them. Some of the lines flourished, and some began to die out. Milan outlived not only his son Milan Walter and his daughter-in-law but also their three children.[71]

As to the causes of death in the area of Harrisville, infectious diseases, both epidemic and endemic, certainly led the list. Some epidemic diseases then prevalent never did hit the area. The great cholera epidemics of 1832 and 1848, so devastating in some places and so greatly feared everywhere, never came nearer than Boston and western Massachusetts.[72] Remoteness did carry some advantages. The great killer was consumption. "No other disease has proved so fatal," wrote Leonard in 1855,[73] and he noted that it killed one out of six persons. Some families were decimated by

67 Leonard, 1855, p. 266.

68 Warren S. Thompson, *Population Problems*, p. 237.

69 For further statistics on the death rate in Dublin, see Appendix 5.

70 *NHS*, Dec. 31, 1874.

71 Leonard, 1920, pp. 445, 787–789.

72 See *NHS*, June–July, 1832, *passim*.

73 Leonard, 1855, p. 266.

this dread disease. Thaddeus and Mary Twitchell, who lived in the northern part of Dublin within the area of Harrisville, saw their seven children die of consumption before any of them were out of their twenties.[74] Leonard concluded that "It has been supposed by some that it is more prevalent here than in other places; but by examining other bills of mortality, we find that an equally large proportion of deaths are ascribed to this disease in many town of New England."[75]

No other disease rivaled consumption, but the number of recorded deaths in Dublin during this same thirty-two-year period due to other infectious diseases were typhoid fever, 30; scarlet fever, 24; dysentery, 18; croup, 15.[76] These figures can only be used in a general way. The town records did not always specify the cause of death, and the causes given must often have been imprecise or erroneous. For example, typhoid fever was frequently confused with typhus. The two diseases were not clearly differentiated until 1837,[77] and the confusion lasted long after. It was probably typhoid that the *Sentinel* reported in the home of the Baptist minister in Pottersville in 1822: "Distressing account of the Typhus Fever, in the family of Elder E. Willard, in Dublin, of five pleasant children, only one survives from a confinement of more than three months."[78] Likewise, "croup" may have meant various respiratory diseases. It may be assumed that the "degenerative" diseases, like cardiovascular disease, also took their toll of those who lived long enough to develop them, but it was the infectious diseases which took the most lives.

Although they seem to have been few in number, there were also accidental and violent deaths. Leonard gives no statistics, but mentions some that occurred in connection with the usually hazardous farm jobs like raising barns, felling trees, and blasting rocks.[79] In Harrisville there were several violent deaths of not unusual variety: a suicide by hanging in 1859;[80] a murder in 1864 when a drunken Irishman, John Grey, beat his wife to death;[81] the unexplained death of a seventeen-year-old housewife in 1870, which was probably the result of an abortion;[82] and the death the same year of an eight-year-old boy named Winn, who somehow got caught in one of the machines in the Harris mill.[83] Sensational though these were, violent deaths were rare in Harrisville; at least, these few are all that have come to light.

For doctors the early residents of Harrisville had to rely on those in Dublin and Nelson. Nelson had its long-time and popular physicians. Dublin had a string of doctors, few of whom remained in town more than a few years.[84] The first resident physician in Harrisville was probably a Dr. Leonard K. Hatch,

74 *Ibid.*, pp. 406–407.
75 *Ibid.*, p. 266.
76 *Ibid.*
77 Felix Marti-Ibanez, M.D. (ed.), *History of American Medicine*, p. 37.
78 *NHS*, March 2, 1822.
79 Leonard, 1855, p. 286.
80 *NHS*, March 11, 1859.
81 *NHS*, June 30, 1864.
82 Census, 1870, Vol. XV, Nelson, Schedule 2, p. 1.
83 *NHS*, March 17, 1870.
84 Leonard, 1855, p. 265.

who according to the census was living there in 1850, aged thirty-three. Despite all the sickness, Levi Leonard mentions that physicians needed an additional income to get along in Dublin.[85] Doctor Hatch did too. He and his wife had ten boarders living with them in 1850, most of them mill workers.[86] His name appeared in some advertisements prescribing "Pepsin" which appeared in the *Sentinel*,[87] and for a while he was also the postmaster in the village.[88] Doctor Hatch may soon have despaired of the village and left, to be followed by other young physicians, but most of the time after 1850 there was probably a doctor scraping along in Harrisville.

Such a state of affairs was not peculiar to Harrisville or Dublin. During the greater part of the nineteenth century, while so much of science was making spectacular advances, medical science was busy tearing down its ancient structure and laying the foundations of modern medicine. In this interregnum medical practice was particularly ineffectual and unconvincing. Doctors and patients alike knew that the old cures like purging and bleeding were useless or worse, but what were the right cures? And causes had to be found before cures. Little wonder that a critical and impatient public left the doctors uncalled and unpaid and turned instead to sure-cure quacks, cults, and patent medicines.[89]

More important for an understanding of this mill town than a purely quantitative examination of the population is some qualitative appraisal. If growth was the key to the former, change is the important feature in the latter. For convenience, such a study can be divided into three periods: 1800 to 1820, a sort of prehistory of the population; 1820 to 1850, during which time the population was homogeneous, almost entirely native-born, and largely under the control of the Harris family; and 1850 to 1870, when the foreign-born became important, a diversity of views increased, and the rule of the Harrises was challenged.

During the earliest period no village existed. "Twitchell's Mills" was as descriptive as it was euphonious. There were settlers, however, within the area of the later town of Harrisville. They were of the same stock and sort as those who had come there in the eighteenth century. Bethuel Harris was probably typical of the leading men in the area in the nineteenth century. He took an active interest in town affairs and displayed a patriarch's concern for the moral welfare of the people. In politics, like most of the people living in Cheshire County, he was a staunch Federalist. As a boy, in 1775, he had seen his father march off to help chase the Redcoats back into Boston. As a youth, he had left Massachusetts during the chaos of Shay's Rebellion. In 1807, he named his

85 *Ibid.*

86 Census, 1850, Vol. XIV, Dublin, Schedule 1.

87 *NHS*, April 30, 1852.

88 *NHS*, June 24, 1880.

89 Richard H. Shryock, *The Development of Modern Medicine*, ch. XIII.

sixth son for Charles Cotesworth Pinckney of South Carolina, one of the last Federalist candidates for President. In 1814, the New England Federalists, bitterly opposed to the war with England, met at Hartford to consider measures of protest or opposition. Twenty towns in Cheshire County sent delegates to a meeting in Walpole, New Hampshire, to choose a delegate for the Hartford Convention. Nelson sent two delegates to this Walpole meeting, and one of them was Bethuel Harris.[90] The Hartford Convention did little except debate, but there was a breath of treason about it that the New England Federalists never lived down. In the hill country of Cheshire County, however, association with the Hartford Convention brought no stigma on Bethuel Harris. During these same years he was also active in the militia and eventually became the major in command of the Second Battalion, Twelfth Regiment, which included the company from Nelson.[91]

Thereafter, as he became more deeply involved in his woolen business, and as the Federalist Party dissolved, Bethuel's political activity ceased. In his late years his principal activity seems to have been the establishment of the new Congregational Church in Harrisville. He never took any active interest in library societies or temperance movements, did not join the Masons, and never sought publicity in the press. Stern, upright, and hard-working, he avoided such embellishments and confined himself to the practical affairs of this world and the next.[92]

In the next period, from 1820 to 1850, nearly all the residents of the village were native-born, and nearly all were associated with the mills. The area of the town of Harrisville was inhabited even more exclusively by the native-born. A few people living in the rural area worked in the mills, but most of them were farmers or had small businesses, such as sawmills, tanneries, or potteries. In the first half of this period the Harrises and their relations formed so large a part of the population that their family may fairly be considered representative of the village population. Later, as the village grew, the Harrises naturally emerged as leaders of the community. Fortunately, a fair amount is known of the residences, occupations, and activities of the second generation of Harrisville's most important family.

The two eldest sons, Cyrus and Milan, left home as young men, but returned to remain in Harrisville and become owners of the woolen mills.[93] The next son, Almon, after several moves settled in Pennacook, New Hampshire, and founded a woolen mill there. The fourth and fifth sons also migrated not to the city but to the West. Lovell, a carpenter as his father had been,

90 *NHS*, Dec. 10, 1814.

91 Bemis MSS, Box 1, Seventeen-page MS by Gen. S. G. Griffin.

92 Cf. Hurd, *op. cit.*, pp. 213–214, including eulogy probably written by Charles C. P. Harris; also portrait of Bethuel Harris in possession of Mrs. Cecil P. Grimes of Pennacook, N.H.

93 Milan spent the last three years of his life with his daughter and son-in-law, J. K. Russell, in Massilon, Ohio.

left Harrisville about 1855 to settle on a farm in Illinois. Years later he returned to spend the last of his life in Milford, New Hampshire. Calmer, a machinist, first moved elsewhere in New Hampshire and then settled permanently on a farm in Minnesota. Both men were in their fifties when they went west to farm. The youngest son, Charles, never moved anywhere. He spent his life in Harrisville, working in the woolen mills and serving as postmaster. Thus, four of Bethuel's sons remained in the woolen business, three at home and one elsewhere in the state, and two went west to farm. Of the six, certainly Milan and Almon would be reckoned successful businessmen.[94]

It is also possible to trace to some degree just what these six sons had in the way of activities and interests. Politically, the Harrises all supported the Whig Party, that advocate of a national bank and a protective tariff. Lovell and Calmer, the two who eventually went west, seldom took an active part in the affairs of the village. With Democracy storming at the gates in 1828, the others might persuade them to sign a petition designed to rally support for John Quincy Adams, but that was about all.[95] Nor did Cyrus, perhaps already wasted by consumption, take an active interest in anything except the church. Of the others, Milan, Almon, and Charles were all active in the church, the Nelson Sunday School Association, the temperance movement, and in the establishment of a Harrisville Engine Company. Milan took an early interest in politics and was several times a delegate from Nelson to the Whig County Convention.[96] His brother-in-law, Abner Hutchinson, was twice chosen to represent Nelson in the state legislature[97] and was also a major in the state militia.[98] None of Bethuel's sons had his interest in the militia, which was by their time something of a joke anyway. Of all the reform movements that were sweeping America in the 1830s and 1840s, only temperance seems to have caught the attention of the Harrises. Few other subjects could so stir Milan Harris's wrath as that of insobriety. There is nothing to indicate that any of the six Harris brothers were actively interested in libraries, lyceums, or the like, and Milan's expressed views on education did him little credit. Their attitude on reform, as it concerned their village, seems to have been simply that mill hands needed to be sober, not erudite.

Milan and his brothers in their roles as leaders of this new community nonetheless command qualified praise. They had built up a flourishing industry and a handsome village. They were respected by their business associates and by the public.[99] As they had only a common school education[100] and little contact with

94 Leonard, 1920, pp. 787–789.

95 *NHS*, Extra, Sept. 19, 1828.

96 *NHS*, Jan. 29, 1835, Jan. 18, 1838, March 4, 1840.

97 Nelson Town Records, Vol. XIV, Town Meetings, 1841–1852 (1845, 1846).

98 Child, *op. cit.*, p. 180.

99 Cf. testimonial letter in *NHS*, Jan. 9, 1857.

100 Child, *op. cit.*, p. 178, says that Milan had only a common school education, and the others probably received no more than that.

101 See Appendix 6.

intellectual currents, their interest in cultural matters was understandably limited. Theirs was a village that was prosperous, attractive, and orderly, a village that did credit to their name.

The 1850s and 1860s saw livelier times in Harrisville. During these years the population and moral climate were changing, tensions developed and sometimes erupted, and the Harrises assumed a more active political leadership at the very time that their leadership was being openly challenged.

The more detailed census schedules, which begin in 1850, make it possible to determine just how the population of Harrisville changed in this period. In 1850, its make-up must have been very much what it had been in years past. The great majority of people in the area of Harrisville then had been born in New Hampshire. The only out-of-staters in any numbers came from neighboring Massachusetts and Vermont. The foreign-born, indeed all of those born outside of New England, were only 2 percent of the total. Most of these were English.

Another ten years brought a modest but significant change in the population. The percentage of native-born fell off slightly. The percentage of residents who were New Hampshire-born fell off considerably more. The foreign-born rose to 7 percent. Among these, the Irish now held first place, the English were next, and the first two French Canadians had appeared in town.

At the end of this period, in 1870, the percentage of native-born residents had declined considerably more, and they had also declined absolutely in numbers. The foreign-born reached nearly 20 percent. If those of foreign parentage, principally the children of these immigrants, are added, the total is a quarter of the population. Again the make-up of the foreign-born had changed. The French Canadians were now the largest group, the Irish were next, and the English ran a poor third.[101] During these twenty years while the total population was rapidly growing the native-born residents steadily declined in numbers, and the foreign-born, in constantly changing proportions themselves, grew from an inconsiderable number to become an important minority.

The census schedules of 1860 and 1870 also show the extent to which the population was dependent on the woolen mills. About 45 percent of *all* the people in the village worked in the mills. For a true appreciation of the importance of the mills there should be added to the mill hands the local farmers who sold wool and firewood to the mills, those who helped to build the new mills that went up in the fifties and sixties, and those who did hauling or other occasional work. The arriving immigrants worked in the mills in ever greater numbers, although some of the women may have worked as domestics and the men at lumbering in the winter and on the roads in summer. Millwork was preferable. By 1870, more than two thirds of all the foreign-born in the area of Harrisville (and they all lived in the village) were working in the mills.[102] The percentage of Harrisville residents who worked in the mills may not have been any higher in this period than it was before the expansion of the 1850s. However, the population was probably more dependent on the mills for its economic well-being than ever before because of the decline in farming and because of the large numbers of immigrants. They had less economic independence than the native-born; there was no ancestral farm to retreat to, and in those days employment notices frequently read "No Irish need apply." There must have been few people in the village, or even in the area of the town, who did not have a vital concern in the prosperity of the woolen mills.

In this period the Harrises who had remained at home became more prominent in local affairs. Milan Harris was sent to the General Court by the town of Nelson in 1851 and re-elected the next year. That year, his brother Lovell was there too, as representative of Dublin. Then, curiously, Milan was defeated in election to that office five times in the late fifties. He was elected state senator twice during the War, and for the rest of the time in this period he was usually a constable or police officer in Nelson.[103] His brother, Charles, was likewise twice elected to the General Court in the late sixties,[104] and his son, Milan Walter, was twice sent there by Dublin.[105] Such a record does not indicate any overweening political ambitions among the Harrises. It is no more or less active a role than would be expected of one of the important families in a small New England town.

The Colonys made their appearance on the local scene in this period. However, they were a Keene family and never had more than one foot in Harrisville. No real comparison can be made between them and the Harrises as leaders of the community. Henry Colony, the president of the Cheshire Mills for twenty-odd years, lived in Harrisville in the handsome house probably built by Cyrus Harris. His brothers and partners, John E. and Alfred T., also lived in the village for some years.[106] Henry made an effort to enter the political scene in Nelson but was defeated in election for the offices of both representative and state senator in 1869 and 1870.[107] His mill hands must have turned out in force for him to have gotten as many votes as he did in Nelson, for the Colonys were, of all things, Democrats. Political affiliation

102 Census, 1870, Vol. XV, Nelson, Vol. XVI, Dublin.

103 Nelson Town Records, Vol. XIV, Town Meetings, 1841–1852, Vol. XIII, Town Meetings, 1852–1876 (sic).

104 Nelson Town Records, Vol. XIII, Town Meetings, 1852–1876.

105 Dublin Town Records, Vol. V, Town Meetings, 1847–1882, pp. 342, 381.

106 Dublin Town Records, Tax Invoices, 1847–1862; 1863–1885.

107 Nelson Town Records, Vol. XIII, Town Meetings, 1852–1876.

then, as well as business rivalry, was to cast the Colonys in the role of leaders of the opposition in this domain of the Harrises.

During the 1850s, a series of incidents occurred which indicated how Harrisville was developing social stresses and strains. Some of these were entirely local in origin, others were local grievances compounded by the conflicts that were then dividing the nation. All of them underline the precise changes taking place in this growing factory village.

The first was a bitter battle of words over the quality of education being dispensed in the little brick chapel that served as a schoolhouse. This developed into a protracted newspaper argument in 1853, which revealed both the growing antagonisms between Harrisville and its parent towns and the special difficulties of the village itself.[108]

The next year the whole country was embroiled in an acrimonious debate over the Kansas-Nebraska Act, which repealed the old Missouri Compromise and opened the way for slavery to be extended into these territories. Dublin and Nelson people were strongly opposed to the extension of slavery. Both towns had given their votes to the Free Soil candidate for president in 1848, and Nelson again did so in 1852.[109] In March 1854, representatives of the towns in Cheshire County met in Keene to oppose the passage of the Kansas-Nebraska Bill. In the seven-man delegation from Nelson were three residents of Harrisville: Milan Harris, Abner Hutchinson, and Mainard Wilson, the blacksmith.[110] This does not mean that Milan had become an abolitionist, as his enemies charged. Rather, he was probably like many other northerners who were first aroused by the threat posed by the Kansas-Nebraska Act and who from that time on actively opposed the extension of slavery.[111] If the furor over the Kansas-Nebraska Act was a fair test, and it would seem that it was, the leading citizens of Harrisville in this period were aware of, and responsive to, the vital political issues of the day.

Another national issue that was of more direct interest to the citizens in Harrisville concerned immigration. The swelling numbers of German and Irish Catholic immigrants who came to this country in the 1840s and 1850s caused a reaction among the native-born Americans that was born of prejudice and fear. Riots occurred in various places, and in the 1850s the nativists organized the American Party, better known as the "Know-Nothings," to enact restrictive legislation. The movement was short-lived, soon being eclipsed by the larger issue of slavery, but

108 See ch. 4, pp. 80–84.

109 *New Hampshire Manual of the General Court*, Vol. I, 1889.

110 *NHS*, Feb. 24, March 3, 1854.

111 The next year Milan Harris signed a petition opposing the extension of slavery but favoring the Fugitive Slave Law. See *NHS*, Jan. 26, 1855.

in the mid-fifties it gained control of several states, including New Hampshire. In 1855, both Dublin and Nelson were caught up in the movement and helped to vote into power the American Party candidate for governor. They gave him as solid support as they had ever given their Federalist heroes.[112] There is nothing to reveal Milan Harris's opinion of the Know-Nothings, but the manner in which he avoided hiring immigrants to work in his mills is suggestive.

In the mid-fifties, there could not have been more than fifteen or twenty people in Harrisville who had come over from Ireland, but if some accounts are credited, these few had a demoralizing influence on the village. In 1856, an account of the Congregational Church in Harrisville pointed out that "in consequence of the increasing number of foreigners, who find employment in the factories, and other influences hostile to the welfare of Zion, the *morals* of the place cannot be said to compare favourably with what they were ten or fifteen years since."[113] Then, about the same time that this was written, Milan Harris sniffed some illegal rum being sold in the village and got involved in another newspaper battle.[114] There were others too who shook their heads and yearned for the old days, evidently the time when everyone in the village had been either a Harris or a Twitchell. One of them, "a prominent and very respectable citizen of Harrisville," did not hesitate to tell the readers of the *Sentinel* of the state of things: "Only last sabbath five persons in the lowest part of the Village were drunk all day, no uncommon occurrence of late, for that once quiet and peaceful Village."[115] Harrisville was changing, and fast, but the point to be noted is that this change was well under way before the foreign-born were there in any great numbers.

Later that year things came to a head. The occasion was provided by the presidential campaign, in which the first Republican candidate for that office, Charles C. Fremont, was defeated by the old Democratic wheelhorse, James Buchanan. Neither candidate was worth getting excited about, but the emergence of the Republican Party, so single-mindedly opposed to the extension of slavery, did introduce a real issue into the campaign and one which bitterly divided the American people.

In Harrisville there was unwonted activity. A Fremont Club was formed there, with Milan Harris as president and Simon Goodell Griffin of Nelson as one of the vice-presidents.[116] In the middle of October the Club held a Fremont rally in the meeting house. The citizens assembled "almost en masse," according to the *Sentinel*. Milan Harris called the meeting to order, the minister

112 *NHS*, March 16, 1855; James D. Squires, *The Granite State of the United States*, I, 315.

113 Robert F. Lawrence, *The New Hampshire Churches*, p. 269. The article on the church in Harrisville is subheaded with the name of "Rev. Wm. S. Tuttle," the minister there at the time.

114 See ch. 4, pp. 94–95.

115 *NHS*, April 4, 1856.

116 *NHS*, Sept. 12, 1856.

of the church offered prayer, and several speeches were made. "Present at the convention were four little children, born in slavery."[117] The zeal that was so characteristic of the new Republican Party was also present, and it all had its effect. It won the Dublin and Nelson voters who had drifted into the ranks of the Free-Soilers and the Know-Nothings, and it brought out men on election day who had not voted in years. Both towns gave Fremont their votes, by margins of three to one and five to one.[118]

With the bitter campaign over and Buchanan safely elected, the Democrats of Harrisville now had their turn to make themselves heard. What they lacked in numbers, they made up for in spirit. For a couple of weeks nothing happened. Then, on Thursday, November 20, there was a "celebration" or, more precisely, a riot. Three accounts of this appeared in the Keene newspapers. The first was Milan Harris's statement of what occurred, and appeared in the sympathetic *Sentinel*:

Mr. Editor: As certain misstatements and falsehoods about myself have been circulated by the rabble which met at Harrisville on the 20th inst. I deem it my duty to set the truth before the public, and leaving off all preliminaries, will come directly to the point.

The order was given by Timothy Colony to remove the gun from its position where it did no harm, down the side of the hill and 'shoot at Milan Harris' factory and the meeting house, and see if they will take an insult!' (It can be proved, I did not hear it.)

The cannon was moved as near and placed in a position to have as great an effect on my mill as was possible. In the lot they were in, it was pointed directly *at* me and my mill, I being in the yard with my men, and discharged.

It broke glass in the mill. It was discharged again, breaking glass as before. I walked up to them and said: 'Those fellows who are firing this gun must pay the damage you are doing in breaking the glass in my mill.' *I did* not touch any *person. I did not forbid*, or ask them not to fire the cannon, or to remove it.

Immediately Tim Colony caught hold of me and talked as none but Tim can talk, and no man ought to repeat. Upon that a mean puppy in the shape of a man, here known as 'Tim Colony's lap-dog,' (alias F. H. K. of Keene) one who is too degraded to live in any civilized society, as I should judge from his conduct on that day, stepped up, laid hold on me, twitched, yerked, hunched and *struck* me several times.

Having stated as much as I propose to do at this time, I will close without giving the facts in relation to 'Capt.' Rynders ordering an assault on the meeting-house; his bounding into the house at the time a religious meeting was being held which had been established from the formation of the church; the mob bursting in the doors after they had been closed

117 *NHS*, Oct. 24, 1856.
118 *New Hampshire Manual of the General Court*, Vol. I, 1889.

by an officer of the house; the breaking up of the meeting, with Tim
Colony ringing the bell by the wheel, cutting the bell rope, striking one,
kicking another, etc.

Yours & c.,

MILAN HARRIS[119]

In the same issue of the *Sentinel* there appeared another letter
to the editor, telling what happened as seen from a different
vantage point. The writer of this lugubrious, slightly droll account
must have been readily identified by local readers at the time,
but his identity is now unknown.

Mr. Editor

It is with reluctance that I appear before the public at this time, and in
this way, but duty demands it; my position in society and my high
standing in the church demand it. The moral, the upright, the peaceful
and church-going citizens of this village and vicinity who look up to
me, who go and come at the sound of my voice they demand it. . . . The
citizens of this place have been expecting for some time (and with reason
too) that something awful was about to take place, as the Ruffian
Democrats here have been very jubilant since the election of Buchanan
by the slavery extensionists. . . . The long looked for event has this day
transpired. . . .

The morning was ushered in with the firing of a cannon stationed upon
a hill a short distance from my palace [sic], and about on a level with my
residing place in the tower. A few guns were fired, and all was still
again. The people here went about their business, believing the law
would protect them in their rights, both civil and religious. As has been
the cusom here ever since the church was formed, the weekly prayer
meeting was appointed last Sabbath to be held this day at 2 1/2 o'clock
P.M. And as it has been my duty for many years to call the people
together for public worship, I did so on this day; the people who chose
came together . . . little thinking there were any in this once peaceful
and law abiding community who would stoop so low . . . as to molest
them in their peaceful worship. But . . . an *edict* has gone forth from the
leaders of the ruffian party . . . that all ministers . . . who dare cry aloud
and spare not the sins of the nation, and of the ruffian party in particular,
must be crushed out. . . .

Church Bell[120]

Several weeks later, there appeared in Keene's Democratic
newspaper, inappropriately titled the *Cheshire Republican*, a glee-
ful editorial comment on the affair:

Mr. Milan Harris of Harrisville . . . has the . . . proclivity for making
himself conspicuous through the newspapers, and dabbling in very small
matters at home.

He has ruled supreme in his little hamlet, until within a short time, and had the pleasure of doing just about as he pleased, besides obliging others to conform to his notions of temperance, abolitionism, and the other godly virtues, until he has come to believe that there is really no other individual over that way but Milan Harris, except it be his son-in-law, Russell. But he is beginning to find out his mistake at last. . . .

Soon after the presidential election some Buchanan boys procured a cannon, and in a proper place commenced firing a salute, which disturbed Milan so that he or his lackeys commenced tolling the bells.

Upon this, the Buchanan boys, we believe, drew their gun in a closer proximity to Milan's premises than would have been done but for this insult, which brought out the gentleman himself, who endeavored to stop the terrible banging about his ears by his usual dictatorial course.

But the Buchanan boys . . . nobly stood by their gun, and perhaps in defending it did pull somebody's nose a little. . . .

The Buchanan boys also went into the church belfry and took their turn at the bell. In this exercise Milan's lackeys also interfered, and in the operation perhaps some of them were more harshly used than they would have been had they kept about their business and not attempted to insult a democratic salute. . . . It would be very natural to suppose that such a result would follow an insult to a democratic celebration by a pack of intolerant, black-republican-know-nothings.

. . . Milan carried a characteristic article to his smut-mill which was so very illiterate and vulgar that no one would stoop to take any notice of it.

There the matter rested several weeks. But Milan was bound for a fight. He had been disturbed by those Buchanan guns, which he could not silence, and failing to provoke a newspaper controversy he and a promising son-in-law determined to try what virtue there was in law.

So on Monday last two of the Buchanan boys were brought before a justice and fined, one six dollars for damaging the son-in-law and the other three dollars for injuries given Milan himself. . . .

So Milan has got his revenge (he's a flaming Christian) though he has been obliged to make a noodle of himself to do it; and the Buchanan boys of Harrisville have the pleasure of knowing that they fired just about as many guns as they pleased without getting down on their knees to Milan Harris, of Harrisville. . . .

BUCHANIER[121]

121 *Cheshire Republican*, Jan. 7, 1857, in C. C. Wilbur (ed.), *The Old Timer*, No. 59, Dec. 1944.

Concerning this affair, the fact that it occurred more than two weeks after the election makes it seem doubtful that this was simply a spontaneous celebration by jubilant Democrats. Too, it seems more than coincidental that the "celebration" took place the same day as the prayer meeting. The article by "Buchanier" was hardly designed to convince the disinterested of the justice of their cause, and it is easy to sympathize with the victims, but what did this riot really mean?

Its origins were complex but discernible. The presidential race was a heated one in Harrisville because the old-time residents had been aroused by a new, zealous political party, and the hitherto negligible Democratic minority was emboldened by its growing numbers and by new and unquestionably vigorous leadership. But there was more to this than simply a healthy interest in politics. The clash resulted from a convergence of developing tensions and antagonisms. One element was the resentment of the lordly position of Milan Harris felt by newcomers, high and low. The *Cheshire Republican* makes that clear. For his part Milan's choice of the word "rabble" reveals something of his attitude. The leader of the demonstration, Timothy Colony, was no "rabble," nor indeed even a resident of Harrisville, but he was the treasurer of the rival Cheshire Mills. The cashbook of the company even includes a modest entry perhaps not unrelated to the event: "Dec. 1, 1856, Paid for Extras at Celebration, $2.87."[122] Enough to pay for the gunpowder; the rest of the combustibles were ready to hand.

Religious antagonisms were also involved. The church was the creation of the Harrises and the rallying point for the like-minded older residents. The political rally held there and the letter of the sanctimonious "Church Bell" attest to that. Those making the demonstration certainly showed no undue respect for the sanctity of the church. The Colonys were Protestants, but very likely some of the others were Irish Catholics who saw in the brick. meetinghouse a new version of that which they had left behind, the hated Church of Ireland. There were native Americans, too, who were either hostile or indifferent to the church. The church was not flourishing, and it was customary to ascribe the reason for this to the "floating population" that worked in the mills.

Whatever the precise origins of the incident, its significance seems clear. Milan Harris's orderly little mill village had gotten out of hand. Things would never again be quite the same. In the largest sense this Buchanan celebration (Democrats may regret that no more worthy occasion could have been found), better than any other incident in its history, marked the passing of the old order and the emergence of a new and different Harrisville.

There would not again be such an incident as this, but violence, lowered morals, and disturbances of the peace were long to be associated with the new village. In 1860, Constable Milan W. Harris posted a notice in the village:

122 CMR, Cashbook "A."

To whom it may concern:

All persons calculating to disturb the peace tonight, or any other night are required to read warning. . . .

Enough of anything is enough, and if their business does not stop . . . and it is necessary to prove who it is that mars other people's property, proof can be had.

It is known who it is that is disturbing the public peace, and the threats that have been made are also known.[123]

"This business," whatever it was, was not aired in the Keene newspapers, but it is safe to conclude that there was some occasion for the posting of this warning.

The Civil War may have served to drain off some of the turbulent spirits in the village, but incidents continued to occur during the sixties. In 1863, there were clashes between Harrisville residents (including a son of Milan Harris) and Dublin authorities who were trying to enforce a local fishing law.[124] The next year the village experienced the only murder in its history. In 1866, one of Milan Harris's employees, given an irresistible opportunity, ran off with a thousand dollars of payroll money.[125] It is only fair to add that he did return to face the music, but empty-handed after a couple of months of living high off the hog.[126] Nor were the Colonys exempt from such troubles; in the late sixties they appear to have been systematically defrauded by some of their employees.[127] Now these events which have come to light do not add up to a Gomorrah. But, if they are seen as the visible seventh of the iceberg, then it may be said that the moral atmosphere of the village was deserving of some of the criticism it received.

When the storm clouds of war finally broke, Dublin and Nelson rallied to the defense of the Union as stoutly as they had fought for Independence. Political differences were, by resolution, laid aside.[128] In 1862, Nelson voted bonuses for army volunteers: $160 for those enlisting in old regiments, $130 for those enlisting in new ones.[129] One hundred and four men, either born in Nelson or living there before the War, served in the Union army. Of these, twenty-nine died in the grim conflict. Nelson could also claim the highest-ranking Union officer from New Hampshire, Major General Simon Goodell Griffin. The women left behind tended the farms and worked unflaggingly in the Ladies Soldiers Aid Society, supplying quantities of bandages and woolen clothing throughout the four years.[130]

The part played by the village of Harrisville was somewhat different. Certainly its residents joined in the common effort.

123 Hall, *op. cit.*, p. 50, says that this notice is in the CMR, "V, Twenty-two packets of bills, commission letters, accounts, etc.," but the writer has not found it.

124 *NHS*, Nov. 12, 1863.

125 *NHS*, Jan. 18, 1866.

126 *NHS*, March 15, 1866.

127 See ch. 6.

128 *NHS*, April 26, 1861.

129 *NHS*, Aug. 21, 1862.

130 Nelson Picnic Association, *Names and Services . . .*; *NHS*, Nov. 21, 1861; cf. Leonard, 1920, pp. 198–219.

Men enlisted and women did war work. In 1863, with new quotas assigned to the towns, the practice of enlisting men outside of town and paying them a bounty was introduced. Milan Harris served as a paid agent for Nelson and spent nearly five thousand dollars in North Lebanon, New Hampshire, paying bounties.[131] Nonetheless, Harrisville seems to have been occupied with making uniforms rather than wearing them. Following the accepted practice, Henry and John E. Colony, and Alfred R. Harris, among others, hired substitutes.[132] Whatever the explanation, no Harris or Colony from Harrisville fought in the Civil War. The mills turned out great quantities of "Government Blue" flannel, the immigrants came in ever greater numbers, and the village grew.

As for the war's consequences, the wealth and population that came with the boom undoubtedly hastened the day when Harrisville would separate itself from Dublin and Nelson. That the war did more than that seems doubtful. The distinctive character and development of Harrisville were foreshadowed in the 1850s. The Civil War had the effect not of introducing new trends or developments but rather of accelerating those already in existence.

131 Nelson Town Records, Vol. XIII, Town Meetings, 1852–1876, Dec. 2, 1863; Hall, *op. cit.*, p. 52.

132 Nelson Town Records, Vol. III, rear of book.

II Abel Twitchell House, ca. 1774
Photo by Jack E. Boucher, New England Textile Mill Survey (N.E.T.M.S.)

III   Bethuel Harris, 1769–
1851
Courtesy of Mrs. Cecil P.
Grimes, Pennacook, N.H.

69

IV   Deborah Twitchell
Harris, 1776–1855
Courtesy of Mrs. Cecil P.
Grimes, Pennacook, N.H.

V M. Harris Company's Upper Mill, ca. 1830–1832
Photo by Jack E. Boucher, N.E.T.M.S.

VI Milldam Canal
Photo by Jack E. Boucher, N.E.T.M.S.

VII   M. Harris Company's Boardinghouse, Sorting House, Storehouse   71
Photo by author

VIII   The Lower Mill, or Granite Mill, 1846–1847
Photo by Jack E. Boucher, N.E.T.M.S.

IX Cheshire Mills' Brick Mill, 1859–1860
Photo by Jack E. Boucher, N.E.T.M.S.

X Cheshire Mills' Boardinghouse, 1851
Photo by Jack E. Boucher, N.E.T.M.S.

XI   Bethuel Harris House, ca. 1819–1821
Photo by Jack E. Boucher, N.E.T.M.S.

XII   Cyrus Harris–Henry Colony House, ca. 1828
Photo by Jack E. Boucher, N.E.T.M.S.

XIII   Milan Harris House,
1833. Persons in picture are
probably Mr. and Mrs. Milan
Harris
Photo by J. A. French,
Keene, N.H., ca. 1870
Courtesy of Earl B.
Spaulding, Nelson, N.H.

XIV   Almon Harris House
Photo by author

XV C. C. P. Harris and
Abner Hutchinson Houses
Photo by Jack E. Boucher,
N.E.T.M.S.

XVI Milan Walter Harris
House, ca. 1852
Photo by Jack E. Boucher,
N.E.T.M.S.

XVII   Cheshire Mills' Tenements, "Peanut Row," ca. 1864
Photo by Jack E. Boucher, N.E.T.M.S.

77

XVIII   Cheshire Mills Tenements on School Street, ca. 1864
Photo by Jack E. Boucher, N.E.T.M.S.

XIX   M. Harris Company's
Tenements, ca. 1868
Photo by Jack E. Boucher,
N.E.T.M.S.

4 SOCIAL LIFE OF HARRISVILLE, TO 1870

The social life of Harrisville did not develop as a series of isolated entities. Every aspect of it was intimately bound to the growth and change of the mill village and was very much under the influence, for better or worse, of the family that was responsible for the appearance of the village.

An illustration of this point lies in the highly important field of education. Because of profound changes in American society, by 1800 formal education in this country possessed a new shape and a new importance. Improvement of the public schools in nineteenth-century New England was real but frequently came in fits and starts, and from 1805 to 1885 it was especially hampered by the district system. While it corrected earlier evils, this law, by dividing small towns into ten or twelve autonomous school districts, resulted in an "atomization" of the educational system in the state. One gains the impression that, then as now, progress was often belated and inadequate.[1]

Locally, Dublin probably provided its children with an education better than average. It had an educational trust fund established by Samuel Appleton of Boston, and the town's learned minister, Levi Leonard, served on the school committee for thirty-odd years.[2] For the approximately 350 students in the town there was an ungraded common school in each of the town's ten school districts. The school tax money was divided among the districts, one third equally and two thirds according to the number of scholars.[3] There was also a "High School" conducted in the autumn of alternate years.[4] Such a school was announced in the *Sentinel* in the summer of 1835:

A school will be opened in Dublin town hall on Monday the 31st of August to continue 12 weeks, under the care of Mr. Richards, of Brown University.

Thorough instruction will be given in the different English branches; also in the Ancient Languages.

Tuition $3.00 per quarter

Board from $1.00 to $1.25 per week[5]

1 George G. Bush, *History of Education in New Hampshire*, pp. 9–53.

2 Leonard, 1855, pp. 258–259.

3 *NHS*, Jan. 20, 1854.

4 *Centennial History of Education in Dublin*, pp. 9–10.

5 *NHS*, Aug. 20, 1835.

Neighboring Nelson's school system was similarly arranged and included a high school which some years at least had both a fall and a spring term, each of twelve weeks.[6]

For Dublin and Nelson such a system may have worked tolerably well. However, the growing village of Harrisville, squarely astride the town line, created something of a problem. In 1840, when Dublin was first divided into school districts, the committee doing the job reported in favor of a half district for the village of Harrisville, but the remainder of the village on the Nelson side would be made into the other half of the district. Of course, by this arrangement the Dublin side of the village received only half as much of the tax money that was divided equally as the other school districts of the town received. At the time the residents of Harrisville did not object to this division, but Nelson did not oblige by forming the other half of the district. What the arrangement was during the 1840s is uncertain. Nelson must have had a school district that included its half of the mill village. The Dublin side, with its half-district allocation of funds, did what it could. The little brick vestry was put to use as a school building and was probably run on some sort of a subscription basis.[7] (Plate XX)

By the end of the forties when the village had enlarged considerably, its residents made an effort to have this half-district rule changed. Dublin shrewdly countered with the proposal to call its half of the village a whole district "on condition that the factory hands who were under twenty-one should not be counted as scholars unless they attended the school." When the villagers had figured this out, they opposed the plan on the grounds that they would not draw as much money as they did before. Dublin, "apparently vexed at their course," let the old system stand and also voted to exclude the factory hands from the count. The Harrisville residents were outvoted, and there the matter rested until the end of 1853.[8]

It was the practice of the County Commissioner of Common Schools, in conjunction with the local school committees, to visit his schools and then make an annual report on their condition. This lengthy report was printed each December in the *Sentinel*. The one that appeared there in 1853 must have opened the eyes of the *Sentinel* readers in Harrisville, for their school came in for unrivaled criticism. It began with a factual description: "The school in district #8, in Harrisville . . . had been in operation 9 weeks, and numbered in all 30 pupils. They were divided into seven classes in Arithmetic, three in Geography, two in Grammar, four in Reading, the same in spelling, and one in

6 *NHS*, Aug. 7, 1839, Feb. 5, 1840.

7 *NHS*, Jan. 6, Jan. 20, 1854. The vestry was 26 by 32 feet and consisted of one large room and two cloak rooms.

8 *NHS*, Jan. 20, 1854. There is some ambiguity in the wording of this account, but this is the writer's interpretation of it.

each of the studies of Writing, Algebra, and History."[9] The
commissioner went on to say that "its record was black enough"
and proceeded with particulars. Deportment was bad. There was
a daily average of four students late and four absent. He could
not speak of the method of instruction in very favorable terms.
The schoolhouse itself was "very inconveniently constructed
inside." After some minor comments, he concluded,

It is true that they have to some extent a floating population in that
district, but no good or valid excuse can be rendered for such a school
as they have had there the present season. . . . I cannot doubt that if they
had looked after their school with one tenth part of the care and interest
that they have after their workshops and other business matters it would
not have been in the condition in which I found it.[10]

The commissioner's candid comments put the fat in the fire
and started a newspaper battle that lasted two months. With
reason, Milan Harris took the criticism personally and wrote a
lengthy answer to the *Sentinel*. It was not very convincing. He
lamely attempted to explain away the criticism of absences,
tardiness, and deportment. He struck back at the visiting officials:
"We were present at the examination, and saw nothing that
looked worse than the sneering and side glances of some of the
committee when a scholar happened to make a mistake in his
recitation. . . ." Then, taking another tack and "allowing the
school was not what it should be," he explained how Dublin
divided its school money, "thus swindling the district out of
their proportion of the school money." He dolefully concluded,
"Well may a district thus used lose their interest and ambition."[11]
The school commissioner wrote a temperate and factual answer
to this, ignoring the many provocative points in Milan's letter.
"I examined the record, and, with one exception, it was the
worst one I had seen in some eighty schools that I had previously
visited. . . . How improvements can be suggested without point-
ing out faults is more than I can tell."[12] On that note he quietly
dropped out of the altercation.
His place was taken by Jonathan K. Smith, chairman of the
Dublin School Committee, who felt called upon to defend
Dublin against Milan Harris's charges. He too wished only to
lay the fact before the *Sentinel* readers. He traced the history of
the school district in Harrisville, allowed as how he thought the
district was now large enough to be counted as a whole district,
and was confident that Dublin would do justice in the case. But,
he maintained, the school committee had always given the village

9 *NHS*, Dec. 23, 1853.
10 *Ibid.*
11 *NHS*, Jan. 6, 1854.
12 *NHS*, Jan. 13, 1854.

its full share of attention, had repeatedly called attention to the condition of the school, and had sought to improve it. "Whether there has been a prejudice there, that has served as a non-conductor of these influences, or a devotion to business so intense as to neutralize them, or whether the influence of the transient population is so powerful as to paralyze all effort I do not decide...." Later he made the point, claiming all present at the time of the examination as his witnesses, that Milan Harris himself had laid almost the whole blame of the school's bad record upon the transient population. If he did say this, Milan changed his plea after reading the school commissioner's indictment. After all, he had to live in the village.[13]

In the course of his argument Smith excused the poor deportment of the scholars by explaining what the commissioner must have had in mind when he wrote that the school was "very inconveniently constructed inside." The brick vestry was built at the edge of the pond and actually projected out over it. The lower floor was a double one with the inch space between the boards stuffed with woolen flocks. Milan Harris claimed, "probably it is the warmest floor in town."[14] It is to be hoped that it was not, for as Smith pointed out:

At the last examination it was cold and windy, and notwithstanding the warmth of the floor, it so happened that the wind came up quite freely through it—woolen flocks and all—and reduced the temperature of everything above the floor so near to zero that there was very little fun in it for teacher, scholars, or visitors. The wood being very poor, the stove could do very little in raising the temperature of the room, and the consequence was that the whole school was obliged to stand crowded about it, probably two-thirds of the time; and yet the deportment of the scholars was as good as could be expected from any scholars under such unfavorable circumstances.[15]

As the battle continued, the participants emphasized what they saw as the main trouble with the school. Milan Harris complained of the half share of revenue that the district received, while it paid "about or quite the heaviest tax of any district in Dublin."[16] When rebuked by Smith for his preoccupation with this aspect of the problem, businessman Harris indignantly retorted,

The gentleman seems to think it is a sin to think about money—that it is quite enough for one class to pay the money and not trouble themselves about the appropriation of it. Well I must confess I had some different ideas about the matter.

13 *NHS*, Jan. 20, Feb. 3, 1854.
14 *NHS*, Jan. 6, 1854.
15 *NHS*, Jan. 20, 1854.
16 *NHS*, Jan. 27, 1854.

I was so *simple* as to suppose that the very *fact of paying* the money created an interest in the spending or appropriation of the same, not to say anything about the *right* in the premises.[17]

Chairman Smith pointed out that Milan Harris had made no retractions in the course of the controversy, despite his contradictions and errors of fact. He concluded that if the school was to be improved, "the improvement must come mainly from the people of the district, rather than removing grievances from abroad."[18] While money was necessary to have a school at all, he added, to have a good school required something more, "viz. a correct idea of what constitutes a good school, and an enlightened zeal in all efforts, public and private, which have a tendency to promote its welfare."[19]

These letters are revealing. The ones written by the County School Commissioner and the Chairman of the Dublin School Committee reflect creditably on those gentlemen and the offices they held. On the other hand, Milan Harris does not come off so well; his own words show him to have been churlish, intemperate, silly, and occasionally less than candid. These letters also describe in concrete terms the educational problems of the village. They provide another example of the discontent of the Harrisville residents over the treatment accorded them by the parent towns. And the letters reveal that all was not well within Harrisville before the arrival of the immigrants, or even Tim Colony. How closely all the criticism struck the mark is indicated by the extravagant nature of Milan Harris's rejoinders.

Although it took time, some good came from all this furor. Everyone concerned must have had some sober second thoughts about the condition of the school. In 1857, Union District Number Eight was established, including all the residents of Harrisville on both the Dublin and Nelson sides of the line.[20] The same year a new schoolhouse was built for the district,[21] a barnlike two-story white frame building set on the high ground of the Dublin side of the village. There was one large room on each floor, and the building had a magnificent view of Mount Monadnock for the teacher, if not the scholars, to gaze at on warm spring afternoons. (Plate XXI) Two years later a new County School Commissioner's report on the district indicated pedagogical improvement as well:

The people of Harrisville have what they deserve: for when a school house, in all its arrangements, is the embodiment of a generous public spirit, a good school is the reward of merit. Perfection we do not expect

17 *NHS*, Feb. 6, 1854.

18 *NHS*, Jan. 16, 1854.

19 *NHS*, Feb. 3, 1854.

20 Nelson Town Records, Vol. III, 1857, in Michael G. Hall, "Nelson, New Hampshire, 1780–1870," p. 45.

21 *NHS*, Jan. 29, 1858.

but it was gratifying to learn that this school has been steadily advancing, and was never so good as now.[22]

As an example of what good teaching meant in those days, the report of the same commissioner on the school in Pottersville is worth quoting:

When scholars, reciting in any study, can take any subject named, first define it as a whole, then give its several divisions, take them up in order, and tell all about them; and when one sits, any other in the class is able to continue the recitation from the precise point where it was left, with scarce a word from the teacher, we are assured they know what it is to study and recite, and have been taught by one who understands what it is to educate.[23]

Before Harrisville became a town, then, there had been substantial progress made in education. Whether it was adequate for the times is another question. The common school had certainly been improved, and ten years after the new school was built, a high school was opened in the village.[24] On the other hand, there were new problems, many associated with the influx of immigrants. French-speaking Canadians came in large numbers. Some of the adults were illiterate. In 1860, there were only five illiterates in the area of Harrisville, less than one percent of its population.[25] Four of these people were Irish-born. In 1870, there were at least twenty-eight illiterates in the area of the town, including twenty-five Canadians, two English, and one Irishman. These amounted to 4 percent of the population, and actually it must have been higher than this. Not one of the twenty-nine foreign-born workers living in Milan Harris's boardinghouse is listed as being illiterate in the 1870 census, a highly improbable situation.[26] Milan Harris never put that kind of a premium on education.

There were other unfavorable factors. Although it never became large, there was an increase in child labor in the mills.[27] The native-born population in the rural areas, drained as it had been for half a century of its most enterprising youths, may well have suffered a loss in its intellectual caliber. Here the historian is on thin ice, but there are indications that this was so.[28] For the half century down to 1870, it would seem doubtful that the population of Harrisville was maintaining, much less improving, its general level of education.

During this time other educational influences and organizations in Harrisville were conspicuous by their absence. Both Dublin and Nelson had formed social libraries in the 1790s. The one in Nelson, though it was reorganized several times, seems to have

22 *NHS*, Feb. 2, 1860.

23 *NHS*, Feb. 26, 1861.

24 *NHS*, Nov. 21, 1867.

25 Census, 1860, Dublin, pp. 65–68, 86–91, Nelson, pp. 60–64. These figures are exclusive of the illiterates in the Dublin Town Farm, which was within the area of Harrisville.

26 Census, 1870, Nelson, Vol. XV, Dublin, Vol. XVI.

27 See ch. 6.

28 See ch. 7.

had an active existence for at least forty years and perhaps much longer. A share in this library cost two dollars, to be paid either in cash or books. Its catalogue shows a heavy concentration on works of history, travel, biography, and philosophy. One article in the society's constitution, apparently a compromise between two schools of thought on the purpose of the library, specified that "no more than a third of the money expended shall be for religious books. . . ." One of the leaders of this library was Amos Heald, Esq. A farmer, carpenter, and mill owner, he had settled in Nelson in 1790 and was for thirty years active in the political, business, religious, and cultural affairs of the town. Another leader was Henry Melville, storekeeper and one-time partner with Milan Harris in the Upper Mill. Although the library had a goodly number of members, at least down to 1835 there is no record of any member of Bethuel Harris's family ever having so much as borrowed a book.[29] There is no evidence either of there ever having been such a subscription library in Harrisville. The town library, when it was finally formed in the late 1870s, succeeded only after considerable opposition.[30]

Of lyceums and similar self-improvement societies there were none. Perhaps the early village was simply too small for these things. The newspapers of Keene and Peterborough contain an impressive number of notices of meetings of such groups, but those places were at an inconvenient distance for residents of Harrisville to go of an evening.

In Dublin, too, there was considerable activity of this sort. During the 1820s, its library society held meetings "for literary purposes" during the winter months. In the next decade a lyceum was formed. At first weekly and later fortnightly the citizens met on Wednesday evenings during the long, dark months from November to March to hear reports, lectures, and debates. In addition the young people of the town formed a "Society for Mutual Improvement," or "Young Lyceum," and held meetings on alternate Wednesdays.[31]

In 1844, the lyceum there was superseded by a Common School Association. For at least ten years this association held meetings during the winter months in the several school districts, two or more in each week. Levi Leonard, who, with Jonathan K. Smith of the great school debate, was a leader in all these groups, said of these school meetings, "Lectures were delivered on educational subjects; and a great variety of topics, relating to the instruction, discipline, and improvement of common schools, were discussed in a free and social manner."[32] Again, there is

29 Record book of Nelson "Philosophical Library," 1808–1824, and "Literary Society," 1824–1835.

30 See ch. 7.

31 Leonard, 1855, pp. 202–203.

32 *Ibid.*, pp. 263–264.

nothing to indicate that any of the Harrises ever took part in these groups, though Milan Harris, and his village too, might have derived some benefit from his attending the meetings of the Common School Association.

The interests of the Harris family were elsewhere and nowhere more than in local religious affairs. Bethuel and Milan were among the founders of the Nelson Sunday School Association formed in 1825. This association, which regularly held school after the Sunday morning service, had in the 1830s as many as 20 teachers and 136 scholars. Both Milan and his brother Almon served as superintendents of this association.[33]

Most of the early residents of Harrisville who belonged to a church attended the Congregational Church in Nelson, but the four-mile trip each way on a Sunday was irksome, and a division was bound to come. It started in the 1830s, when the villagers began to hold religious meetings on Sunday evenings, probably in some of the Harris homes, and then organized a Sunday School, which at first met in one of the mill storehouses.[34] About 1838, Bethuel Harris proposed erecting a structure more suitable for religious purposes, one which might also be used as a schoolhouse. The "vestry" as the little one-story brick building came to be called, was finished in 1840. Its cost was about one thousand dollars, of which Bethuel Harris contributed over two thirds.[35] Whether to economize on real estate or not, the vestry was built at the edge of the pond and extended out over it. The gabled end faced the village, and the roof was embellished with a "stylish, carpenter-Gothic cupola."[36] However frigid its interior may have been in winter, the exterior of the vestry was handsome indeed and in complete harmony with the rest of the village.

Many years later Charles C. P. Harris wrote that "at the time this vestry was built no one had supposed that a church would be organized in this place for years."[37] Since the vestry on which the Harrises spent a thousand dollars would have been completely inadequate as a church building, this statement of Bethuel's youngest son may be taken at face value. However, the villagers had not reckoned with the fragmenting forces inherent in American Protestantism. According to a contemporary account the break with the church in Nelson and establishment of a church in Harrisville was precipitated by a controversy over baptism. An evangelist was making great efforts to draw off Nelson people to a Baptist Church in Munsonville, a small factory village in the northern part of the town of Nelson. The Nelson minister rose to the challenge and preached on the subject

33 Nelson Sunday School Association, Records, 1825–1839.

34 Robert F. Lawrence, *The New Hampshire Churches*, p. 267.

35 D. H. Hurd (ed.), *History of Cheshire and Sullivan Counties, New Hampshire*, p. 217.

36 Ada Louise Huxtable, "Progressive Architecture in America: New England Mill Village, Harrisville, New Hampshire," *Progressive Architecture*, Vol. 38 (July 1957), pp. 139–140.

37 Hurd, *op. cit.*, p. 217.

of baptism. The congregation then chose sides, with most of the Harrisville residents supporting the Nelson minister, Mr. Ballard, and his position in favor of "sprinkling." The faction favoring immersion may have been strong, for after a time the Harrisville people decided to form their own church and tried, vainly, to induce Mr. Ballard to be their minister. Without deprecating Harrisville's lively appreciation of the importance of this theological question, it seems clear that this controversy only precipitated a break that must have come sooner or later. The real cause was simply the distance from the town of Nelson and the divergent life of the mill village.[38]

The church in Harrisville was organized in 1840. Bethuel Harris and about thirty others petitioned the church in Nelson for letters of dismission that they might organize a church in the village. Of these thirty, eleven were Harrises or married to Harrises.[39] The petitioners also requested the Nelson church to join with them in calling an ecclesiastical council for the purpose of organizing the new church. Their former brethren complied with this petition, and the nearby churches at Swanzey, Troy, Antrim, and New Ipswich, New Hampshire, and one in Warwick, Massachusetts, were invited to serve on this council. For some reason the Congregational Church in Dublin was not invited. The council met, and the new church was duly instituted at services on September 22, 1840. Thirteen members were added at this time, making forty-four in all.[40]

As the pretty little vestry was clearly inadequate for a church, a new building was soon erected. Bethuel Harris again took the lead in this and, in fact, underwrote the construction. He proposed that the church should raise what it could to defray the expense, and he would supply what might be lacking. This turned out to be about three fifths of the total cost of $3,500.[41] For this sum the villagers got a handsome church. It was built of the same red brick as the rest of the village, had a gleaming white pedimented gable inspired by the Greek Revival, and was crowned by a graceful white spire. The church was placed just back from the canal, which served as a reflecting pool for its lovely facade and excellent lines. (Plate XXII) It is impossible now to tell much about its interior. It held 350 people[42] and was well lighted by three large windows on either side of the church. Milan Harris donated an organ and employed an organist.[43] Finally, the completed church was dedicated in August 1842.[44]

In the course of the next few years old Bethuel made other contributions. He presented a permanent fund of $1,200, and he gave each of his children a slip, or pew, in the church. Just behind

38 Lawrence, *op. cit.*, pp. 267–268.

39 Bemis MSS, Box 6, XXXII, 3, "The Church in Harrisville."

40 Hurd, *op. cit.*, p. 218. For a list of the charter members of the church, see Bemis MSS, Box 6, XXXII, 3.

41 Hurd, *op. cit.*, p. 218.

42 Hamilton Child, *Gazetteer of Cheshire County, New Hampshire*, p. 181.

43 *Ibid.*

44 Hurd, *op. cit.*, p. 218.

the church, on what is called Harrisville Island, he provided for a cemetery and presented each of his children with a lot there.[45] However, of these two important village institutions provided by Bethuel Harris, only the cemetery was destined to flourish.

Virtually no records of the Harrisville church now exist, but its history can still be traced. For perhaps ten years this Evangelical Congregational Church at Harrisville[46] showed vigor and growth. The first minister had been hired in 1841 and was paid an annual salary of $400.[47] He may or may not have received the use of a parsonage too; by the 1850s the church was providing one for its minister.[48] The Reverend Mr. Whiton served the church from 1841 until his death five years later. His salary did not compare with that which, for example, the minister in Dublin received,[49] but, small as it may now seem, it was probably more than the mill workers were earning, and there is no reason to suspect he died of malnutrition. In this first decade there were "two extensive revivals," resulting in fifty-odd "hopeful conversions," but about half of these converts soon moved away.[50] In the late forties there were some sixty members, the Harrises themselves amounting to about a fifth of the total.[51]

Before long, however, the church's progress faltered. That much seems beyond doubt; the important question was why. In later years the cause would be attributed (along with much else) to the large number of Roman Catholic immigrants working in the mills, but the church was on the downgrade before the immigrants arrived. A contemporary history of the Congregational Church in New Hampshire explained that "the manufacturing character of the population, making for much temporary residence, operated unfavorably to the growth and prosperity of the church."[52] And Charles C. P. Harris wrote in 1886, "Virtually this church has been a missionary church, many having come here to labor in the mills, and, after being here a time, united with the church; afterwards making their residences at other places, they asked and received letters of dismission to other sister-churches."[53] This may all have been true enough, but as the explanation for the church's stagnation it is unsatisfactory. The piety of the early mill girls in Lovell was always commented upon, and they were certainly only in temporary residence there. There was more than this to the decline of the Harrisville church.

Charles A. Bemis, who as a boy lived in Pottersville in the 1850s, wrote a historical sketch of the church in Harrisville, and he too testified to its early loss of vitality:

45 *Ibid.*, p. 214. Bethuel may have provided for only a family burial plot. Years later the town of Harrisville bought the lots from some of the sons.

46 *NHS*, Jan. 5, 1842.

47 Lawrence, *op. cit.*, p. 268.

48 See *Map of Cheshire County, N.H.*, 1858.

49 Levi Leonard received $600 in 1820, his first year in Dublin. See Leonard, 1855, p. 179.

50 Lawrence, *op. cit.*, p. 270.

51 "A Confession of Faith and Form of Covenant of the Evangelical Congregational Church in Harrisville, N.H."

52 Lawrence, *op. cit.*, p. 269.

53 Hurd, *op. cit.*, p. 219.

Although the church began with great enthusiasm, was composed of strong and earnest men and women, was the only church in the community, ministered to large and loyal families, had its flourishing Sunday School and prayer meeting, somehow it failed to grow, to gather in the young people as was desired. . . . The children in the Christian homes were growing up but experienced no spiritual awakening and manifest no disposition to come into the church. Before ten years had passed the tide apparently had turned. Each year the losses exceeded the gains and the church was slowly losing ground.[54]

Bemis made no reference to the floating population. His point that the young people were not coming into the church was quite something else.

Bemis also provided some statistics on the church. He reported that in 1855 the church had nineteen male and thirty-six female members, that its benevolences totaled eighty dollars, and that it had an average congregation of ninety people. The figure for the "average congregation" seems large for a village of less than 200 people, but it may have been largely a question of what else to do of a Sunday in Harrisville. The disparity of the sexes among the members and the meagerness of the benevolent contributions tell their own story. Money was raised by circulating subscription papers once a year, hardly a method calculated to bring the mill workers to the effective support of the church.[55] It was in fact only the fund left by Bethuel Harris and a slightly larger one left by a Miss Chloe White that kept the church in repair and paid the minister.[56] Most of the ninety people in the church on a Sunday morning must have been riding free.

In point of fact the floating nature of the population seems no more nearly the whole explanation for the condition of the church than the tax inequities were for the condition of the school. Rather it seems that a large part of the village population utterly lacked any religious zeal, and the church was not likely to so inspire them. The red brick church was not the result of any community effort but was established by the mill owners and, in large part, run by them. It may well be suspected that the pulpit was used to fulminate against the evils of the Democratic Party, the political party that must have had the allegiance of a good many of the mill workers. The attack on the church during the Buchanan "celebration" was an indication of just how alien it was to some of the villagers. To give the church some real vitality and popular support, the Harrises might better have hired that immersing evangelist in Munsonville of whom they so strongly disapproved. He might have been able to use the vestry-over-the-water to good advantage. More important, he might

54 Bemis MSS, Box 6, XXXII, 3, "The Church in Harrisville."

55 Lawrence, *op. cit.*, p. 269.

56 *Ibid.*

well have stirred the interest of the floating population of mill workers. In the largest sense it would appear that the prompt decline of this Congregational Church was one of the earliest evidences that the paternalism of the Harrises was not the answer to the needs of the growing village.

By way of comparison the other church in the area of Harrisville was a Baptist Church in Pottersville. The first mention of it was in an article of the Dublin town meeting of 1784, when the "Baptist Society" petitioned to be excused from paying the salary of the Congregational minister. The petition was granted by the town, and the next year the Dublin Baptists organized themselves into an independent church. At its beginning there were thirty members of the society, evenly balanced between men and women. For the first dozen years they held their meetings during the winter in private homes and during the summer in a barn.[57]

In these years when Dublin and Nelson were still growing, the Baptist Church in Pottersville prospered. In 1794, Elijah Willard came to the village from Fitchburg, Massachusetts, to teach school. That same year he was ordained as pastor of the Baptist Society. "His was the longest and most successful pastorate the church ever enjoyed."[58] Three years later the society built its first house of worship, "after the usual style of those days."[59] The main building was forty feet by thirty, had a porch at each end, and was located on top of a hill, "as was then customary."[60] Finally, in 1815, the church petitioned the legislature "to be incorporated into a body politic," the petition being signed by Elijah Willard, Amos Heald of Nelson, and five others, in behalf of the church and society.[61]

Among their records in these early years are some interesting letters, which provide considerable insight into this Baptist Society. One of these, dating from about 1802, was a letter of excommunication:

The Baptist Church of Christ in Dublin to Brother Ezra Morse. This is to inform you that we have a command from our Lord Jesus Christ to withdraw from every brother that walks disorderly [.] We think it has been made evident to us by a number of witnesses that you have thus walked.–

1st In saying you could buy any man with a Glass of rum which we think is speaking evil of our fellow men which when the Chh made it a matter of labor as we thought in faithfulness and we thought it was proved you denied and concluded the witnesses were mistaken.

2nd by your unchaste conduct towards your Neighbors wife which we think is a heinous crime according to Job 31, 11—1st Cor. 7, 2, and

57 Hurd, *op. cit.*, p. 215.
58 *Ibid.*
59 *Ibid.*
60 Leonard, 1855, p. 193.
61 *NHS*, Dec. 23, 1815.

many other passages, and when it became a labour in the church and you undertook to confess your fault in the matter it appeared you kept back part, and saying you had no evil design or that you did not mean any hurt appears to us an inconsistency.

For all which we view it necessary to withdraw the hand of Fellowship from you untill God by his Grace shall return you again by repentence. By order of the Chh.

William Bank, Chh Clerk[62]

As there were other letters of a similar nature, it is not surprising to find that the membership of the society fluctuated considerably. The trend, however, was for the membership to decline. In the ten years following 1818, only five new members were admitted. By 1829, when Pastor Willard resigned and the church was reorganized, the whole number of admissions had been 227. Of these, 66 were dismissed to other churches, 29 were excluded, and 58 died. How many of the remaining 74 were in communion with the church in 1829 is unknown, but the reorganized church that year had 44 members.[63] An account of the church's history admits that in 1829 a successor to the Reverend Mr. Willard was sent by the State Convention, "the church being in a very low state. . . ."[64] Elijah Willard was "highly esteemed and dearly beloved by the church and by the people generally,"[65] but his pastorate was too long for the good of the church, even though his long tenure may have been as much the result of the church's decline as its cause.

In the records of the next twenty years, until it was again reorganized in 1849, it is difficult to read any improvement in the condition of the church. During this period the society had no less than ten ministers.[66] In 1842, the brethren sent to the regional Baptist Association a doleful letter rehearsing their failures. Looking to the future, it concluded,

Our peculiar location, and the churches which have risen up around us prevent our even anticipating the measure of prosperity we once enjoyed . . . but we have reason to think we should enjoy greater prosperity than we do, if we unitedly and fully come up to the Lord against the Mighty. But in consequence of being so few and feeble, we sometimes feel, however improper, almost, if not quite discouraged. . . .[67]

The church also reported that it had a membership that year of fifty-five people and that these contributed a total of ninety-eight dollars in benevolences.[68]

The condition of the church was probably not helped when some of its members took up with Millerism. The leader of this

62 A book of Records for the Baptist Church of Dublin, 1785–1857.

63 Leonard, 1855, pp. 191–192.

64 Hurd, *op. cit.*, p. 215.

65 *Ibid.*

66 Leonard, 1855, p. 192.

67 A book of Records for the Baptist Church in Dublin, 1785–1857.

68 *Ibid.*

sect won wide attention and frenzied support when he predicted that March 21, 1843 was to be the date for the Second Coming of Christ. In Pottersville, the Millerites held meetings in the Baptist Church during 1842 and early 1843, but when the appointed day came and went just as any other day, "the excitement soon subsided." Probably it was not mere coincidence that a week later the church dismissed its minister.[69]

The next year the Baptists did manage to erect a new meeting-house or rather remove the old one to a new location, one town lot to the westward, where they fitted it up in "modern style."[70] As was the case in the church in Harrisville, the Baptist Church rented its pews. All but a few of the pews in the remodeled church were subscribed to that year by twenty-four members who paid amounts ranging from four to fifty dollars.[71] However, the new meetinghouse did not bring the hoped-for revival. The succession of ministers continued, and the church was again reorganized in 1849, a procedure that may have been the ecclesiastical equivalent of voluntary bankruptcy in business.

For the next two decades there is little available evidence, but the church seems to have held its own. At the time of this second reorganization, "sixty members renewed their covenant obligations."[72] Within a few years the membership was reduced to forty-four, the same number as in 1829, but by the end of the sixties the membership may have risen.

In these last years before Pottersville was included in the new town of Harrisville, there was occasional mention of the Baptist Church in the columns of the *Sentinel*. In 1858, there appeared an account of a festival held at the church on the Fourth of July. Jonathan K. Smith, the corresponding chairman of the Dublin School Board, was President of the day. He and others made "interesting remarks upon patriotic and national subjects (slavery not excluded)." The Declaration of Independence was read, the church choir sang, and the ladies of the church provided a collation which the diplomatic writer of the account described as "ample."[73]

At the society's annual meeting in 1867, the members voted to move their church once again to a new location and remodel it, "and also to dispense with the services of the pastor while repairing the house. . . ."[74] The reason for these removals is unexplained; perhaps they were the simplest way to get rid of unwanted ministers. The records also show a plan of the new meetinghouse and the names of the subscribers to the slips, or pews. Most of the twenty-eight subscribers lived in Pottersville, but at least one was a Harrisville resident. He was Zophar Willard,

69 Leonard, 1855, pp. 192, 194.

70 *Ibid.*, p. 193.

71 A Book of Records for the Baptist Society in Dublin, 1815–1903.

72 Hurd, *op. cit.*, p. 215.

73 *NHS*, July 16, 1858.

74 Hurd, *op. cit.*, p. 216.

grandson of the Reverend Mr. Willard and enterprising business-man in the mill village.[75] On the basis of such meager evidence it would seem that the Baptist Church in 1870 was at least no worse off than it had been for a number of years.

In appraising the history of the Baptist Church and comparing it with its neighbor in the mill village, it is important to remember that, unlike the Congregational Church in Harrisville, this church depended upon a steadily declining community. Nor did it have the support of a family as wealthy and powerful as the Harrises. Nonetheless, the church was probably as strong num-erically in 1870 as it was a half century or more before when the population had been at its peak. When the church went downhill, the members humbly took the blame onto themselves rather than point to external causes, and there was always a small group ready and able to reactivate the society. Among its members, early and late, were vigorous and enlightened citizens like Amos Heald, Jonathan K. Smith, and Zophar Willard. If there was little reason for self-satisfaction in the Baptist Society, a compari-son with the Congregational Church in Harrisville in the same period must be, on the basis of existing evidence, distinctly favorable to the church in Pottersville.

Another aspect of the mill village's social life that received staunch support from the Harrises was the cause of temperance or, more precisely, total abstinence. Not that they began the movement, even locally. Temperance was one of the hardiest of all reform movements in the thirties and forties, and it was a popular cause in the hill towns of Cheshire County. Nelson formed a Tem-perance Society in 1829. The next year it had eighty-two members, and not a single Harris was among the officers.[76] By the early forties nearly half the population in Dublin and Nelson belonged to the local temperance societies, and it was claimed that no stores in either of the towns sold ardent spirits.[77]

By this time the Harrises recognized the importance of temperance and joined in the crusade. It would have been only natural if they had been converted to the cause by the single argument of the reduced working efficiency of drinkers, but their motives were probably broader than this. Milan Harris was active in the county total abstinence movement by 1842, and at the end of the decade he was elected President of the Washington Total Abstinence Society, for Cheshire County.[78] Almon and Charles C. P. Harris were also active in promoting the movement locally.[79] The one Harris of whom it may be suspected that he did not share the family viewpoint on abstinence was old Bethuel. His name is missing from the list of those active in the movement,

75 A Book of Records for the Baptist Society in Dublin, 1815–1903; Child, op. cit., p. 179.

76 NHS, Nov. 12, 1830.

77 NHS, March 2, 1842, April 12, 1843.

78 NHS, March 2, 1842, Feb. 22, 1849.

79 NHS, Feb. 14, 1844.

and one report on temperance in Nelson in 1843 indirectly
admitted the blemish on the family escutcheon: "The village of
Harrisville . . . contains about 15 brick buildings, mostly large,
all of which, except one were erected without the assistance of
rum, and that was built 24 years since."[80] The dates point directly
at Bethuel Harris's great house, but, whatever effect the rum
might have had on the workers, the house itself did not suffer.
After nearly a century and a half it still stands straight and firm,
without a hint of its dark origins.

However popular the temperance cause was in Dublin and
Nelson, the residents of Harrisville were plainly of two minds on
the subject. This is revealed by another of those newspaper
battles that tell so much of the town's history. The whole thing
was rather oblique and mysterious, but the libel laws being what
they were, one cannot much blame the writers. It all began in
March 1856 with a brief but journalistic account of the store in
Harrisville, whose proprietor was selling goods at from 20 to
30 percent discount and attracting a large trade thereby.[81]

The next week the same notice appeared in the *Sentinel* again,
but then with a neat twist it declared that the crowd was there to
save from conflagration the occupant of the store, "who on the
eleventh of March voted the *Loco foco* ticket, for the avowed
purpose of selling rum in Harrisville." The letter concluded with
the remark that on the previous Sunday five persons in the
village had been drunk all day, and that this was now no uncommon occurrence in Harrisville. Although the letter was
unsigned, the editor added that "the above sent us by a prominent
and very respectable citizen of Harrisville, is published as an
advertisement."[82] An unusual advertisement, it must be allowed.

The occupant of the store, one D. W. Clement, was not the
man to let such a charge go unanswered, and two weeks later the
whole letter was reprinted, together with Mr. Clement's
comment:

In your paper . . . appears the above . . . without any signature, said
to have been sent you for publication by a prominent and respectable
citizen of Harrisville. The said citizen (he cannot claim any title to
respectability, for no respectable person would descend to so scurrilous
and lying a communication for publication) is no other than Milan
Harris, who from his position in business, the community had a right to
expect truth and decency but he has not displayed either. He says . . .
'the occupant of the Bradley Store voted the loco foco ticket for the
avowed purpose of selling rum in Harrisville.' *It is false in every particular.*
Now if Harris wishes to slander the people of the village of Harrisville,
let him come out over his own signature, if he wishes to hear some part

80 *NHS*, April 12, 1843.
81 *NHS*, March 28, 1856.
82 *NHS*, April 4, 1856.

of his own history, which might not be so agreeable, and which he might not wish to be made public, and furthermore the church to which he belongs would be doing him a benefit if they would caution him not to bear false witness against his neighbor, and advise him to attend to his own business.[83]

After printing this, the editor of the *Sentinel* indicated he would like to drop the matter, but Milan Harris had to have the last word. He wrote another letter, lamely explaining that he had only transmitted the accusing letter from its writer to the *Sentinel* and that he believed the facts in it to be true. He concluded by hinting darkly,

I apprehend there is something behind what has come out, which instigated the writing of the article entitled 'Fire & etc.' judging from report some 'hole' too *foul* and *dark* for my limited conception of language to describe, therefore I will not make the attempt but hope the authorities may find the way into it.[84]

With this final insinuation by Milan Harris, the newspaper battle inconclusively ended, but, whatever its issue, it is not hard to believe that if Clement still kept the village store at the end of that year, he was probably in the front ranks of those who carried out the Buchanan "celebration."

Although like other reforms temperance as a national movement was eclipsed by the crusade against slavery and its aftermath, in Harrisville the supporters of the movement continued to be active. In 1869, the village instituted a new temperance society, "Beacon Light Lodge No. 93, International Order of Good Templars."[85] It soon had eighty-five members.[86] Needless to add, the Harrises and their relations were active in this new society. Although this temperance group at times enjoyed a considerable measure of public support, just as in the case of the church it had an official air about it, and it was very likely a partisan issue between the Harris faction and their opponents, the old residents and the new.

The Harrises also fostered organizations that the village residents viewed with more unanimous approval. Such a one was the Harrisville Engine Company. This was incorporated by act of the state legislature in 1839, under the names of one Alpha Knight and Bethuel, Milan, Almon, and Charles C. P. Harris.[87] The Harris mills represented an important investment to protect, but still it seems that this was a very progressive step for so young a community. The Harrisville Engine Company soon boasted of an efficient engine, for in 1845 the *Sentinel* told of its

83 *NHS*, April 18, 1856.

84 *NHS*, April 25, 1856.

85 *NHS*, May 6, 1869.

86 *NHS*, Aug. 26, 1869.

87 *Laws of New Hampshire*, 1839, c. XII.

timely arrival, saving the barn, if not the house, of Chauncy
Barker, who lived at a distance of a mile and a quarter from the
village.[88] By the 1850s and 1860s, the woolen mills had their own
systems for fire protection, apparently very up-to-date and
efficient. These consisted of force pumps attached to the mills'
waterwheels. These pumps, with hydrants and hose, were said
to "possess the power of 4 or 5 engines worked by hand."[89]
The village continued to have its share of fires,[90] and some strange
ones too,[91] but it could not be said that the Harrises had not
acted promptly and effectively to protect the mill village from
this menace.

Another organization of this period in which Bethuel Harris
was much involved was the militia. The state had what was
probably a reasonably efficient militia totaling three divisions
early in the nineteenth century. It was in its best condition in the
decade or so after 1812. Thereafter it underwent a steady decline,
until the system was virtually abandoned in 1851.[92]

The fate of the militia company in Nelson was much the same
in miniature. General Simon Goodell Griffin, who was long active
in the militia, left some valuable notes vividly describing it in
these years. According to the general,

The Packersfield [Nelson] Co. made a fine appearance, particularly about
the time when Saml. Scripture was captain . . . and Bethuel Harris
Ensign. It turned out then to the number of 120 rank and file, with a
handsome uniform of blue faced with buff, which was rich and showy.
The annual special training, when the co. was called out for drill, were
gala days for the town. . . .[93]

In addition to this company there was organized around 1820 a
Nelson Rifle Company. This was composed of about fifty young
men who were especially fond of military exercises, and was
considered something of a "crack" outfit in the militia.[94]

The general, for whom the militia was part of his childhood
memories in Nelson, went on to tell how the training days for
these companies were great attractions for all the boys in town
and how the regimental muster, held in the fall, was the red-letter
day of the whole year. These musters seem to have resembled
county fairs, for the general recalled attractions such as gingerbread, candy sticks, sweet cider, pop beer, and raw oysters, which
"with the side shows and other contrivances to pick away our
four-pence-ha'pennies and nine pences [meant] we usually came
home with empty pockets." But the real attraction of these days
was the muster itself:

88 *NHS*, Sept. 10, 1845.

89 CMR, Letter Book
"A," Feb. 16, 1852; *NHS*,
July 2, 1868.

90 *NHS*, April 26, 1861,
April 17, 1862.

91 See ch. 6.

92 C. E. Potter, *Military
History of New Hampshire*,
p. 394; Bemis MSS, Box 1,
Seventeen-page MS on
militia in Nelson, by S. G.
Griffin; *NHS*, Oct. 9, 1819.

93 Bemis MSS, Box 1,
Seventeen-page MS on
militia in Nelson, by S. G.
Griffin.

94 *Ibid.*

To go to those [regimental] musters, usually held at Troy, we used to start from here in the middle of the night so as to be on the field before sunrise, and not lose a moment of that precious day. . . . There would be company drills in the morning. . . . Then came the battalion drill and the review of the regiment by the General attend[ed] by a gorgeous staff, each one of whom we believed to be equal to a Field Marshall, followed by a sham fight in the afternoon. . . . Those sham fights were very good imitations of real battles. . . . [and] they were a great deal safer and more enjoyable to look at than some real ones we saw later in life. . . . And then, when the sun went down, with heads full of the excitement and wonders of the day,—and some of the older ones possibly with something stronger,—we would start for home, and the roads and little murmuring streams as we passed sleepily along were all vocal with the strains of that martial music which had charmed our ears all the day long. . . .[95]

Although the Twelfth Regiment in review may have looked like the Corps of Cadets to him then, the militia must have been well past its best days when young Simon Griffin first saw it. In 1833, the *Sentinel* urged a change in the militia law on the grounds that it could not "put the militia in a more uneficient condition than they are at present."[96] By this time most of those who served in the militia did so only because they were required to by law; they "had no pride in their appearance or efficiency as soldiers." In fact, in these later years the drill days were turned into something of a burlesque, as Griffin described what was by then known as the "Old Slam Bang Co.":

The law required that every enrolled soldier should appear on parade armed and equipped, and each was warned a certain number of days in advance . . . to appear at the trainings and musters with his musket in good condition, with ram-rod and bayonet, knapsack, canteen, and cartridge box, with two spare flints, and a priming with a brush. These last were for the old flint locks. They wore no uniforms, and being required to appear with all this paraphernalia of war, as the military spirit declined toward the last of those years, and the duty became irksome, they frequently took the liberty, for the sake of having a little fun, of wearing all sorts of ludicrous things, and making themselves as ridiculous,—like the Antiques and Horribles of our present 4th of July parades. There was no law against such merriment, and the officers were powerless to prevent it, unless it was carried to great extremes.[97]

So much for America's tradition of citizen soldiers. Aging Bethuel Harris, who had been Major of the Second Battalion in the militia's heyday, must have had some wry comments on such goings-on. His practical-minded sons escaped from the whole

95 *Ibid.*
96 *NHS*, June 13, 1833.
97 Bemis MSS, Box 1, Seventeen-page MS on militia in Nelson, by S. G. Griffin.

silly business, for by provision of its charter the members of the Harrisville Engine Company were exempt from military duty in the militia. Other residents of the mill village must have done what the law required in the way of service, but there may have been even less enthusiasm for it in this manufacturing community than in the rural areas. Of course, observing it all from the sidelines was a different matter, for whatever its military value the entertainment value of the militia seems to have been beyond question.

Recreation and leisure-time activity were not in the mid-nineteenth century the major problems they are today. When work in a woolen mill took ten or twelve hours a day, six days a week, the rest of the time pretty well took care of itself. Nonetheless, there were opportunities for diversion. Church affairs, elections, literary societies, and the like, provided much of the organized activity. In addition there were the usual outdoor activities, varying with the season, available to the population of a small village. The rough countryside, the ponds and the streams gave ample opportunities for hunting, fishing, swimming, and skating. The editor of the *Sentinel* occasionally mentioned such recreation: "This is the season for Sleigh Rides. We counted 53 single Sleighs and one large double one, on Wednesday, from Dublin, as they gracefully moved through Central Square, after the company had taken dinner at Whitney's."[98]

Of holidays there were only a few. Christmas was little celebrated even well into the nineteenth century,[99] and, to judge from the columns of the *Sentinel*, it took a very definite second place to Washington's birthday. The Fourth of July was regularly commemorated in some towns but not in others. There is no mention of any such occasion in Nelson before 1842,[100] but Dublin seems to have regularly celebrated the day. Levi Leonard described the "usual form" of this, as it was in 1839:

The Dublin Grenadier Company paraded; a procession was formed, and marched to Centre Meeting-house; where a prayer was offered by Rev. James Tisdale, the Declaration of Independence read by Dr. Albert Smith, of Peterborough, and an address delivered by Rev. L. W. Leonard; after which, the procession again formed, and proceeded to the American Hotel, in front of which, under an arbor, a dinner was provided by Mr. Joseph Morse.

In the afternoon of the same day, there was a temperance celebration, with an address by John Preston, Esq., of New Ipswich.[101]

As in this instance, patriotic holidays were frequently used by different groups to boost their cause, whether it was the Whig

98 *NHS*, Jan. 19, 1837.

99 "The annual celebration of Christmas which was scarcely known in most of the rural districts of New England thirty-five years ago has grown to be almost as f . . . an institution as Thanksgiving. . . ." (Illegible word) *NHS*, Jan. 2, 1884.

100 Bemis MSS, Box 1, Seventeen-page MS on militia in Nelson, by S. G. Griffin. Griffin said he believed this was the first occasion when the old Nelson picnic grounds were used for a public gathering.

101 Leonard, 1855, p. 268.

Party, temperance, or the antislavery crusade. In comparison with the holiday celebrations of twentieth-century America, the striking feature then would have been not the fewer number but the modest scale of the festivities. The *Sentinel* might contain a list of new books available at a local store for Christmas giving, but the paper did not swell to ten times its normal size every December. The Reverend Mr. Leonard's Fourth of July address would have dealt thoughtfully with the blessings of liberty and avoided the demagogic histrionics of a later era. The parade would have been short and unimpressive to modern eyes. In short, commercialization of holidays hardly existed, and people were more easily entertained.

The growing village of Harrisville came to have its own social occasions, and although most of these were part of a later era, some took place in the 1860s. When Milan Harris added a new woolen mill to his plant in 1867, the dedication was properly celebrated by a "concert and ball." The only remaining evidence of this event is a dance program, but it tells a good deal. The occasion began with a musical program, with a grand march followed by quicksteps, polkas, and waltzes, all of which ended with the playing of the "Battle Cry of Freedom." Then came the dancing, with time out for a supper. The music was provided by "The Peterborough Cornet and Quadrille Band." Contra dance followed quadrille, and the band played such special numbers as "Hull's Victory," "Money Musk," and "Durang's Hornpipe." Whomever the program belonged to spent most of the evening dancing with the feminine members of the Colony, Griffin, and Hutchinson families, but it is hard to believe that the dedication was not a celebration for the whole village to enjoy.[102]

At the end of that same year, a local correspondent sent to the *Sentinel* a lengthy account of how Christmas had been celebrated in Harrisville: "As Christmas time drew near, the people of this village gladly united in a festival for the promotion of good feeling and social sympathy which a common table and Christmas tree are calculated to produce. . . . On Wednesday evening last the capacious schoolroom was thronged by all classes of citizens." There were songs by the choir, recitations, and a dinner for all. There was also a Santa Claus and a Christmas tree, and presents were given to the children. The choir gave its leader, Milan W. Harris, two silver goblets and a salver. There were presents for the two school teachers and the minister, and "the young ladies of the Mills" presented the Reverend C. M. Palmer with an armchair. "The company separated with the general feeling that it was good to spend an evening in social concourse and be made

102 "New Mill Dedication," Jan. 10, 1867, Dance program in possession of John J. Colony, Jr.

sharers in the happiness which such a festival should always
bring."[103]

103 *NHS*, Jan. 2, 1868.

Behind the saccharine sentiments, one sees here the self-
conscious effort of a small and heterogeneous community to
promote unity and good feeling. Whatever its success and however shallow the efforts may have been, it represented a more
promising approach to the village's social problems than simply
mourning for the good old days when the Harrises spoke only to
Twitchells. These two occasions also suggest that, at least by the
late 1860s, Harrisville was developing a social life distinctively
of its own kind and making.

5 AGRICULTURE, MINOR BUSINESS, AND THE RAILROAD, 1800–1900

The woolen mills have been the central fact in the economic life of Harrisville. However, there are other aspects that need to be considered. There were farms in the vicinity before any of the woolen mills were built, and later, as the mill village grew, other business and industry developed. Much of it was adjunctive to the woolen mills; some of it was largely independent. All of it was subject to change, and the economic life of the area altered considerably during the course of the nineteenth century. In 1900, though the town was already past the peak of its growth, a variety of enterprises were still flourishing.

Farming in Harrisville shared the same fate as most of New England agriculture, and its history is in sharp contrast to the rise of the woolen mills. In the very early years of the nineteenth century, farming predominated as a means of livelihood and, if not characterized by luxurious living, was at least buoyant and self-confident. One writer has called the period from 1790 to 1830 "The Age of Self-Sufficiency" in northern New England,[1] and this applied to towns like Nelson where "everything that was needed to sustain life could be produced on the farm, with the exception of a few small imports and the need for currency to buy professional service."[2]

Such self-sufficiency required a mixed type of farming. Henry Melville, writing about 1823, reported that in Nelson "the principal articles of produce are beef, pork, butter, and cheese. Of grain there is not any more raised than is required for the use of the inhabitants."[3] Sometime later Levi Leonard wrote that in Dublin the main crops were corn, oats, barley, and potatoes.[4] Within the area of Harrisville the crops must have been much the same. Milan Harris always retained an interest in farming. At an agricultural fair in Nelson in October 1855, Milan had on exhibition "12 varieties of apples, potatoes, and garden vegetables."[5] In a town like Harrisville a manufacturer could still live close to the soil.

Also, from colonial days to the present the farmers of New

1 Harold F. Wilson, *The Hill Country of Northern New England*, p. 15.

2 Michael G. Hall, "Nelson, New Hampshire, 1780–1870," pp. 75–76.

3 Letter from Henry Melville to John Farmer, Concord, N.H., ca. 1823.

4 Leonard, 1855, p. 121.

5 *NHS*, Oct. 5, 1855.

England have tapped the sugar maples in the snows of early spring, and the quantity of sap gathered in some years must have been tremendous. In 1864, the *Sentinel* reported that "the amount of sugar made in Nelson the past season is estimated . . . to be over 30,000 pounds." In Dublin the same year 110 persons made an aggregate amount of nearly 55,000 pounds.[6] At an estimated price of fifteen cents per pound at that time,[7] maple sugar was an important cash crop for the farmers of Dublin and Nelson.[8]

The soil around Harrisville is neither the best nor the worst. One gazetteer in 1885 reported that it was "generally good and yields fine crops,"[9] a rosier statement than the facts warranted. A greater handicap to farming than the quality of the soil was the rough, rocky, and uneven terrain. The soil could be improved when farmers took the trouble to manure their fields,[10] but little could be done about the boulder-strewn hillsides with granite ledge lying inches below the surface.

Some of the farmers were progressive and experimental. Jonathan K. Smith of Dublin raised a variety of grains, including wheat. In 1838, he produced a crop equal to forty-one bushels per acre,[11] a figure he could have quoted without apology to any wheat grower in the country. Nonetheless, the farm of J. K. Smith must have been exceptional. In general the farms of Dublin and Nelson could not rival those in the West, as those towns' steady decline in population indicated. Levi Leonard listed the names of over one thousand adult males, most of them with families, who emigrated from Dublin before 1853.[12] Many of them may have gone to Lowell or Boston, but their destination did not alter the effect on the farms about Harrisville. "By 1830, the buoyancy of life on the uplands had worn away. . . ."[13]

One enterprise that gave a temporary reprieve to many farms in northern New England was sheep raising. The success of this industry rested on three factors. The first was the American woolen industry's growing need for an adequate domestic supply of good quality wool.[14] The second was the fact that the rock-strewn hillsides of northern New England were better suited to grazing than cultivating, and the cold climate induced sheep to grow a heavier coat than in regions where the climate was milder. The third factor was the introduction of the Merino breed of sheep from Spain in 1809. This famous breed had been jealously kept from export by the Spanish government, but the disorders of the Napoleonic Wars allowed the American consul in Lisbon, William Jarvis, to purchase a flock and send it to Boston in 1809.[15]

6 *NHS*, June 16, 1864.

7 *NHS*, May 21, 1868.

8 *NHS*, April 25, 1883.

9 Hamilton Child, *Gazetteer of Cheshire County, New Hampshire*, p. 175; Leonard, 1855, pp. 122–123, gives the data from several chemical analyses made of the soil in Dublin.

10 Leonard, 1855, p. 121.

11 *NHS*, Jan. 25, 1838.

12 Leonard, 1855, pp. 418–423.

13 Wilson, *op. cit.*, p. 26.

14 Arthur H. Cole, *The American Wool Manufacture*, I, 82.

15 Wilson, *op. cit.*, pp. 75–76.

In New England the Merino was an immediate success. In the next six years upward of 24,000 of these animals arrived in America.[16] Merino rams were advertised for sale in the *Sentinel* as early as July 1810.[17] So popular did sheep raising become in New England that before long it was described as a "mania,"[18] reaching its peak in the years between 1830 and 1845. After that, the declining tariff rates on imported wool and, more important, the increasing production of western wool, meant the eventual end of large-scale sheep raising in northern New England, despite a brief revival of the industry there during the Civil War.[19]

The rural area around the village of Harrisville shared in this "mania." The first Merino sheep in Nelson were added to some native flocks there in 1830.[20] In 1836, more than six thousand sheep were pastured in Nelson, more than in any other town but one in the county.[21] By the end of the decade the total number of sheep in the town had increased 200 percent, and most of this increase came in the small flocks of between fifty and one hundred head.[22] It was in these years that the huge sheep barns were erected, the cleared land pushed to its farthest extent, and the stone walls, which one encounters today running meaninglessly through dense forest, built to mark off pastures and meadows.

Hall writes that in Nelson "the boom during the Civil War was, if anything, more exaggerated than that during the late thirties and early forties."[23] One explanation for this was that by the sixties the busy Harris and Colony mills provided a ready market for large amounts of wool. But Hall also points out that the end came more quickly this time too. After the war the market was glutted with woolens, and increased importations of raw wool from South America and Australia meant disastrous competition for the sheep growers of New England.[24] By 1871, there were only fifteen hundred sheep in Nelson.[25] Some sheep raisers might survive, and a few might even prosper. The Cheshire Mills continued to buy local wool in small amounts for many years.[26] However, by 1870 the production of wool as a staple commodity was a thing of the past. "Long lines of sheep were driven across the State and into Massachusetts to the slaughter houses at Brighton. . . . The great sheep barns slowly fell into ruin; the hill pastures grew back into brush and forest; and the days of sheep farming as a major enterprise came to an end."[27]

One authority on the region has written that the years between 1870 and 1900 were for the rural areas of northern New England the period of their greatest distress. "In these years . . . the shock of the widespread desertion of farms and the pronounced decline

16 James D. Squires, *The Granite State of the United States*, I, 232.

17 *NHS*, July 28, 1810.

18 Squires, *op. cit.*, I, 232.

19 Wilson, *op. cit.*, pp. 76–88.

20 Bemis MSS, XLIII, p. 2, in Hall, *op. cit.*, p. 80.

21 C. Benton and Samuel Barry, *A Statistical View*, pp. 14–16.

22 Hall, *op. cit.*, pp. 81–82.

23 *Ibid.*, p. 83.

24 Wilson, *op. cit.*, p. 86.

25 Board of Agriculture, *Annual Report*, I, 324–325, in Hall, *op. cit.*, p. 83.

26 CMR, General Journal, 1872–1882.

27 Squires, *op. cit.*, I, 237.

in rural population, with their social and economic consequences, stunned the hill country." He goes on to say that though in the years to follow the situation was far from satisfactory, the region had by that time built up sufficient resistance to maintain its agricultural way of life in spite of adverse conditions.[28]

Although this may have been true of the region generally, it is doubtful that it applies to Dublin and Nelson. Allowing for the loss incurred by the formation of the new town of Harrisville in 1870, the census figures indicate that the two parent towns suffered a greater population decline in the early decades of the twentieth century than they did in the last of the nineteenth.[29] Indeed, Nelson in the 1870s was even able to clear itself of the large debt incurred during the war years[30] and, together with Dublin, continued to show considerable vitality. Various factors helped to maintain these towns during the late nineteenth century, including the popularity of Dublin as a summer resort for the very wealthy, the small industry in Nelson, and the woolen mills in Harrisville.

The farming population in Harrisville fared much the same as their neighbors in Dublin or Nelson. Although the farmers formed a minority in the new town, they were a considerable group. According to the Census of 1880, about 40 percent of the town's population lived outside the mill village, there were fifty-eight farms in the town, and farmers and farm laborers made up about 23 percent of the town's work force. After the woolen mill workers, farmers were the most numerous occupational group.[31]

Even more than their neighbors across the town line, the farmers of Harrisville benefited from the mills. In the last analysis, it was the mills that brought the railroad, a means of shipping as advantageous for the farmer with cash crops as for the manufacturer. The sale of wool to the mills by the local farmers has already been mentioned. Sheep raising continued, though in ever decreasing numbers. There were 612 sheep in town in 1874, 434 in 1885, and 210 in 1900.[32] In addition to wool the woolen mills also bought locally great quantities of cordwood. For heating its plant and doing its processing, the Cheshire Mills alone burned about a thousand cords of wood annually. In 1885, the company paid out nearly $2,400 on wood accounts to a score of local people.[33] Cordwood was a dependable cash crop, if a cheap one, and it could be harvested and marketed during the slack winter months.[34] Finally, the mill village with a population of over five hundred people in 1880[35] was a ready market for locally raised meat and farm produce. In fact the

28 Wilson, *op. cit.*, p. 95.

29 See Appendix 4.

30 *NHS*, Sept. 23, 1875.

31 *NHS*, Aug. 12, 1880; Census, 1880, Harrisville, pp. 11–28. Since the Census of 1880 is the last one open for inspection, this line of inquiry ends with that year.

32 Harrisville Town Records, Tax Invoice, 1874; *HAR*, 1886, 1901. In these years attacks on flocks by dogs got to be so troublesome as to discourage sheep raisers further. See *NHS*, Oct. 18, 1882.

33 CMR, Cashbook "C," 1882–1889, pp. 113–157.

34 *NHS*, Feb. 7, 1883. "Owners of teams are waiting patiently for snow, as probably there are fifteen hundred cords of wood to be hauled to Cheshire Mills and private families in the village." *NHS*, Jan. 15, 1896.

35 Census, 1880, Harrisville.

great proportion of the sheep raised in Harrisville may in this period have been mutton sheep, as this was the trend in the state at large.[36]

In the last decades of the century local farmers made efforts to change with the times. Many turned to dairying.[37] The number of cows in Harrisville varied greatly through the years and fell precipitously after the Panics of 1873 and 1893, but there were 224 cows there in 1886 and about the same number at the end of the century.[38] Butter, milk, and cheese could be sold not only in the mill village but in Keene and elsewhere, and the railroad provided quick transportation to these markets. In one year, 2,742 cans of cream, worth nearly four thousand dollars, were sent by railroad from Harrisville to a milk company in Wilton, New Hampshire.[39] The town sought to encourage the dairy industry by granting tax exemptions to creameries and cheese factories.[40]

The farmers of Harrisville made other efforts to keep up with new methods of agriculture and to protect their way of life. They experimented with silos,[41] and they organized a Grange in Pottersville.[42] This organization, though it was more active in promoting social affairs than anything else, sponsored lecturers from the "United States Experimental Station" at Hanover, New Hampshire, on such subjects as stock feeding.[43]

In conclusion, it appears that Harrisville farming underwent the same long decline that it did throughout northern New England. Nonetheless, the farmers put up a stout defense of their rocky hills and had their plight eased by the simultaneous emergence of the industrial village of Harrisville.

Overshadowed though they were by the woolen mills, several minor industries existed in the area of Harrisville. Many of the individual companies were very small or very short-lived, or both, and there would be little point in trying to trace them all individually. Some of the industries, however, made a sufficiently large contribution to Harrisville's economy to warrant description.

One of the earliest industries, begun in Pottersville in 1795, was the manufacture of brown earthenware.[44] There was a local clay of excellent quality,[45] and during the years of the Napoleonic Wars when trade with Europe was disrupted, the industry throve. At one time there were eight or ten pottery shops in the vicinity of the village.[46] In the early years of the nineteenth century when hard money was seldom seen in these hill towns, this brown earthenware served as a form of currency. "Farmers in the vicinity of the potteries were glad to exchange their surplus

36 Wilson, *op. cit.*, p. 187.

37 *Ibid.*, pp. 191–194.

38 Harrisville Town Records, Tax Invoice, 1874; *HAR*, 1886, 1891, 1896, 1901.

39 *NHS*, Dec. 21, Dec. 28, 1887, Oct. 31, 1888.

40 *NHS*, March 7, 1883.

41 *Ibid.*

42 *NHS*, Dec. 17, 1884.

43 *NHS*, Oct. 30, Nov. 6, 1889.

44 It was begun by one David Thurston. Child, *op. cit.*, p. 177; D. H. Hurd (ed.), *History of Cheshire and Sullivan Counties, New Hampshire*, p. 217.

45 Child, *op. cit.*, p. 177.

46 *Ibid.*

products for it. They carried the ware to various parts of this and adjoining states and exchanged it for cash or for such articles as were needed. . . ."[47] After 1815, the industry slowly declined. It could not compete with the cheap English whiteware and tinware and was virtually driven from the market.[48] As late as the Census of 1860, there were still four men living in the village who listed their occupation as "potter," but another ten years and they were gone.[49]

An essential minor industry or trade for any community in the eighteenth and nineteenth centuries was a blacksmith shop. Harrisville's first antedated the village itself by a good many years, for Jason Harris's blacksmith and trip-hammer shop, built about 1778, was the second building erected along Goose Brook.[50] Probably his shop was too remote to get much trade; at any rate Jason Harris soon gave it up and moved into Nelson village.

Some years later there was another blacksmith shop built on the same site, this one belonging to Mainard Wilson.[51] It was an attractive brick building. Built out over the stream, it stood three stories high and had its waterwheel directly under the shop. Wilson's shop, or stand, did more than blacksmithing. This was not unusual, for a skilled blacksmith often made and repaired all sorts of tools, fixtures, and simple machinery. In 1850, Wilson employed five men in his shop at a cost of $150 a month for wages. He ran seven turning lathes by waterpower and made factory machinery which accounted for three fourths of his income.[52]

In 1859, this shop was bought by the Cheshire Mills.[53] They used it, and still do, to make machinery parts and repairs needed in the woolen mill. The building is still standing, but, because of its use by the Cheshire Mills, it has not been kept in as fine a state of preservation as have many other buildings in the village. By 1877, if not before, another blacksmith stand was in operation just behind the home of Milan Harris.[54] The village probably never had more than one stand at any one time but, with about two hundred head of horses and oxen in the town during the last quarter of the century,[55] was always able to support one blacksmith.

The most important minor industry in Harrisville was the manufacture of woodenware, and there were a number of shops. The area offered in abundance the two essentials of waterpower and timber. The first such enterprise was started in 1838 along Goose Brook in the eastern part of Harrisville.[56] This shop turned out washboards, clothespins, mop handles, and, like all such shops, a certain amount of sawed lumber.[57] Another shop was

47 Hurd, *op. cit.*, p. 217.

48 *Ibid.*

49 Census, 1860, Dublin, pp. 87–91; Census, 1870, Vol. XVI, Dublin, pp. 19–20.

50 Leonard, 1855, p. 273.

51 *Ibid.*

52 Census, 1850, Vol. XIV, Dublin, Schedule 5.

53 Cheshire County Records, Book 198, p. 60, in CMR.

54 C. H. Rockwood, *Atlas of Cheshire County, N.H.* (1877). This stand was operated by one George Tufts, and later by Robert MacColl, until about 1928.

55 *HAR*, 1886, 1901.

56 It was started by George Handy and Nathaniel Greeley. Leonard, 1855, p. 275; Hurd, *op. cit.*, p. 211; *Map of Cheshire County, N.H.* (1858).

57 Leonard, 1855, p. 275.

soon started on the same stream just below the mill village. In 1850, this shop employed five men at a monthly cost in wages of $110 and was manufacturing boxes and various kinds of lumber.[58] Further to the west, in Pottersville, there was still another wooden-ware shop which, beginning in 1849, manufactured much the same sort of thing.[59] A half mile north of Harrisville Pond, Joel Bancroft had built a sawmill and woodworking shop, which in time became quite an extensive little plant. Together with five or six houses, it actually constituted a small subvillage and had the singular name of Mosquitoville.[60]

In the years after Harrisville became a separate town, the woodenware industry continued to prosper for a time. The water-power was more fully exploited by the construction of new dams and reservoirs.[61] In 1880, there were seven dams along Goose Brook within the town lines.[62] Also, the decline of farms left greater acreage for growing timber, and vast quantities were used. One shop, owned by Zophar Willard, consumed an average of a thousand cords of wood annually.[63]

In 1880, there were twenty-two men in Harrisville who worked in the town's four woodenware shops,[64] and five years later the number of workers had doubled.[65] Some of the shops claimed to have the most modern machinery available for the making of clothspins, pail handles, trays, shingles, laths, and boxes.[66] The woodenware industry was probably on the decline by the end of the century. However, in the late nineties the Winn family began a chair factory that operated successfully for a number of years.[67]

Of commercial enterprises probably the first was a store in Pottersville. It was in the house built by Levi Willard, son of the Reverend Elijah and father of Zophar, and was in operation for some years before 1838. "The principal stock of a country store in those days was New England rum, molasses, and salt fish."[68]

In the mill village a store was built about 1838, possibly by Cyrus Harris.[69] This building still stands, close by Abel Twitchell's house overlooking the gorge and the mills. It was built of red brick, two-and-a-half stories high, with the gabled end at the front. In general style it closely resembled the other buildings erected in the same period. If Cyrus Harris did build it, he had little or nothing to do with running it as a store. In 1842, one William Hyde in Harrisville offered his business and stock (but apparently not the building) for sale "on account of ill-health."[70] Two years later another notice appeared in the Sentinel, advertising a "New Store and New Goods at Harris-ville." The storekeeper added a note that reveals another home

58 It was run by Thaddeus Perry Mason. Census, 1850, Vol. XIV, Dublin, Schedule 5; Leonard, 1855, p. 276.

59 Leonard, 1855, p. 276.

60 Map of Cheshire County, N.H. (1858).

61 Hurd, op. cit., p. 212.

62 NHS, June 24, 1880.

63 Ibid.

64 Census, 1880, Harrisville.

65 Child, op. cit., p. 177.

66 Child, op. cit., p. 177; NHS, June 24, 1880, Oct. 18, 1882.

67 Leonard, 1920, p. 578; Fisk and Wadsworth, Map of Dublin, N.H. (1853 and 1906); cf. NHS, May 17, 1899, Jan. 3, 1900. The Winn's factory superseded Amos Perry's saw and box mill.

68 NHS, Aug. 18, 1886. The location of the store was supposedly No. 10, Lot 21, Range X, in Dublin where the Baptist parsonage later stood. See Leonard, 1920, pp. 663–664.

69 Hurd, op. cit., p. 213.

70 NHS, Jan. 26, 1842.

industry of the nineteenth century: "Palm leaf will be furnished to Braiders, and good hats taken in exchange for Goods, at fair prices. . . ."[71]

The operation of the store continued to change hands every few years, and its storekeepers included the D. W. Clement who tangled with Milan Harris in 1856. Who owned the store during these years remains a question. It is not likely to have been Milan Harris. The Cheshire Mills owned it in 1860,[72] but the Colonys apparently did not run, or wish to run, a "company store" in the usual sense of the word. When the store came on the market, they sometimes took it over for a period of time and sold, leased, or rented it as they could.[73]

Among the records of the Cheshire Mills is the ledger of one "Ebenr. Jones, Esq." who kept store in Harrisville briefly in the late 1850s. A sampling of the accounts of Milan Harris and Henry Colony in 1858 reveals interesting data on both the items they purchased and the prices they paid:

71 *NHS*, May 1, 1844; cf. Leonard, 1920, p. 579.

72 CMR, "April Letters, 1860," Homer S. Lathe to Mr. Colony, April 2, 1860.

73 Interview with John J. Colony, Jr., Jan. 13, 1962.

74 CMR, Ebenr. Jones Ledger, p. 6.

75 *Ibid.*, p. 3; cf. *Historical Statistics of the United States*, p. 124, Series 101–112.

76 *NHS*, Oct. 13, 1870; Fisk and Wadsworth, *Map of Dublin, N.H.* (1853 and 1906).

77 C. H. Rockwood, *Atlas of Cheshire County, N.H.* (1877).

78 *NHS*, Nov. 10, 1881.

Henry Colony[74]				*Milan Harris*[75]		
1 1/4#	Mackeral .08	.10		1/2	Bush. Salt	.38
4	pocket knives	.67		16#	Sugar	1.44
1	Bbl. Extra flour	7.00		1	pr. shoes	.58
1/4#	peppermints	.05		5	doz. crackers	.25
1	Bar soap	.21		4#	nails	.18
1#	Figs	.16		1	reader	.42
2	doz. crackers.	.10		2#	coffee	.25
1	Pr. suspenders	.13		1	gal. molasses	.38
1#	Black tea	.50		10	Lard	1.40
6#	Lard .14	.84		1	file	.10
2	prs. shoes	1.33		44	window glass	.12
4#	Rice	.24				

Even from this bit of evidence it is clear that it was the Colonys, not the Harrises, who were given to frivolities.

As the mill village expanded in the post-Civil War era, its commercial life became more active. Other stores appeared,[76] though the red brick one was the principal store and the one that has remained in business down to the present. A map of 1877 showed a livery stable and, alas, a saloon.[77] A chatty little column in the *Sentinel* in 1881 indicated the increasing variety of commercial enterprise in Harrisville: "A. W. Sprague of Keene intends to open a store here this week, over S. Earle's barber shop. Millinery rooms have been opened in the house of F. Jewett."[78]

There was a hotel in the village by 1860. It was in the house later owned by Zophar Willard, near the western end of the village, and its sign read simply, "Harrisville Hotel."[79] It was of modest size and may not have long survived. In the late sixties another hotel was opened opposite Willard's house.[80] Recently it has been remodeled into an attractive dwelling, but for many years the rambling old frame building was in a state of dilapidation. Even in its early years the building must have added little to the appearance of the village. Its very name, "The Union Hotel," suggests that it was but another of the thousand such dreary places across the land that served the needs of a traveling public with a minimum of comfort or beauty. In 1881, the hotel was bought by a Charles Blake, of Jaffrey, and shortly after reopened as the "Nubanusit House."[81] The new name suggests some aspiration for elegance, if not its achievement. (Plate XXIII)

The old register of the hotel is gone, and information about its operation is scarce. The main building had sixteen rooms, in which at the turn of the century there were "neat white enameled bedsteads and furnishings to match. . . ."[82] For its custom the hotel depended largely on summer or permanent boarders and on traveling men. About 1900, there were usually ten or fifteen guests staying there.[83] Mr. Blake also kept horses for hire, but it is difficult to see how the operation of the hotel itself could have showed much profit over the years.

If the hotel itself was dreary, the same could not be said of all its guests. Couples of doubtful character and relationship stayed there,[84] the colorful Dr. Byrnes lived there for years,[85] and there are reports that prostitutes were sometimes brought into the hotel from cities like Manchester and Lowell. It is local opinion that at the turn of the century Charles Blake was making more money from the sale of illegal (and watered) whiskey than he was from the hotel itself.[86] Whatever the means, the Nubanusit House managed to keep its doors open well into the twentieth century.

No innovation was longer debated or deemed to be of greater economic consequence to Harrisville than the coming of the railroad. Many remote parts of the country saw their economic salvation in the steam locomotive, and the region of the Monadnock highlands was no exception. As early as 1835 when the railroad "fever" was first running high[87] but before there was a single piece of track in the state of New Hampshire, a local committee planned a line that was to run from Brattleborough, Vermont, to Nashua, New Hampshire. The most difficult stretches of the route were the highlands in Dublin, but the committee assured the public that these could be easily surmounted by the use of inclined planes on each side of the heights.[88]

79 *NHS*, Dec. 27, 1899. In 1899 this first hotel was reopened, probably for summer visitors.

80 *NHS*, Oct. 13, 1870.

81 *NHS*, Nov. 10, Nov. 24, 1881.

82 *NHS*, March 14, 1900.

83 Interview with Ernest Blake, son of Charles Blake, Aug. 1960.

84 *NHS*, April 8, 1880.

85 Cf. Newton Tolman, *North of Monadnock*, "Rum, Pneumonia, and Dr. Burke."

86 Interviews with old-time residents, Aug. 1960.

87 *NHS*, Sept. 10, 1835.

88 *NHS*, Aug. 6, 1835.

Just the previous year Pennsylvania had opened a canal and railroad route across the state by devising the famous Portage Railway. In that system, inclined planes and stationary engines were used to pull the canal boats and railroad cars over the Allegheny Mountains.[89] It was very likely this ingenious, if uneconomical, precedent that inspired the New Hampshire railroad enthusiasts to see the inclined plane as an easy way to surmount heights only half the altitude of the Alleghenies.

This early burst of enthusiasm for a railroad across the state failed, and the plan was never tried. No further mention of such a route appeared in the *Sentinel* for another fourteen years. In this interval the railroad had pushed up the Merrimack and Connecticut River valleys. When amid much celebration the tracks reached Keene in 1848, local enthusiasm was renewed for a line to connect the eastern and western parts of the state.

The next year the *Sentinel* mentioned the possibility of a railroad to run from Keene, through Harrisville, to Concord.[90] Meetings were held in various towns to discuss the relative advantages of various routes. In reporting a number of these in June 1851, the *Peterborough Transcript* said that "the meeting at Harrisville was respectably large and much sympathy for the project is felt in that quarter."[91] However, the enthusiasm again proved ephemeral, and nothing more was to be heard of such a line for still another decade. Without doubt the growing mill village of Harrisville favored such a railroad, but it was not yet strong enough to have any real influence in the matter.

The 1860s saw Harrisville boom and the prospects for the railroad revive. These were undoubtedly helped by the fact that the state government was now firmly in the hands of the industrialists and their agent, the Republican Party. The first step came in 1864 when the state granted a charter to the Manchester and Keene Railroad.[92] The next year a survey of the route was made. The section planned between Harrisville and Keene was altered to obtain an easier grade,[93] but there was never any doubt that the road should go through Harrisville. This was not only because of the importance of the mill village, and the fact that people there wanted the railroad, but also because Harrisville was the lowest place at which to cross the summit of the highlands between the Merrimack and Connecticut Rivers.[94] However, after thirty years of talk many people found it difficult to take the project seriously. Among the accounts of the Cheshire Mills is a skeptical notation: "Nov. 17th 1865 pd. to M. Harris   Paid $150.00 for Survey of Manchester and Keene Railroad. Will probably be refunded if the railroad is

89 Edward Channing, *A History of the United States*, V, 15–16.

90 *NHS*, April 26, 1849.

91 *Contoocook Transcript* [*Peterborough Transcript*], June 25, 1851.

92 *NHS*, Sept. 18, 1879.

93 *NHS*, May 9, 1867.

94 *NHS*, Oct. 13, 1870.

ever built. Don't think it ever will be. . . ."[95] With a technically feasible route laid out, the railroad corporation was formally organized in 1868, and despite his earlier skepticism Henry Colony was one of the original directors.[96] A company was formed to build the road,[97] but then came new trouble.

The construction company (and nothing has come to light about who were its leading figures) asked for a two-hundred-thousand-dollar gratuity from the towns along the route to build the twenty-five-mile section between Keene and Greenfield, New Hampshire, where the tracks of another railroad had already reached.[98] Such a proposal was a common way of financing construction in that extravagant era of railroading, and, despite opposition, most of the towns along the route voted amounts varying from $2\frac{1}{2}$ to 5 percent of their valuations.[99] In Dublin and Nelson, however, a battle ensued, which finally ended with Harrisville's "secession."

The question whether to vote the gratuity must have been the leading topic of conversation in those two towns during 1869. It was mainly the farmers who were in opposition.[100] Back in 1835 they might have favored the plan, but in the intervening years they had seen their farms decline, and they knew it was the experience of some other agricultural towns that the arrival of the railroad did not check, but only intensified, the flight of the young people from the farm to the city.[101] The New Hampshire farmers' economic decline only served to increase a natural conservatism. And they noted that the arguments for the railroad stressed the fact that it was needed in order to develop fully the waterpower so abundant between Peterborough and Keene. This had little appeal to the farmers, who saw no reason why they should pay higher taxes to help the manufacturer. Thus the promises that the railroad would cause "this section of the state to bloom as the rose"[102] left their grim, weather-beaten faces unmoved. The bitterness that the Nelson farmers felt toward the Harrisville element, determined to have the railroad one way or another, was expressed by some local bard in a newspaper published in Nelson in March 1870:

The Latest Excitement

Not by John G. Whittier

If you will give attention,
     I'll try and tell you how
A portion of our citizens
     Are making quite a row

95 CMR, General Journal, 1858–1866.

96 NHS, Sept. 18, 1879.

97 Hurd, op. cit., p. 212.

98 NHS, June 11, 1874.

99 Hurd, op. cit., p. 212.

100 Hurd, op. cit., p. 212; Hall, op. cit., p. 53.

101 James W. Goldthwait, "A Town That Has Gone Downhill," The Geographical Review, XVII, No. 4 (October 1927), p. 539.

102 NHS, Oct. 13, 1870.

Old Nelson was good enough
   To live in, so it seems,
'Till the subject of a railroad
   Burst in upon their dreams.

This Railroad would cost "a pile,"
   But on having it they were bent,
And all they asked the town to do,
   Was to vote them Three Per Cent.

But Old Nelson didn't see it,
   And then some of them found
They couldn't live here any longer
   They must have another town.

There once was a great rebellion,
   Within this land, you know;
Don't this seem something like it,
   When they say they are bound to go?

The South seceded on a pretext
   Of oppression and abuse.
And in this Town rebellion,
   They make the same excuse.

They calculate to draw the line
   Between the old town and the new,
To accommodate the voters
   Who on Election day are TRUE.

And now, it remains to be seen
   Whether this aimed–at separation
Shall end as did those efforts
   To divide our dear old Nation.[103]

103 *The Nelson Clarion,*
March, 1870.
104 Dublin Town Records,
Vol. V, Town Meetings,
1847–1882, p. 381.

The issue of the gratuity came to a head late in 1869. In Dublin a special town meeting was held on November 9. The town records tell what happened:

On the motion that the town raise a gratuity of 3% of its valuation to be assessed at such time & in such manner as the town may hereafter deem expedient, to aid in the construction of the Manchester and Keene R. R., to be expended between the towns of Keene and Peterborough— 105 voted in the affirmative & 85 in the negative, which motion was lost; the law requiring a 2/3 rds vote[104]

The element in favor of the railroad did not give up easily. Another special town meeting was held in December, but, though the margin was closer (124 to 82), the subsidy again failed to pass.[105] Finally, when at the next regular town meeting in March 1870 the same article came up, the town voted to postpone the matter indefinitely.[106]

In Nelson it was much the same story, but the defeat of the measure was more decisive. A special town meeting was called in November, and the article of principal interest read, "To see if the town will vote a gratuity of five per cent of the valuation of said town, or any less sum than five per cent, to the *Manchester and Keene Railroad Corporation*, to assist and encourage the building of a railroad between Keene and Greenfield. . . ." When the article came up, the Nelson people "voted not to raise three per cent of the valuation of the town to aid the Manchester and Keene Railroad, 56 voting in favor of the gratuity, 67 against."[107] Nelson was clearly against voting the gratuity, and no further ballots were ever taken. The embattled farmers of Dublin and Nelson had won their fight, but it was a Pyrrhic victory.

The Harrisville people had sat through many frustrating town meetings in their day, but the defeat of the railroad was too much. For thirty-odd years they had seen the towns of Dublin and Nelson vote down their petitions for roads and shortchange them on their school money. The Harrises and Colonys felt a special grievance. The town line so ran through the village that their residences were on the Nelson side and their woolen mills on the Dublin side. This had been purely accidental, but it may have allowed some opportunity for double taxation. At any rate the two leading families *felt* that they were being taxed exorbitantly and with no legal residence in Dublin could do little about it.[108] Since Harrisville had been growing in wealth and numbers while both Dublin and Nelson continued to decline, people in the mill village were inclined to go it alone. The exasperating defeat of the gratuity by the stubborn farmers convinced them it was time to act.

The expedition with which Harrisville got its independence did credit to its leading citizens and may also have been a testimonial to the efficiency of the railroad lobby in the state legislature. Although the names of the prime movers in the "secession" barely show in the records, they must have included Milan Harris and Henry Colony. Relations between the two families were hardly cordial, but they could act together when it was to their advantage. A petition to be constituted a separate town was drawn up and signed by "every voter in the village, with few

105 *Ibid.*, p. 382.

106 *Ibid.*, p. 384. The determination of Harrisville to "secede," if necessary, was evidently foreseen. Dublin, by a vote of 114 to 70, decided not to oppose a division of its town.

107 Nelson Town Records, Vol. XIII.

108 *NHS*, June 24, 1880.

exceptions."[109] Since nearly every voter depended on Milan Harris or Henry Colony for his livelihood, this near-unanimity of opinion is not surprising.

The petition was then presented to the June session of the New Hampshire Legislature. The Journal of the House for that session states, "By Mr. Scott, of Keene, petition of Milan Harris and 157 others, to constitute the town of Harrisville from a part of Dublin and Nelson."[110] Curiously, the petition itself was not printed in the Journal of the Legislature, according to the usual procedure. Nor is there any record of the petition in the Office of the Secretary of State in Concord where it should be on file. Whatever the explanation for this may be, a move by the representative from Peterborough to table the matter was defeated. Instead, the petition was approved and a charter granted to the new town, effective July 2, 1870.[111] The whole affair is like a chapter lifted out of Winston Churchill's *Coniston*. Harrisville was indeed "railroaded" into existence!

Freed of their fetters, the railroad advocates in the new town proceeded with the business at hand. Some delay was to be expected while Harrisville got organized, but finally, on August 10, 1872, a special town meeting "almost unanimously" voted a gratuity of 5 percent for the railroad.[112] This amounted to $14,500.[113] Just two days earlier a glowing prospectus for the railroad appeared in the *Sentinel*. It advised the public that the subscription books were open for the capital stock of the railroad, gave details, and urged support for the venture by purchase of the stock. Dividends were expected to be not less than 6 percent a year. It also mentioned the possibility that the new line might be leased to some other railroad (before it was even built). The prospectus was signed by the seven directors of the Manchester and Keene Railroad, including Milan Harris and Henry Colony.[114] Among the names of the other early directors of the corporation were General Simon Goodell Griffin and Ezekiel A. Straw, who despite his unlikely name was governor of the state.[115] It was a list of names well calculated to help overcome any obstacles in the way of the enterprise.[116]

Not by any means, however, were the company's difficulties over. The business depression that hit the country in 1873, and bankrupted Milan Harris, kept the project at a standstill. Only in 1876 was a permanent survey completed and actual construction begun. "More than one thousand men are at work on the Railroad between Greenfield and Keene. . . ." reported the *Sentinel* in August of that year.[117] The laborers were paid $1.25 a day.[118] It may be assumed that they earned their pay, but

109 *Ibid.*

110 *Journal of the Honorable Senate and House of Representatives of the State of New Hampshire*, June Session, 1870.

111 *Laws of New Hampshire, 1870*, XLIV.

112 Hurd, *op. cit.*, p. 212.

113 *NHS*, June 11, 1874.

114 *NHS*, Aug. 11, 1872.

115 *NHS*, May 29, 1873, June 11, 1874, Sept. 18, 1879.

116 The Cheshire Mills apparently invested in the railroad corporation. The mills' financial records show "assessments" between 1876 and 1878 totalling more than $15,000. See CMR, General Journal, 1872–1882.

117 *NHS*, Aug. 10, 1876.

118 *NHS*, July 20, 1876.

whether they got it is another question, for the construction company failed and so did its successor.[119] Probably the contractors had not duly appreciated the difficulty of grading those hills of granite. It has been said that this road had the steepest grade of any railroad east of the continental divide.[120] Insufficient capital may also have helped cause the failures, for some towns were not paying the gratuities they had voted. Finally, however, in 1878, a single-track railroad was "nominally completed."[121] Harrisville was rewarded for its faithfulness with three depots: one in the east part of town, one at the southern edge of the village, and one in Pottersville, now officially called "West Harrisville."[122] At long last the railroad had come to Harrisville.[123]

There is no record of any celebration when the railroad was completed, no "Driving of the Golden Spike," no flag-bedecked first train through, thronged with dignitaries. Probably everyone was exhausted. Either that or else they had now such doubts about the wisdom of the whole venture that a celebration seemed inappropriate. Well they might have had their doubts, for after the completion "a single train was operated over the line for some months and then discontinued because it did not pay operating expenses."[124] The *Sentinel* reported that an injunction had been put upon the railroad by its own employees in February 1879, "as they had not been paid off since the road commenced running." Consequently, it continued, "no trains run over the road now. The banks along the road have caved in so that the road is not in a running condition."[125] Resumption of service was promised in June,[126] but if any trains did run that year, it was not for long.

It was all pretty discouraging for the people in Harrisville, and there were the jibes of their neighbors in Dublin and Nelson to endure. Their discouragement is apparent in the local news columns of the *Sentinel*: "The M. & K. railroad is quiet under the snow—nothing moving except the boulders in the cuts."[127] "Another rumor is abroad that the M. & K. Rr. is to 'start.' We wish it would start and never come back; perhaps then some suitable conveyance would be provided for travellers."[128]

Partly responsible for the trouble were the towns that were trying to "weasel out" of paying their gratuities. A Dublin correspondent with the *Sentinel* referred to this with all the self-satisfaction of one who had known better:

People here are listening to hear the whistle of the engines on the M. & K. Rr., now the court has ruled Hancock must pay her gratuity.

119 Hobart Pillsbury, *New Hampshire . . . A History*, II, 469; Hurd, *op. cit.*, p. 213.

120 Interview with John J. Colony, Jr., Jan. 13, 1962; another source specifies a gradient of 1,100 feet in 14 miles. See *Becco Echo*, Vol. 4, No. 2 (June 1953).

121 Pillsbury, *op. cit.*, II, 469.

122 Hurd, *op. cit.*, p. 213. The name apparently never was popular, and it was changed to "Chesham" in 1886. See *NHS*, Oct. 20, 27, Nov. 24, 1886. A Mr. George B. Chase gave the village its new name. *NHS*, Mar. 7, 1900.

123 The original rails were iron. These were replaced with steel rails during the summer of 1887. See *NHS*, June–Sept., 1887, *passim*.

124 Pillsbury, *op. cit.*, II, 469–470.

125 *NHS*, March 6, 1879.

126 *NHS*, June 19, 1879.

127 *NHS*, Jan. 22, 1880, "O.C."

128 *NHS*, Feb. 12, 1880, "Z."

. . . It begins to be seen that the road will not make all the towns rich through which it passes, but such an expectation was not reasonable, and the disappointment will not release payment of gratuity.[129]

Even in Harrisville it was not until 1883, eleven years after the gratuity had been voted, that the town directed the selectmen to borrow money to pay the balance due, about five thousand dollars.[130] The payment of the gratuities undoubtedly was a great help in keeping the line in operation.

The railroad resumed service in September 1880, and apparently there were no major disruptions of service thereafter.[131] Four passenger trains ran daily, a morning and an evening train each way. Two more trains were sometimes added during the summer season.[132] Naturally, the complaints of the citizenry continued about the schedule, the mail service, the shabby station in West Harrisville (now Chesham), and the delays. "It is vexatious to go to the station and wait an hour, more or less, for a train. . . ."[133]

More important economically than the passengers was the freight the line carried. By the mid-eighties this was heavy and still increasing.[134] Two freight trains ran daily,[135] and it is easy to imagine what they carried: raw wool, coal, soaps and dyes for the woolen mills to be exchanged for bales of finished woolens destined for the great woolen market of Boston; general merchandise for the small stores along the route; woodenware and lumber from the numerous mills between Peterborough and Keene; farm produce and particularly dairy products.

Probably as a result of the power struggle among the several railroads in the state as well as its own financial difficulties, the Manchester and Keene in its first nine years of service was operated successively by the Nashua and Lowell Railroad, the Connecticut River Railroad, the Concord Railroad in conjunction with the Boston and Lowell Railroad in 1884, and finally, in 1887, by the Boston and Maine Railroad.[136] Since "no dividends were ever paid on the original stock" and the net loss for operating the line between 1884 and 1895 was over one hundred thousand dollars,[137] it would hardly seem that the M. & K. was worth fighting over.

The railroad did not fulfill the rosy promises that had been painted for it. It was a source of much frustration to the traveling public, and it must have been a grievous disappointment to any stockholders who really expected those annual 6 percent dividends. Nonetheless, it gave Harrisville better communication with the "outside world" than it had ever had before. Indeed it

129 *NHS*, April 22, 1880, "p."

130 *NHS*, March 21, 1883, "O.C." The method by which the town financed this gratuity is nowhere spelled out. An "assignment" of the gratuity voted was held by Governor Hale, and it was with him that the Harrisville selectmen settled. A local person wrote in the *Sentinel* that the Governor had generously written off the interest due on the note "on account of his good treatment by some of our good citizens in the canvas last November." See *NHS*, March 28, 1883.

131 *NHS*, Sept. 9, 1880.

132 *NHS*, Jan. 24, June 13, 1883; Hurd, *op. cit.*, p. 213.

133 *NHS*, Dec. 1, 1881, "O.C.," May 22, 1889, May 14, 1890.

134 Hurd, *op. cit.*, p. 213.

135 *NHS*, June 13, 1883, "O.C."

136 *NHS*, June 11, 1884; Pillsbury, *op. cit.*, II, 469–470.

137 Pillsbury, *op. cit.*, II, 469–470.

was the best possible answer to its transportation problem before the coming of the automobile. In 1890, it was estimated that in the course of a year more than five thousand people took or left the train at Chesham Station alone.[138] The railroad must have amply repaid Harrisville for its gratuity. Certainly it was a decided economic gain for the new town and may have been one reason why the Harrisville area maintained itself relatively well during the last decades of the nineteenth century at a time when so much of northern New England was suffering depopulation.

The last two decades of the century saw the beginnings of a new industry in the vicinity of Harrisville, one that would eventually become the most important in the area. This was the coming of tourists and summer residents. This development had its origins in the new wealth and leisure of the urban middle class after the Civil War, combined with rural New England's natural beauty, climate, and declining population. Dublin had attracted summer visitors for many years. As early as 1846 a summer boardinghouse was established there, and "by 1879 ten summer boarding houses were filled to overflowing each season."[139] However, Dublin was known principally for the very wealthy summer residents who built palatial villas there and entertained such notable guests as Lord Bryce, Mark Twain, and President Taft.[140]

Eventually, some of this influx of summer people into Dublin spilled over into Harrisville, though with exceptions they were not the very wealthy. More important than the proximity of Dublin was the completion of the railroad. There is virtually no mention of summer visitors to Harrisville in the pages of the *Sentinel* before 1881, but from then on they are one of the frequently mentioned items of local news during the spring and summer months.

These visitors took up various kinds of residence. Some built or rented lakeside cottages. The first of these were built in 1885 on Breed Pond, just a mile from the Chesham Station.[141] They were so popular that in 1888 one writer to the *Sentinel* said, "we think we can safely predict that there will be fifty cottages erected around the lake within five years."[142] In later years this development spread to the banks of Harrisville Pond and North Pond, both ideal for the purpose. Other summer people bought rundown farms from discouraged owners. In some cases these summer people had come from the area originally, gone to the city, and were now returning to recapture their youth.[143] Still others camped out along the shores of the ponds. The great majority of summer visitors during the eighties and nineties,

138 *NHS*, May 14, 1890, "O.C."

139 Leonard, 1920, pp. 605, 607.

140 Leonard, 1920, pp. 611–613. Winston Churchills' novel, *Mr. Crewe's Career*, is said to be laid in Dublin.

141 *NHS*, July 8, 1885.

142 *NHS*, Aug. 29, 1888, "O.C."

143 *NHS*, May 19, 1881.

however, must have been what were known locally as "city boarders." Some stayed at the Nubanusit House,[144] but most of them boarded with private families.

The residents of Harrisville saw the importance of this new industry and sought to give it encouragement. Every season the *Sentinel* told of people who were making alterations and additions to their homes to attract summer boarders.[145] The town improved the camp sites,[146] and when the owners of summer cottages asked for a new road, the selectmen recommended it, and the town voted it without a fraction of the traditional opposition.[147] As early as 1888 the State Fish Hatchery was stocking Silver Lake[148] with white perch.[149] Chesham was so much revived by the summer people and the railroad that in 1887 a new general store and fish market were opened there.[150] Special trains were run for the benefit of the tourists,[151] and one enterprising resident operated a small steamboat on North Pond.[152]

A noteworthy event in this new development was the establishment of Marienfeld Camp in Chesham in 1899, supposedly one of the first boys' camps in the country. The camp had originally been established near Milford, Pennsylvania, by Dr. C. H. Henderson of Pratt Institute. After three years of operations he bought a two-hundred-acre farm on the west side of Silver Lake and moved the camp to this new location.

In a full-column article the *Sentinel* described the new camp, the program, and the dress of the campers, who ranged in age from eight to twenty-two. "They wear the least clothing that the weather permits; no hat and preferably no shoes and stockings, or else sneakers." Garments of blue were preferred, firearms and tobacco were not allowed, and there was a corps of instructors in attendance. "The object of the camp," the *Sentinel* went on to explain to its readers, "is to offer a thoroughly wholesome outdoor life for boys during summer and to add the spiritual and mental elements needed to produce a sound balance of character. The main business of the boys is with sunshine, fresh air and exercise, with opportunities for study in preparing for college examinations or making up more advanced work."[153] The camp was well received by the town and had a long and successful career.

Just how much this new industry based on summer visitors amounted to in the latter part of the nineteenth century is difficult to determine. However, even as early as 1884 there were fifty or so "city boarders" who spent the month of August in Harrisville,[154] and the number of summer visitors seems to have steadily increased after that.[155] Shortly after the turn of the

144 Always known locally, and less elegantly, as "Blake's Hotel."

145 *NHS*, July 12, 1882, May 9, 1883.

146 *NHS*, July 12, 1882.

147 *HAR*, 1891.

148 The efforts at improvement included a craze for renaming the local bodies of water. Goose Brook became Nubanusit River, Harrisville Pond (originally Brackshin Pond) became Lake Nubaunsit, and Breed Pond (originally Pleasant Pond) became Silver Lake. Such pretensions were unfortunate for historical and aesthetic reasons, and the town would do well to restore the original names.

149 *NHS*, Oct. 17, 1888.

150 *NHS*, Oct. 26, 1887.

151 *NHS*, June 13, 1883.

152 *NHS*, April 11, 1888, "O.C."

153 *NHS*, July 26, Aug. 23, Oct. 18, 1899.

154 *NHS*, Aug. 6, 1884.

155 Neighboring Nelson was slower to attract the summer people, perhaps partly because of an unhospitable attitude on the part of the residents. However, before the end of the century it too was having its appearance and its economy affected by the summer tourist or resident. See Bemis MSS, Box 6, XXXVI, a, 20.

century the value of the nonresidents' property in Harrisville amounted to 25 percent of the town's total valuation.[156] In 1900, the summer tourist industry in Harrisville probably ranked third in importance, behind woolen manufacturing and farming. But it held the greatest promise for the future.

No account of Harrisville's economic life in these years would be complete which failed to suggest the vital atmosphere of the community. Even in 1870, before it reached its peak, the *Sentinel* called Harrisville "the most lively business place of its size known in this section of the state."[157] Certainly it was a place of activity, bustle, and noise. It was peopled with busy industrialists, tradesmen, teamsters, salesmen, mill hands, and their large families. Its sound was a medley of clattering looms, rumbling trains, tolling factory bells, and always the rushing waters of Goose Brook.

156 *Invoice and Taxes of the Town of Harrisville, 1906.*

157 *NHS*, Oct. 13, 1870.

6 THE MILLS OF
HARRISVILLE, 1861–1900

The Harris Mills, 1861–1882

By 1860, Harrisville was fast outgrowing the patriarchal control of Milan Harris, but he could certainly regard the future of his woolen mills with confidence and cautious optimism. The recent Panic of 1857 had not hurt him. In fact it allowed him to expand, for when Harris and Hutchinson failed he was able to take over most of their land and buildings. Thereafter, Milan's plant consisted of the Upper Mill, the Middle Mill, dye house, storehouses, a brick boardinghouse, and the two all-important dams on Goose Brook.[1] Helping Milan to run the business were several members of his closely knit family: his two sons, Milan Walter and Charles Romanzo, his brother, Charles C. P. Harris, and his son-in-law, J. K. Russell. At the beginning of the 1860s, the concern was in good shape, produced quality woolens, and enjoyed a high reputation in the industry.

The decade of the 1860s was one of prosperity and expansion for the mills. The main stimulus was the Civil War, which brought great new demands for woolen cloth. With this in prospect, Milan Harris in June 1861 secured a charter of incorporation from the state legislature. In the act the corporation was called the "National Mills," but it was never referred to as such. The names of the incorporators were Milan Harris, Milan W. Harris, Joseph K. Russell, Alfred R. Harris, and George L. Wright. The last named was for many years Milan Harris's boss carder.[2]

Milan may have been partly prompted to take this step of incorporation by the law's provision for limited liability, especially after the experience of 1857. More important probably was the need for broader control and capital to expand and modernize. The Middle Mill, the first one built in the village, was obsolete. Milan Harris had extended the length of his Upper Mill and added a tower at the front to house the stairs, but ground space in that narrow, rocky gorge was limited. Expansion was clearly indicated, but it would be a major undertaking.

During the war years a large part of the productive capacity of the Milan Harris mill was devoted to making cloth for uni-

1 *Map of Cheshire County N.H.* (1858).

2 *New Hampshire Laws, 1861*, ch. 2542.

forms. As early as August 1861, the *Sentinel* reported that "Milan Harris and Co., of Harrisville, are manufacturing the cloth for the uniforms of the New Hampshire volunteers, and, we are glad to learn, are making a good article. The cloth is a grey doeskin, called cadet mixed, is three fourths of a yard wide, and costs less than 80 cents a yard."[3] The only other mention of fabrics made in these years are notations on the weavers' payroll records. These indicate that the company also produced flannels, cords, tweeds, and cassimeres.[4]

Among the very few surviving records of the Harris mills is a payroll book covering the years from 1861 to 1867.[5] The entries in it were not made with future historians in mind, and some of them defy explanation.[6] Nonetheless, it is a valuable record and provides a dimension to the knowledge of Milan Harris and Company that would otherwise be lacking.

In 1860, the company had employed an average of thirty male and thirty female workers.[7] Although all the company's employees may not have been listed in this payroll book, it indicates that during the Civil War the average number of employees rose by 10 or 15 percent. The year 1861 had its ups and downs, but from mid-1862 to early 1866 the work force probably did not fall short of sixty. Peak employment was reached in 1864, when the mill had an average of sixty-nine workers on the payroll. No seasonal pattern of employment appears in the record.

Considering the boom times that the war brought, it is a little strange that employment at the mill did not increase more than it did. Maybe the company was working to full capacity and no further expansion was possible without disrupting production. Yet even in 1864 a number of the operatives were not working full time. There may have been unavoidable bottlenecks in production, but this would seem an inadequate explanation. On balance, it would appear that though the war years were unquestionably good ones, they may have fallen short of being boom years for the Harris woolen mill.

The long lists of names in faded ink also permit some analysis of the company's work force. Although the company reported an equal number of male and female operatives in the Census of 1860, this record shows that during the first year of the war the male workers were much in the majority. As would be expected in wartime, from the middle of 1862 until the fall of 1865 the female hands either equalled or slightly exceeded the males in number. Thereafter, the men were again in the majority.[8]

In the division of work between the sexes the Harrisville mills must have paralleled the industry generally during the last half

3 *NHS*, Aug. 22, 1861.

4 "M. Harris Payroll Book," Sept. 1864.

5 "M. Harris Payroll Book," Jan. 1861–June 1867. Unfortunately, this book was sold in the auction of an estate, and its whereabouts is now uncertain.

6 Entries like "Discount on Wages Exclusive of Board," during 1861.

7 See Appendix 3.

8 This pattern was in line with the proportion of men and women in the woolen industry nationally. Cf. Edith Abbott, *Women in Industry*, p. 361, Appendix B.

of the nineteenth century. Supervising and mechanical work was done almost entirely by men. Wool sorting, scouring, and carding operations were distinctly men's work. Spinning was primarily a man's job. Weaving was a process that was done by both men and women. During the war most of the weavers were women. Later, with the influx of the immigrants from Canada, the percentage of male weavers rose until they were sometimes in the majority. Among the finishing processes, burling and mending were done by women, and dyeing, pressing, napping, and shearing were done by men.[9]

The names in the payroll record show that during these years the old-stock families still predominated among the workers. Beal, Farwell, Heath, Russell, and Yardley were among the most frequently listed. There are a sprinkling of Irish names, such as the Winns. Irish names are not always identifiable as such, but it is doubtful if the Irish exceeded 10 or 15 percent of Milan Harris's work force during the war years. There were almost certainly some English immigrants working there too. Somewhat notable is the absence of French-Canadian names. Not one of them is to be found on the payroll of Milan Harris, though by 1867, when the book ends, the neighboring Cheshire Mills were employing at least eight or ten French Canadians.

It is a little startling to discover that until October 1865 the workers at the Harris mills were paid quarterly. This must have been a convenience to Milan Harris and also saved him the expense of short-term loans to cover his payroll, but for his workers it meant a long time between paydays. Advances were undoubtedly made and credit allowed the workers by the local stores and tradesmen. In the largest sense these quarterly paydays are a measure of the simplicity and subsistent nature of the workers' lives and of the economic power of the mills. After October 1865, bolstered by the profits of the war years, the company paid its workers monthly. Also, since the Cheshire Mills had long paid its workers monthly, Milan Harris was probably under pressure to do likewise. Another feature, more curious than significant, is that some of the wage rates were figured in shillings, though of course the pay was in dollars. This was the case with several workers in 1861, none of them immigrants, and it was still occasionally done in 1866.

The following table is based on this payroll record. It shows the daily wage rates for some of the jobs in the mill and the computed daily earnings for the spinners and weavers, who were on piecework:[10]

9 Arthur H. Cole, *The American Wool Manufacture*, II, 107–110.

10 Asterisks denote positions held by the same individual from 1861 to 1867.

Position	1861	1862	1863	1864	1865	1866	1867
O'seer, Weave Rm.★	9/	1.67	1.75	2.50	4.00	4.00	4.00
Weavers (Avg.)	.72	—	.74	1.00	1.12	.96	.95
O'seer, Card Rm.★	1.88	1.88	1.88	2.25	2.50	5.00	5.00
Spinners (Avg.)	1.23	1.16	1.82	1.51	1.75	—	—
Carpenter★	1.25	1.25	1.25	1.75	2.00	2.00	2.00
Watchman	1.00	1.00	1.00	1.33	1.50	1.75	1.75

11 Cf. *Historical Statistics of the United States,* p. 90, Series 578–588; CMR, Payroll.

12 "M. Harris Payroll Book," 1861–1867.

As one basis for comparison, it is interesting to note that for two years, beginning in the spring of 1865, Milan Harris and his son Milan Walter each appear to have drawn $66.66 a month. This was less than a couple of their overseers were drawing. It is to be noted that there were few appreciable wage increases until 1864 and that it was the overseers who won the largest raises. The payroll does not show if and how the pay rate of the spinners and weavers increased in these years, but it does show that in 1863 the weavers were paid from two to five cents a yard for the cloth they wove on the mill's narrow Crompton looms. The wages Milan Harris paid were competitive with those paid by the Cheshire Mills. There, and in the woolen industry generally, wages rose steadily but only slightly until the last year of the war. Then, larger gains were made.[11]

The earnings of the workers depended not only on their wage rates but also on the number of days they worked in a month. In this there was sometimes considerable variation. Naturally, of those listed the watchman's employment was the most stable; he worked every night or at least six nights a week. The overseers and maintenance hands like the carpenter worked twenty-four or twenty-five days in a month. The spinners and weavers, on the other hand, often fell far short of full-time employment. The few entries that show this information suggest that even in the middle of the war years the weavers might only work eleven to fifteen days a month. The spinners' employment varied widely, sometimes as few as ten or as many as twenty-seven days a month. One explanation for this could be that there was an imbalance in the mill's machinery, that is, too few cards to keep the jacks and looms supplied. Since it was the spinners and weavers whose work was tied to production, the fact that some months they worked less than half time has its implications not only for their earnings but, as already mentioned, for the company's as well.[12]

In the years between 1866 and 1874, Milan Harris's company underwent reorganization, expansion, and finally, collapse. In 1866, a third interest in the concern was sold to General Simon Goodell Griffin, lately returned to Keene after distinguished service in the war. Exactly what the general did is not clear, but he was certainly not just "window dressing," and he probably put considerable capital into the company.[13] Letters in the Bemis collection give him the title of "Agent for Milan Harris Woolen Company." In 1870, the company was again reorganized to include one H. A. Gowing, in addition to Milan Harris, his two sons, and General Griffin.[14]

Both of these reorganizations were probably prompted by the need for additional capital, for, belatedly, Milan Harris had undertaken to expand. In 1867, he dismantled the old Middle Mill and replaced it with a new and larger one which with its machinery cost nearly $75,000.[15] The new mill was in existence only fifteen years, but pictures of it survive. The family resemblance to the other mills in the village was clear. It was built of the same red brick and had the same granite sills, the broken cornice, and square tower as the other mills. Nevertheless, there were differences. It was bigger, having four floors and a loft. The distinguishing roof windows of the earlier mills were replaced by the conventional skylights. And the high tower was capped, not by a cupola, but by an ornate cornice and a fenestrated mansard roof. (Plate XXIV)

As to the company's operations in these years, its production consisted of "Moscow Beavers" and "Tricuts," in shades of blue, brown, black, and dahlia. The former fabric was a copy of a German cloth; the latter was described as a "thick woolen." They were both quality cloths. The *Sentinel* reported the Tricuts to be "superior, in fabric and finish, to anything of the kind ever made in this country, fully equal to most of the French and German tricuts."[16]

Construction of the New Mill, as it was called, considerably increased the company's production. The number of spindles and looms operating in 1870 was more than double the number reported in 1860, and the company's annual production rose from 90,000 to 150,000 yards in the same decade.[17] Though it did not have the capacity of the Cheshire Mills, it probably could produce as much as the average-size woolen mill in the country.[18]

Somewhat curiously, Milan Harris did not put Crompton broad fancy looms into the New Mill,[19] though they had now been on the market for ten years and had the advantage of greater speed as well as greater width to recommend them.[20] Nevertheless, this lag in modernization held true for what Cole calls "the representative woolen mill" in 1870,[21] so it merely suggests that Milan Harris was no longer the progressive manufacturer that he had been in the 1820s.

13 *NHS*, July 2, 1868, Aug. 6, 1884.

14 This man probably lived in Dublin at one time and may have been the same man who was a partner in the Boston factoring house of Gowing, Grew, and Company. It would explain some things if he were, for these agents owned the Harris mills in later years. Dublin Town Records, Vol. V, p. 390.

15 *NHS*, July 2, 1868.

16 *NHS*, Oct. 13, 1870.

17 See Appendix 3.

18 Cole, *op. cit.*, II, 220–221.

19 Census, 1870, Vol. XVI, Dublin, Schedule 4.

20 Cole, *op. cit.*, I, 313.

21 *Ibid.*, II, pp. 220–221.

Of the other operations of the company, it is known that they bought wool from the Boston firm of Nichols, Parker, and Dupee, and perhaps also from a "Dodge Brothers and Company,"[22] but who their agents were, what their profits were, and much else—all remains unknown.

The expansion of the Harris mills brought a commensurate increase in the number of workers. The *Sentinel*, in 1868, reported "91 employees on the payroll besides nearly 20 others otherwise employed."[23] Without counting the extra help, which was probably engaged in freight hauling, wood cutting, and tending the boardinghouse, this total represents a 50 percent increase in the work force from the immediate post-war years.[24] To accommodate the new workers, Milan Harris built a row of tenements behind his mill[25] and also a huge white frame boardinghouse across the street from these tenements.[26] This building was four stories high and had an ell on one end. On the wall inside the front door were pasted the rules and regulations, designed to promote safety, order, and the good reputation of the company. Smoking was forbidden, boarders were not permitted to have visitors in their rooms without permission from the boardinghouse master, and both male and female boarders were required to keep to their respective ends of the boardinghouse after nine o'clock in the evening.[27]

The Census of 1870 provides information about those then living in the frame boardinghouse and affords an opportunity to compare the company's workers at that time with those it employed a decade earlier. In 1870, there were sixty-two people living in the boardinghouse. Fifty-six of them worked in the mill. This was more than half the company's total number of operatives. Of those living in the boardinghouse, six were born in Canada, seventeen in Ireland, four in England, and one each in Germany and Italy. The total number of foreign-born was twenty-nine, or nearly half the boarders. This was quite a change from 1860 when there were only two immigrants living in the company's brick boardinghouse. Milan Harris had reluctantly hired the Irish in preference to the French Canadians, but the expansion of his mill had compelled him to hire both.[28]

The Census of 1870 also provides information about the sex and age of the workers at the Harris mills. That year the company reported employing an average of fifty men and forty women,[29] as compared with an equal number of each a decade earlier. The ages of the workers who lived in the boardinghouse also show a contrast to those in 1860. While the workers in the years before the Civil War had almost all been in their late teens or early

22 CMR, Milan Harris & Company, "Drafts," 1866–1870.

23 *NHS*, July 2, 1868.

24 "M. Harris Payroll Book," 1861–1867.

25 See ch. 3, p. 48.

26 The boardinghouse was torn down some years ago, but it may be seen in the background of the picture of the New Mill (Plate XXIV).

27 Interview with John J. Colony, Jr., Jan. 13, 1962.

28 Census, 1870, Vol. XV, Nelson, pp. 10–12.

29 Census, 1870, Vol. XVI, Dublin, Schedule 4.

twenties, there was among the boarders in 1870 a significantly higher proportion of workers in their late twenties, thirties, and forties. These facts suggest that not only were the workers changing in their national origins but there was now in the Harrisville mills, as elsewhere, a more permanent body of workers than had been known earlier in the woolen mills of New England.[30]

Further proof of this lies in the increased number of children employed in the mills. In 1870, the Milan Harris Woolen Company reported employing, in addition to the ninety adults, an average of eight children and youths. To the government, this category meant boys who were sixteen or younger and girls who were fifteen or younger.[31] Four of these youths have been identified. They ranged in age from twelve to fourteen, and all four were born in Canada.[32] This was not exceptional. The indigent French Canadians were known as "conspicuous offenders" in evading the school laws to put their children to work in the cotton and woolen mills of New England.[33] Indeed, it appears that child labor, after being nearly extinct in Harrisville in 1860, was revived by the influx of French Canadians.

Nonetheless, too much should not be made of the matter. In the whole town of Harrisville in 1870 there were only fourteen youths who met the official description of child laborers, and their median age was fifteen and a half. They amounted to 8 percent of all the mill workers in the town.[34] Another ten years' time would find that child laborers in the town numbered twenty, had a median age of fifteen, and constituted nearly 11 percent of the mill workers.[35]

In the woolen industry generally, due to the development of more automatic machinery and to increasing public pressure, the trend was toward ever fewer child laborers. In 1869, they amounted to about 12 percent of the industry's work force, and by the end of the century this figure had declined to about $5\frac{1}{2}$ percent.[36] In Harrisville the number of child laborers, about average in 1880, probably declined thereafter. The local mills' divergence from the national pattern is at least partly explained by the fact that the immigrants arrived in Harrisville later than they did in the larger manufacturing centers.

In the fall of 1873, the country was severely hit by a financial panic. The subsequent depression was keenly felt in Harrisville, and in the fall of 1874 the Milan Harris Woolen Company failed.[37] There was no mention of it at the time in the *Sentinel*, perhaps out of consideration for Milan Harris. Nor is there anywhere any explanation given for the failure of this company so long in

30 Census, 1870, Vol. XV, Nelson, pp. 10–12; cf. Abbott, *op. cit.,* pp. 143–145.

31 Census, 1870, Vol. XVI, Dublin Schedule 4.

32 Census, 1870, Vol. XV, Nelson, pp. 12, 14; Vol. XVI, Dublin, pp. 3, 4.

33 Iris S. Podea, "Quebec to 'Little Canada': The Coming of the French Canadians to New England in the Nineteenth Century," *New England Quarterly*, Vol. XXIII, No. 3 (Sept. 1950), p. 373.

34 Census, 1870, Vol. XV, Nelson, pp. 12, 14; Vol. XVI, Dublin, pp. 3, 4.

35 Census, 1880, Harrisville.

36 Cole, *op. cit.,* II, 103–104.

37 Hamilton Child, *Gazetteer of Cheshire County, New Hampshire*, p. 178; Bemis MSS, Box 1, LV, J. K. Russell to C. A. Bemis, April 28, 1908.

business, so well regarded, and so lately in a thriving condition. Old Milan Harris had been in woolen manufacturing for over fifty years, had survived two earlier panics, and his company was said to be the oldest woolen manufactory in New England. Why then did he fail?

To begin with, in 1874 Milan Harris was seventy-five years old and ailing.[38] Yet, instead of having his work easier in his old age, the opposite was true. His plant was larger than ever, but with whom could he share the management and responsibility? His eldest son, Milan Walter, had died in August 1873.[39] Of the younger son, Alfred Romanzo, practically nothing is known. The only time his name appeared in the *Sentinel* was when he was arrested and fined ten dollars for violating the Dublin Pond Fish Law,[40] and he eventually went to California.[41] Milan's other partners, General Griffin and H. A. Gowing, had other business interests and probably spent little time in Harrisville. Milan's brother, Charles C. P. Harris, had stopped working for him in 1866.[42] Joseph K. Russell, his son-in-law and trusted overseer, on whom Milan counted heavily, had "seen a better country" and, chary of a partnership in a family concern, had gone to Massillon, Ohio.[43] Just at the time when he needed help the most, old Milan Harris seems to have been left very much alone.

Moreover, Milan Harris had overextended himself. After the Civil War, the woolen industry was left with an overexpanded capacity and inventory that greatly depressed the market.[44] Milan Harris, furthermore, had not expanded until 1867, a time when he might better have been retrenching. The New Mill had barely started to produce when the boom ended.

Finally, it is possible that the Manchester and Keene Railroad had something to do with the failure. One of the causes of the Panic of 1873 was that huge amounts of capital had been invested in railroad construction, tieing up those funds with little prospect of immediate returns. It may have been a similar story in Harrisville. The subscription books of the Manchester and Keene Railroad were opened in August 1872, and Milan Harris, as one of its directors, may have invested heavily in the venture.[45] Indeed, it would have been strange if he had not been one of the leading financiers of the new railroad. If he did invest heavily in this unprofitable venture, it would have weakened the financial position of his company at the worst possible time.

Thus it was that the Milan Harris Woolen Company failed. Old Milan Harris saw his fortune swept away[46] and his mills pass into the hands of others. For the man whose industry had

38 Bemis MSS, Box 1, LV, J. K. Russell to C. A. Bemis, April 28, 1908; Leonard, 1920, p. 788.

39 Leonard, 1920, p. 788.

40 *NHS*, Nov. 12, 1863.

41 Leonard, 1920, p. 788.

42 "M. Harris Payroll Book," 1861–1867.

43 Bemis MSS, Box I, LV, J. K. Russell to C. A. Bemis, April 28, 1908, Box 3, LV, J. K. Russell to C. A. Bemis, June 9, 1908.

44 Cole, *op. cit.*, I, 376–388.

45 *NHS*, Aug. 8, 1872.

46 *NHS*, Aug. 6, 1884.

been so largely responsible for the town's growth, it must have been a bitter loss.

The Harris mills were to run again, but things were never the same, and the men who ran the mills were not of the same stripe as the man who had built them. Perhaps as a result of the bankruptcy, the property came into the possession of Gowing, Grew, and Company, a commission house with offices in Boston and New York.[47] During the worst of the depression the mills remained shut. Then, in 1875, they were leased to a woolen manufacturer in Gilsum, New Hampshire,[48] but after only a year he disposed of the lease to the firm of Colony, Craven, and Company.[49]

The three members of this firm were Fred Colony, a son of Henry Colony; Zophar Willard, a native of Pottersville who owned several small businesses in Harrisville; and one Michael Craven, who was to be the principal figure in the operation of the Harris mills after Milan's bankruptcy and retirement. Not a great deal is known about Craven. His reputation with the Colony family has not been good, and that speaks of more than just prejudice. He came to this country from Ireland, married a Rhode Island girl, and moved to Harrisville from some place in Massachusetts during the early seventies. In 1880, he and his family lived in the big frame boardinghouse that belonged to the mill. Craven must have had some experience in woolen mills, but when Colony, Craven, and Company was formed he was still only thirty. For a young immigrant from Ireland, Craven would seem to have done well for himself.[50]

Virtually nothing is known about operations at the mills during this phase of the company's short and troubled existence. Just one item appeared in the *Sentinel*, but it is of interest because it describes one of the few strikes that have occurred in Harrisville:

The help employed by Colony, Craven, and Co., to the number of thirty, are on strike; not a strike for higher wages, but for a reduction in time, being required to work fifteen hours every other day—or from 5 a.m. to 9 1/2 p.m., or leave their job. The company has orders for goods to such an extent that they feel obliged to take this course to meet the demand, and are willing to pay for all the labor performed.[51]

In addition to trouble with their operatives, the partners had trouble among themselves. A bitter personal animosity developed between Fred Colony and Michael Craven, which finally led to the dissolution of the firm three years after it was formed.[52] Fred

47 CMR, C. & W. Copybook "C," p. 347.

48 *NHS*, Nov. 11, 1875.

49 *NHS*, June 24, 1880.

50 Census, 1880, Harrisville, p. 16.

51 *NHS*, June 19, 1879.

52 CMR, C. & W. Copybook "C," pp. 103–104.

Colony withdrew (taking the funds of the company with him so Craven said[53]) and refused to have anything further to do with his erstwhile partner. This left the company's affairs in disorder for several months,[54] but then it was reorganized as "Craven and Willard" and resumed operations in January 1880.[55]

Much more is known about the company during the last three years of its existence. The lease from Gowing, Grew, and Company expired in 1880, and its renewal was occasion for considerable correspondence between Michael Craven and the commission house. Craven objected to three new clauses in the lease. One related to the possibility of the lessee subletting the mill. The commission house wanted a veto right on this, but Craven argued that if he was to be so restricted, then the commission house should be willing to redeem the lease should such a situation arise.[56] The owners also wanted some provision about preferred creditors, but Craven won that point, arguing, "If we was to sign such a lease and our creditors to know it we could not get one dollar on our name."[57] The third point concerned whether rent should be charged during any period when low water in the brook should force the mill to lie idle.[58] These new clauses inserted by the owners suggest they had no great confidence in Craven. There is no knowing just how these issues were settled, but in any event the lease was renewed.

Gowing, Grew, and Company's concern about their property was understandable, for they had just spent twenty thousand dollars improving the plant.[59] Most important, the old narrow looms were sold[60] and replaced with thirty-six new "Broad Crompton pickfinder looms," purchased from George Crompton in Worcester.[61] Other new machinery to match was also installed.[62] The spinning machines had already been improved, although they were not the most modern. Until about 1870, the woolen industry "was still employing the 'hand jack,' a machine which could be operated only under the guidance of a skilled workman."[63] These jack spinners, many of them foreigners, had a notorious reputation in the industry for arrogance and disorderly habits. Their unreliability and the flawed yarn that they sometimes produced were a great hindrance to the efficiency of the industry. The manufacturers' desire to eliminate this bottleneck prompted several innovations in this country and abroad, based on the technology of cotton manufacture. Here, there appeared about 1870 several attachments that converted the semiautomatic jack into a fully automatic machine. This development meant the jack spinner could be replaced by a less skilled and presumably less intractable operative. The Harris mills

53 CMR, C. & W. Copybook "D," pp. 121, 364.

54 CMR, C. & W. Copybook "C," pp. 103–104.

55 CMR, C. & W. Copybook "C," p. 116.

56 Ibid., pp. 358–359, 372–373.

57 Ibid., pp. 358–359.

58 Ibid., 358–359, 372–373.

59 Ibid., p. 75; NHS, Jan. 22, 1880.

60 CMR, C. & W. Copybook "C," p. 44.

61 Ibid., p. 192. These broad looms bought by Craven and Willard had a speed of sixty "picks" (a single transmission of the shuttle) per minute, not very fast for the broad looms of 1880. See CMR, C. & W. Copybook "C," p. 227; Cole, op. cit., I, 313.

62 NHS, Jan. 22, 1880.

63 Cole, op. cit., II, 88.

adopted some of these devices for the ten spinning jacks which they ran to make filling yarn but retained the hand jacks to spin warp thread.[64] More efficient and more flexible than these "self-operators" was the English "spinning mule," introduced in this country about 1865 and quickly adopted by progressive manufacturers. The adoption of the self-operator and the spinning mule were important in the woolen manufacture for breaking the hold of the jack spinners on the industry, improving the quality of the yarn, and adapting to automatic machinery the last important process in woolen manufacturing.[65]

The plant of the Harris mills (Craven and Willard continued to use on occasion the good name of the Milan Harris Woolen Company[66]) now represented a considerable increase in the amount of capital, and the owners saw to it that their property was well covered with insurance. The plant included the old Upper Mill, the New Mill, picker house, boiler house, dye house, storehouses, woodsheds, boardinghouse, and tenements.[67] The New Mill was insured for $18,000, its machinery for $36,000, and its stock for $30,000.[68] Policies were placed with the American Mutual Fire Insurance Company, the Mill Owners Mutual Fire Insurance Company, and the Arkwright Mutual Fire Insurance Company,[69] but unfortunately it is not known just which policies applied to the New Mill.

The fabrics made by Craven and Willard were mainly beavers, worsted goods, and cassimeres. Their total annual production ran to about 200,000 yards,[70] somewhat more than Milan Harris turned out ten years earlier. The mill was limited to making piece-dyed goods, as opposed to stock or wool-dyed.[71] The company bought wool, some of which came from Montevideo, from several wool dealers in Boston.[72] Besides wool, Craven and Willard also bought cotton and cotton yarn, which they used in the warp of some of the fabrics they wove.[73] In an attempt to reduce costs, they urged the yarn manufacturers from whom they bought to use part cotton waste in making the warp thread.[74] From time to time they also purchased and used a considerable amount of shoddy, or recovered wool fiber.[75] The use of these substitute materials does not necessarily signify anything improper. Shoddy was first used on a large scale by the industry during the Civil War.[76] Before 1880, the woolen industry, driven by close competition and a limited wool supply, was using nearly as much substitute material, such as shoddy or cotton warps, as it was new wool.[77] However, these materials could not be used without having an effect on prices, as Craven and Willard were to discover.

64 CMR, C. & W. Copybook "D," pp. 399, 416, Copybook "C," p. 219.

65 Cole, op. cit., II, 88–90; Letter from James C. Hippen, Curator, Merrimack Valley Textile Museum, Aug. 2, 1968.

66 CMR, C. & W. "D," p. 392.

67 NHS, June 24, 1880, Dec. 20, 1882.

68 NHS, Dec. 20, 1882.

69 CMR, C. & W. Copybook "D," p. 294.

70 NHS, June 24, 1880.

71 CMR, C. & W. Copybook "C," p. 103.

72 Ibid., p. 199, Copybook "D," p. 79.

73 CMR, C. & W. Copybook "C," pp. 38, 69, 80, 103–104.

74 Ibid., pp. 112, 113.

75 CMR, C. & W. Copybook "C," pp. 8, 147, Copybook "D," p. 684. C. & W. paid eight to ten cents a pound for the shoddy.

76 Cole, op. cit., I, 315–316.

77 Ibid., II, 69–71.

The mill used a variety of methods to sell its goods. At the beginning of 1880, Craven and Willard received a number of inquiries from different commission houses, but they declined all offers.[78] Gowing, Grew, and Company owned the mill and had spent a considerable amount improving it, and while the lessees were under no obligation to send their goods to that commission house, Craven pointed out virtuously that "as a matter of principle and justice" the mill owners should have the first chance to sell their goods.[79]

By the middle of the year Craven was having his doubts about the arrangement. The commission house had apparently sold a large order of beavers at a low price, forced Craven and Willard to buy wool at high prices to cover this order, complained about making advances to pay for the wool, found fault with the quality of the fabric produced, and then apparently reneged on the contract![80]

Mutual recriminations followed. The agents complained about poor quality fabrics. The manufacturers complained about slow sales, below-market prices, and lack of information about just what fabrics the agents wanted.[81] The arrangement was clearly not working, as one letter from Craven to the commission house indicates:

I have tried pretty hard not to have any trouble and hard words with our commission house and fully started with the belief that your interests would be ours and ours yours so far as having the mill do well   But the the way our goods have been handled the past year it has made me sick Just think of mill idle 3 months and then have some 20,000 yds on hand unsold   And this writing is painful to me but am compelled to do it And must frankly say do not think our goods are showen or worked on as they ought to be.[82]

The next month, writing to another agent, Craven was still fuming: "I wish I could make goods so Buyers could smell them way out to Oregon then the commission men could surely find no trouble in disposing of them."[83]

By the fall of 1881, Craven and Willard were resorting to another method of disposing of their goods, selling directly to the clothier and the department store. In letters to Messrs. Nathan Levi and Company, in Rochester, New York, and to Messrs. Jordan Marsh in Boston, Craven offered Moscow Beavers in quantity for from $1.65 to $1.80 "net cash 30 days," giving them the benefit of the commission and guarantee they usually had to pay their commission house.[84] Even when Craven and Willard used this method of selling their woolens they ran into stiff

78 CMR, C. & W. Copybook "C," pp. 42, 158.

79 Ibid., p. 75.

80 Ibid., pp. 414–415.

81 CMR, C. & W. Copybook "D," pp. 303, 455, 488, 504, 449–551.

82 Ibid., p. 540.

83 Ibid., p. 563.

84 Ibid., pp. 885, 899, 933, 904, 920.

competition, for many other manufacturers had the same idea.[85]
In fact, this shift from selling through the commission agents to
selling direct to the clothing manufacturer and the department
store was one of the major changes in the woolen industry's
marketing system in the late nineteenth century.[86]

As if these difficulties were not enough, Craven and Willard
also had trouble over water. This was nothing new in Harrisville.
Occasionally, the mills had had to shut down briefly on account
of low water.[87] The Harrises and Colonys had disputes over the
use of the waters of Goose Brook.[88] Farmers whose lands bordered
the ponds objected to the flooding caused by the mills raising the
water level behind their dams.[89] And people below the mills
objected to their brook being polluted by the strong vegetable
dyes and waste waters from scouring which all three mills
dumped into the brook.[90] Nonetheless, Michael Craven may
have felt that he received more than his share of water troubles.

Everyone agreed that it was most unusual for the waters of
Goose Brook to fail, but fail they did in the fall of 1880.[91] The
water shortage forced Craven and Willard to close down the
night before Thanksgiving, and they did not resume full-time
operations until three months later.[92] Woolens in process could
not be completed, and business was lost. A large number of
operatives left town to seek employment elsewhere, and for
those who stayed it was a bleak winter.[93]

The drought finally ended in February, but the waterwheels
had not been turning a month when Craven and Willard,working
extra hours to make up for lost time, found themselves in fresh
trouble. On March 25, 1881, Michael Craven wrote to Gowing,
Grew, and Company's New York office to explain it: "The
writer just returned from Boston Are going into the Law
Business with our friend Henry Colony he has sued us for
$11000 damages for using this water. . . ."[94] Craven advised the
company's creditors that should it lose the case it would have
ample means to pay all debts and explained the suit was brought
"to recover damages from us on the ground of our running our
Mills extra time when the Cheshire Mills did not want to use the
water."[95]

The case dragged on for many months, and during this time
when they were trying to get their goods out, Craven and
Willard were prevented from working overtime and from
running their waterwheel except in regular and customary hours,
because friend Henry Colony had them served with an injunction
to prevent it.[96]

As the mill had been shut down for three months late in 1879

85 *Ibid.*, p. 970.

86 Cole, *op. cit.*, II, 136–146.

87 *NHS*, Feb. 8, 1877.

88 CMR, C. & W. Copybook "D," p. 456.

89 Cf. Dublin Town Records, Vol. V, p. 358.

90 Interview with John J. Colony, Jr., Feb. 16, 1961.

91 Cf. *NHS*, July 2, 1868, Oct. 13, 1870.

92 *NHS*, Dec. 2, 1880, Feb. 24, 1881.

93 *NHS*, Dec. 16, 1880.

94 CMR, C. & W. Copybook "D," p. 454.

95 *Ibid.*, pp. 457–468.

96 *Ibid.*, pp. 874, 898, 902.

and again the next winter because of the water shortage, labor recruitment was a major concern for Craven and Willard in 1880 and 1881. When running full time, they employed 130 workers.[97] They used a variety of methods to find help. Some workers could be hired right in Harrisville. From their operatives they heard of others looking for employment. These they wrote to come for interviews and offered to pay expenses if no agreement was reached.[98] They ran advertisements in such newspapers as the *Springfield Republican*, the *Boston Journal*, and the *Providence Journal and Bulletin*,[99] specifying precisely what kind of help they wanted and didn't want:

Worcester Spy
Please insert the following in daily 6 times.
> Wanted
> A few first class fancy woolen weavers on Crompton Broad Pickfinder Looms on white work. Steady employment and good pay. No rummies need apply
> > Craven & Willard[100]
Harrisville, N.H.
June 21st, 1880

The company drew its help from a wide variety of places in all the New England states, from Sanford Corner, Maine, to Stafford Springs, Connecticut, but it drew most heavily on woolen mill workers living in Massachusetts.[101]

The most frequently stressed qualification for working at the Craven and Willard mills was sobriety, as indicated by these two letters to prospective employees:

I must be frank with you and say that you have habbits that are not for your interest or your employers and should you continue these habbits it would be useless to come to work for us. . . .[102]

We have a first class carder in every sense of the word but his habbits are such that we cannot depend on him   he is an intemperate man and the first time he goes on a spree he gets through working for us.[103]

For Craven and Willard, as well as for the other mills in Harrisville, drunkenness and the resulting absenteeism was a major problem.

In their search for steady, dependable workers, the company preferred to hire family help whenever possible.[104] This does not mean that Craven and Willard employed children on any scale, for their role in the woolen mills was quite limited. There was probably some advantage in wage rates for the company to hire

97 *NHS*, June 24, 1880.

98 CMR, C. & W. Copybook "D," p. 169.

99 CMR, C. & W. Copybook "C," pp. 290, 397ff.

100 *Ibid.*, p. 397.

101 CMR, C. & W. Copybooks "C" and "D," *passim.*

102 CMR, C. & W. Copybook "D," p. 149.

103 *Ibid.*, p. 499. Cf. p. 585.

104 *Ibid.*, p. 210.

a family with several grown daughters who were weavers or boys who could work in the spinning or card rooms. More important, family members proved to be more dependable as workers than single men or women.

However, Craven was handicapped in this policy by the limited number of tenements he had to offer.[105] So he was also alert for individuals who showed industry and ambition. In answering letters from those who showed promise, he sometimes grew expansive:

105 *Ibid.*
106 *Ibid.*, p. 923.
107 *Ibid.*, p. 291.
108 CMR, C. & W. Copybook "C," p. 262, Copybook "D," p. 292.

Mr. G. D. Rice   Oct. 11th 1881
          Lowell Mass
Dear Sir
          Your letter to our advt for a Boss Weaver and Designer. . . . . We have engaged a man. . . . However like the tone of your application well and if you half as anctious to get a start in this World as you pretend you are and do not feel above your business as there is lots of young men who say they would like a start etc. etc. but feel above their present worth and are afraid to soil their hands Such folks want nothing to do with as they would be no benefit to themselves or anyone else Now if you think it worth you may come and see me Immediately Would like to have a talk with you and if you mean what you write can make quite a man of you
                              Yours Respt.
                              Michael Craven[106]

Finally, in regard to hiring practices there is the interesting matter of hiring workers employed by other woolen mills. It seems that some mills had a "gentleman's agreement" not to hire each other's help, or at least not without the worker first getting a release from the mill he was leaving. Despite all the ill will between the rival companies in Harrisville itself, there may well have been some such understanding between them, born of self-interest. The proximity of the mills would have encouraged compliance. In letters to operatives living in other towns, Craven sometimes simply declined to hire the person: "I do not want to hire any of Mr. John S. Collins Help as it might make hard feeling with Mr. Collins and you are probably well aware he is quite sensitive on that point."[107] In other letters Craven specified that the operative would have to "work a notice" before he hired him.[108] Just how long this might be is not mentioned; it may have been two weeks or a month.

Nevertheless, this kind of "gentlemen's agreement," like other agreements between manufacturers, was probably honored more in the breach than in the observance. In March 1880, Craven wrote a letter to the head of a mill in nearby Marlborough, charging him with stealing one of his weavers, and concluding,

I wrote you a short while ago that We should not hire any help from your Mill untill they brought a Writing to the affect that you was willing to have them come hear to work and I presumed you would be redy and willing to do the same by us I will now say the agreement on our part is withdrawn and you may have the privilege to hire any and all the help from our Mill You see fit to and We shall do the same   Hoping this will meet your approval, I am. . . .[109]

Concerning wages and hours the information is scant. The contents of some letters Craven wrote to a man in Gilsum, New Hampshire, suggest that upon request he occasionally provided certain other woolen manufacturers with information about his company's wage rates.[110] Letters to job applicants sometimes mentioned the rate paid. However, the total information about the wages paid by Craven and Willard in 1880 and 1881 is small and easily summarized:[111]

Position	Daily Rate
Overseer, Spinning Room	2.00
Overseer, Weave Room	3.00
Weavers	1.50–2.25 (avg.)
Cropping Shear Operator	1.00
Second Hand, Finishing Room	1.50
Second Hand, Card Room	1.25
Carpenter	2.25

These wage rates were roughly competitive with those paid by the Cheshire Mills. The workweek in the Harrisville mills of 1880 was still a long one, probably just as long as it had been half a century earlier. In the Craven and Willard mills it was sixty-eight hours, and it was probably the same at the Cheshire Mills.[112]

The occupants of Craven and Willard's boardinghouse indicate the make-up of their work force. According to the 1880 census, there were living there, in addition to Michael Craven and his family, the boardinghouse keeper and his family, and forty-six workers. (They paid three dollars a week for board.[113]) Ten of these forty-six were women, all but one of them single. The majority of those living in the boardinghouse were in their twenties.

The data on the national origins of those living in the boarding-house suggest that Craven and Willard had some other preferences

109 CMR, C. & W. Copybook "C," pp. 175–176.

110 Ibid., p. 219, Copybook "D," p. 240.

111 CMR, C. & W. Copybooks "C" and "D," passim. Weavers were paid at the rate of .096 per yard plus .01 yard bonus for weaving an average of sixteen yards of perfect cloth for each working day in the month. Spinners were paid, per hundred run, .33 for filling thread and .59 for warp thread, but what this meant in terms of a daily wage is not clear. Cf. CMR, C. & W. Copybook "C," p. 219, Copybook "D," p. 240.

112 CMR, C. & W. Copybook "D," p. 127; Cf. CMR, Payroll, Sept. 1882, Pat Whalen in Finishing Room.

113 CMR, C. & W. Copybook "C," p. 433.

in their hiring that were not mentioned in their letters or adver- tisements. Of the forty-six workers in the boardinghouse, thirty were native-born, nine were Irish, two were French Canadian, and five came from various other European coun- tries.[114] The striking feature is the near-absence of the French Canadians at a time when they were not only the largest foreign- born group in the town but also constituted 16 percent of the total population of Harrisville.[115] This avoidance of the French Canadian immigrants by Craven and Willard was in sharp contrast to the policy of the Cheshire Mills but consistent with the earlier policy in the Harris mills.

It was also consistent with the general pattern of hiring within the woolen industry. Cole points out that woolen manufacturers had a preference for native-born operatives or at least a dislike of the new arrivals among the immigrants. He suggests that data showed these new arrivals to be less stable in their habits and that manufacturers were hesitant about encouraging the settlement of large bodies of such immigrants in the small communities where the mills were often located.[116] However, as the case of Harrisville illustrates, not all woolen manufacturers agreed on these matters.

It is unfortunate that the surviving Craven and Willard records do not go beyond November 1881, for they might shed light on the sudden end of the Harris mills, an end not devoid of mystery and sinister implications. Looming large in the meager evidence is a brief article that appeared in the *Sentinel* in April 1882:

An attempt was made Monday evening to burn the 'new mill' occupied by Craven and Willard. A wool sack saturated with kerosene oil was thrust into the main gears in the basement, which are boxed, and set on fire. The fire was discovered before much damage was done. As yet there is no clue to the perpetrators of the dastardly act[117]

The *Sentinel* made no further mention of this episode, but late in November it reported very low water again for the mills in Harrisville.[118] Then a month later it reported the destruction of the New Mill by fire. The newspaper gave no particulars, but the destruction was complete. A photograph taken after the fire shows only the blackened walls left standing.[119] The news- paper's correspondent from Harrisville lamented that this was a severe blow to the prosperity of the town, as indeed it was. At the time of the fire there were 138 persons on the company's payroll. Bitterly he pointed out that the owners of the mill would not suffer, as there was a total of $84,000 insurance on the building and its contents.[120]

114 Census, 1880, Harrisville, p. 16.

115 See Appendix 6.

116 Cole, *op. cit.*, II, 115– 116.

117 *NHS*, April 26, 1882, "L."

118 *NHS*, Nov. 22, 1882.

119 Photograph in possession of John J. Colony, Jr.

120 *NHS*, Dec. 20, 1882.

In Harrisville, rumors and suspicions of arson must have abounded.[121] Did not the attempt to burn the mill the preceding April alone cast suspicion? The suspects for arson might include some disgruntled employee or, more likely, some representative of the owners or Michael Craven himself. After all, these mills were said to be "nearly fireproof," they were furnished with the very best of fire-fighting equipment,[122] and, with this single exception, the record of all the Harrisville mills in coping with the hazard of fire was excellent. Moreover, the waters of Goose Brook were running low. Might not the fear of another winter without waterpower, the kind that the previous year had caused Michael Craven to write so many bitterly complaining letters, have pushed Craven or the owners to take this extreme step? It would not have been the first such fire.

Whether the fire was accident or arson, the result for the town was the same. Within a couple of weeks the company shut down entirely. Bids were taken for the hundred tons of junk metal lying in the wheel pit of the burnt mill, the old Upper Mill was boarded up, most of the company's operatives left town in search of work, and all was strangely quiet.[123] The celebrated Milan Harris Woolen Company was no more.

The Cheshire Mills, 1861–1900

In sharp contrast to the vicissitudes of the Harris mills is the story of the Cheshire Mills from 1861 to the end of the century. Uninterrupted operations and continuity of family control and management provide a large degree of unity for this period of the company's history. Fortunately, the company records are sufficient for the reconstruction of its history.

Undoubtedly a principal factor in the survival and prosperity of the Cheshire Mills was the quality and continuity of its management. During this long period, in fact for over fifty years, the company was under the direction of two very capable presidents. After sharing control with his brothers during the 1850s, Henry Colony in 1862 assumed the combined offices of president, treasurer, and clerk of the corporation.[124] After he removed from Harrisville to Keene, about 1870,[125] though his son Frank H. Colony may have had the actual supervision of the plant,[126] Henry retained all the principal offices until his death in 1884. Thus, Henry was the dominant figure in the company for most of its first thirty-four years, and as the *Sentinel* said in his obituary, "it was to his skill and ability as a business man that the success of this large manufactury has been largely due."[127]

121 Even today, eighty years later, it is still said in the town that it was arson.

122 *NHS*, July 2, 1868.

123 *NHS*, Jan. 3, 24, 1883.

124 "Records of the Cheshire Mills," Feb. 20, 1862.

125 *NHS*, July 23, 1884.

126 *NHS*, March 7, 1883.

127 *NHS*, July 23, 1884.

Henry Colony's successor was his youngest brother, Horatio.[128] This brother had quite a lively and varied career before he became president of the Cheshire Mills. Born in 1835, and perhaps enjoying some favor as Josiah Colony's youngest son, Horatio did not go directly into business like his brothers but instead went to the Albany Law School. He graduated from there in 1860, was admitted to the New York and New Hampshire bars, and practiced law in Keene until 1867, when he became a partner in the Faulkner and Colony Woolen Mill in Keene.[129]

Horatio had political as well as business interests. He was a delegate to the Democratic National Convention in 1868. It met in Tammany Hall and was hardly inspiring,[130] but it must have taught Horatio Colony something about politics. When Keene became a city in 1874, Democrat Horatio Colony scored a major political upset by getting himself elected the first mayor of that traditionally Federalist-Whig-Republican stronghold.[131] Then as if to prove it had not been a mistake, he ran again and was reelected the following year. He ended his political career with a term in the state legislature.[132]

Horatio Colony was a Mason, a Knight Templar, and a director of various local institutions and societies.[133] He had an impressive bearing and a pleasant sense of humor. In 1884, he was forty-nine, married, and the father of three children. Thus, he came to the Cheshire Mills a man of considerable education and experience.

On Henry's death in that year, Horatio purchased his share in the company from his brother's heirs.[134] Then, a meeting of the corporation in September chose Horatio to fill the offices of president, treasurer, and clerk.[135] His annual salary in the early years seems to have been two thousand dollars,[136] and that was, of course, considerably augmented by his share in the company's profits. Shortly after the turn of the century, Horatio gave up the important position of treasurer to his son, John J. Colony,[137] but until that time the direction of the company's affairs was largely in his hands.[138]

Under the direction of Henry and Horatio Colony, the Cheshire Mills had a profitable record. Within the family it is said ironically that the company has not made money since the Civil War. At any rate there is no question it made money then. The minute book of the corporation shows that from 1863 to 1868 the company paid dividends totalling $5,287.50 on a share of stock having a par value of $1,000[139] and a net asset value of perhaps $2,500.[140] In addition to these dividends, there must have been undistributed profits and profits that were put back into

128 There was also a half brother, Josiah D. Colony, twenty years younger than Horatio.

129 *Granite Monthly* (Nov.-Dec. 1917), p. 231.

130 David Muzzey, *The United States of America*, II, 20–21.

131 *NHS*, April 16, 1874.

132 *Granite Monthly* (Nov.-Dec. 1917), p. 231.

133 *Ibid.*

134 *Cheshire Republican*, Special Trade Edition, Aug. 18, 1899.

135 CMR, "Records of the Cheshire Mills," Sept. 9, 1884.

136 CMR, Cashbook "C," 1882–1889, p. 235, Dec. 31, 1887.

137 CMR, "Records of the Cheshire Mills," 1904.

138 *Cheshire Republican*, Special Trade Edition, Aug. 18, 1899.

139 The capital stock of the Cheshire Mills has always consisted of the forty shares originally issued and divided equally among the four Colony brothers.

140 Based on a capital investment of $100,000 in 1870. See Appendix 3.

the business. Lacking the figures for the company's sales in those years, only a rough estimate can be made of the profits. In 1870, the Cheshire Mills reported the value of its production (and presumably its sales) to be $200,000.[141] Assuming that this figure would not have been lower during the boom years of the 1860s, the company's profits in those years must have been at least 20 percent. Such profits would not have been exceptional at that time in the woolen industry.[142]

Those extraordinary years are the only ones before the end of the century for which the company records show the actual amount paid out in dividends. In many years the annual meeting simply voted to divide the surplus of the previous year among the stockholders. In other years the minutes merely say that the meeting voted to accept the treasurer's report.

Profits and dividends there must have been during most if not all of these years. Nonetheless, a goodly share of the profits was always put back into the company. In 1899, the meeting voted to carry the balance to the reduction of the construction account, and the next year, after voting a dividend of $500 a share, the meeting did the same thing.[143] Whatever the dividends may have been, it was popular opinion, occasionally showing up in the pages of the *Sentinel*, that the owners of the Cheshire Mills through the years had been rewarded for their efforts by "large pecuniary gains."[144]

With the aid of these profits the plant of the Cheshire Mills was expanded and improved in this period. No new mills were erected between 1861 and 1926, but at the turn of the century the company's two mills contained a total of nine cards, forty-four broad looms, and four thousand spindles.[145] In terms of the company's relative size within the industry, these figures meant that though the Cheshire Mills did not rival the mammoth woolen mills like those of the American Woolen Company, it was considerably larger than the average woolen mill in the country at that time.[146] But, if not among the largest, there is little doubt that for efficiency and general condition the Cheshire Mills rated among the best of American woolen mills. A contemporary sketch depicts an idyllic scene which, though not literal, may be suggestive of the mill's appearance in these years. (Plate XXV)

The improvement and expansion of the Cheshire Mills' power system in this period illustrates the high efficiency of the plant. After experiencing several power failures, due either to drought or overuse of the water supply, the Colonys installed in 1884 a reserve steam engine which generated 120 horsepower. This was

141 See Appendix 3.

142 Cf. Cole, *op. cit.*, I, 380. Cole shows that the dividends paid by other woolen manufacturers rose from 9 percent in the 1850s to over 27 percent in 1865.

Of course, there were wartime taxes to be paid. There was a 5 percent tax on incomes over $600 and an excise tax on manufactured goods, including woolens, that ranged from 3 to 5 percent ad valorem. In most instances, the manufacturer's tax was simply passed on to the consumer. On the face of it, it would not seem that these taxes could significantly alter the picture of profitable wartime production at the Cheshire Mills. Cf. Harry E. Smith, *The United States Federal Internal Tax History from 1861 to 1871*, chs. III, IX, *passim*.

143 CMR, "Records of the Cheshire Mills," 1860–1900, *passim*.

144 *NHS*, July 2, 1868, Feb. 9, 1887.

145 *Cheshire Republican*, Special Trade Edition, Aug. 18, 1899. For a complete listing of the company's machinery, see Appendix 9.

146 Cole, *op. cit.*, II, 209–210, 220–221.

capacity enough to run all their machinery.[147] This engine was run by burning great quantities of cordwood and, in later years, coal.[148] The wisdom of this installation was proved later the same year, when water again ran low and the mill could switch to running its machinery by steam.[149]

Ordinarily, the machinery was driven by water turbines. As the plant expanded, new and improved models were installed. Thus, when the brick mill was built in 1860, a ninety-horsepower water turbine was put in that utilized a twelve-foot fall of water.[150] And in 1866, the original Fourneyron wheel in the main mill was replaced by another wheel that utilized a twenty-eight-foot fall of water.[151]

Most impressive was the power system constructed in 1888 at the dam on the site of the burned New Mill. Here, about midway between the head of the ravine and the main mill, the Colonys installed a turbine wheel and enclosed it in a frame wheelhouse. Then they ran an iron pipe, or penstock, five feet in diameter, all the way from the head of the ravine to the turbine. By bringing the water this distance, they utilized a thirty-foot fall and produced the greatest amount of power ever developed at one place on the stream. To convey the power from the wheelhouse to the mill, a frame tower was built over the wheelhouse, and a rope transmission cable was run from the turbine to the top of the tower, around a wheel, and then through the air to the main driveshaft of the Granite Mill, some three hundred feet downstream. The rope cable itself, spliced to make it endless, ran quintupled between the powerhouse and the mill and was about 4,600 feet long. This intricate system, after a new turbine was installed in 1900, gave the Cheshire Mills about half again the amount of power the company then used.[152] The installation of this "rope drive," conveying power such a distance, was considered quite a feat at the time. The device was used for sixty years until the company converted to electric power in 1948. By then it was an antique oddity to be marveled at by historians and engineers.[153]

From time to time the company made other additions and purchases of property. In 1887, they quietly bought the property of the Milan Harris Woolen Company,[154] an anticlimatic end to the long rivalry between the two companies. It was widely hoped that the Colonys would rebuild the burned mill, but this was never done. The other buildings were repaired[155] and used for storage or in other capacities. In addition to the tenements built in the sixties, the Colonys bought other houses in the village to ensure accommodations for their expanding work force.

147 *NHS*, Feb. 6, 1884; Child, *op. cit.*, p. 176.

148 CMR, Copybook "B," pp. 940–941.

149 *NHS*, Nov. 26, 1884.

150 Census, 1870, Vol. XVI, Dublin, Schedule 4. This wheel was replaced in 1881 by a 30″ Standard X.L.C.R. water turbine. Cf. Humphrey, *op. cit.*, p. 15; *NHS*, July 14, 1881.

151 This was a Boyden-type wheel, made by John Humphrey of Keene. Cf. Humphrey, *op. cit.*, p. 15.

152 The wheel installed in 1900 was another 30″ X.L.C.R. turbine wheel. Cf. *NHS*, Nov. 28, 1900; Humphrey, *op. cit.*, p. 15; CMR, Copybook "B," pp. 890–892.

153 *Christian Science Monitor*, Magazine Section, July 27, 1946; *KES*, Aug. 1, 1950.

154 *NHS*, Feb. 9, 1887.

155 *NHS*, Aug. 31, 1887, May 30, 1888, "O.C."

Just after the turn of the century the company was renting to its employees a total of twenty-eight dwellings.[156] To protect its waterpower system and to effect other economies, the Cheshire Mills purchased water rights and wood lots in Harrisville and Nelson. By 1900, the property in Harrisville owned by the company was extensive: a few years later it had a total valuation of $96,100, which was about 23 percent of the total valuation of property in the town.[157]

Like prudent manufacturers, the Colonys were careful to see that their plant was fully covered by fire insurance. They even increased the coverage when they temporarily stocked an unusual quantity of wool because of market conditions.[158] Similarly, the underwriters kept a careful watch on conditions at the mill and occasionally called for improved conditions.[159] In 1901, the plant seems to have been covered by one blanket policy written by the Boston Manufacturers Mutual Fire Insurance Company, for which the Cheshire Mills paid an annual premium of $1,272.[160] The mill was a good risk. When small fires did occur, they were quickly extinguished by the company's own fire-fighting equipment.[161]

During the latter part of the nineteenth century the business operations of the Cheshire Mills (principally purchasing, manufacturing, and marketing) were carried on in a fashion that was in large degree simply a continuation or elaboration of methods used in earlier years. However, the mill records do show certain new features and methods in the last part of the century, and these deserve attention.

At least as late as the mid-eighties, some wool was still purchased from local sheep raisers.[162] The bulk of it, however, seems to have been domestic wool purchased either in the West or through dealers in Boston. In the late eighties, the company bought from one western dealer considerable quantities of medium-grade wool raised in Colorado and Missouri.[163] In 1900, Horatio Colony was still buying wool from dealers in such places as South Dakota,[164] but he found it was frequently advantageous to buy these western fleeces in the Boston market,[165] the most important wool center in the country.[166] Under depressed market conditions, wool bought there was often cheaper than in the West, and also he could be more certain of the evenness of its quality and the accuracy of its weight. On this score Horatio's admonitions to western wool buyers do not compare with those his brother Timothy wrote in the 1850s, but the problem was apparently still much the same as then:

156 See Appendix 7.

157 *The Invoice and Taxes of the Town of Harrisville, N.H., Taken April 1, 1906.*

158 CMR, Copybook "B," pp. 950–951.

159 *Ibid.*, pp. 858–859. Also, directly after the destruction of Milan Harris's New Mill, the Cheshire Mills had a new insurance appraisal made. See Appendix 9.

160 *Ibid.*, pp. 950–951, 971.

161 NHS, April 3, 1887. This equipment was augmented in 1899 by the erection of a 15,000 gallon water tank, supported on a high tower near the mill. NHS, Sept. 27, 1899.

162 CMR, Cashbook "C," pp. 27, 95, 101.

163 CMR, Copybook "B," pp. 218–219, 399–400, 403–404, 562–564, Letters to "Mess S. Bienenstok & Co."

164 *Ibid.*, pp. 903–906.

165 *Ibid.*, pp. 813–814, 847–848.

166 Cole, *op. cit.*, I, 273.

167 CMR, Copybook "B," pp. 853–854.

168 Census, 1870, Vol. XVI, Dublin, Schedule 4; Child, *op. cit.*, p. 176.

169 Cole, *op. cit.*, II, 164–172.

170 *NHS*, Oct. 13, 1870; CMR, Copybook "B," pp. 893–894.

171 *NHS*, Oct. 14, 1885.

172 CMR, Copybook "B," pp. 305–306.

July 6th 1900

A. C. Kibling Esq.

Dear Sir:—

I wrote you . . . in relation to the shortages of wool. . . . In your invoice running from No. 12 to 60 there is 78 lbs. shortage. This we did not intend to make a claim on. . . .

Great pains should be taken on weights. In some small towns it is almost impossible to find an accurate pair of scales, and they should be tested before using. In this last invoice I find quite a number of bags 3 lbs. short. . . . We make no claim but you must use all precautions possible.

Very truly yours

Horatio Colony, Treas.[167]

As to how much wool the Cheshire Mills bought in these years, the few statistics available show that in 1870 the company used 300,000 pounds of domestic wool and that fifteen years later it used double that amount.[168]

The woolen industry experienced a period of hardship and severe readjustment in the half century following 1870. This was due in large part to the dynamic growth of the rival worsted branch of the industry but also to the effects of tariff changes and rising standards of consumption. For the woolen industry as a whole "the period was one of readjustment under pressure, rather than one of any considerable decline," but the market for certain fabrics, such as cassimeres, satinets, and especially flannels, did decline. On the other hand production of rough woolens, like tweeds and cheviots, increased. Overcoatings, cloakings, and especially woolen dress goods increased so greatly as to help tide the woolen industry over this period of stress.[169]

The production of woolens at the Cheshire Mills was only partly in line with these general trends. During the last forty years of the century the company continued to turn out great quantities of flannels. "Government Blue" and "Cadet Mix" flannels were made during the Civil War. In the years that followed production included standard items like plain red, blue, orange, and scarlet flannels, blue and indigo mixed flannels, and twilled scarlet flannels.[170] In the mid-eighties, the mill was turning out 100,000 yards of flannel cloth a month.[171] The quality of these cloths must have been good for the Colonys to have continued selling as much as they did in a declining market.

At the same time the company did manufacture special items in response to changing demands of the market. In the eighties the mill made quantities of "Stanley Suitings," a trade name for a type of flannel suiting.[172] At the end of the century, in addition to the standard flannels and some heavy flannels made for the

Government,[173] it made cheviots, dress goods, and "plaid backs," a double-woven fabric used for coats.[174]

The Cheshire Mills continued to market its woolens through the agency of a commission house. Until 1871, the company sold through Faulkner, Kimball, and Company, of Boston, the same house the Colonys had started with twenty years before.[175] Thereafter, the records show that the mill sent its goods to a Faulkner, Page, and Company, of New York and Boston, which may have been the same house reorganized.[176] Not counting the small quantities of woolens sold directly from the mill[177] and with the exception of one other agent used briefly during the nineties,[178] the mill's entire output was sold through this commission house. Not that other agents did not seek their business, for they did, and Horatio Colony did not neglect to mention this to Faulkner, Page, and Company when he was seeking orders: "We have only 8 or 10 days work for our fancy looms, and you must send us orders without delay. Can you not send us orders for several hundred pieces of fancy cheviots? We are offered plenty of orders from other houses, but refuse them all, as we consider ourselves under your charge and direction."[179]

Relations between the Colonys and their commission house were generally open and cordial. Horatio Colony sought advice from the agents, as Henry Colony had done before him. A letter he wrote in November 1887 asked what kind of fabrics and in what quantities they could make for stock the next year and also what orders they might have for special items.[180] However, if the agent determined what fabrics should be made, the manufacturer had the greater say in regard to prices. Horatio Colony's letters to his agents contained many instructions regarding what items should or should not be held, what prices should be gotten, and what prices might be asked, if necessary, to clear the goods.[181] Of course sometimes the arrangements went awry:

I note the sale of 18 pcs. of Fibiline coverts at 60 cts. 7 off. . . . I thought the understanding was that *sacrifice sales* were to be first reported and assented to. As these goods were not particularly . . . they should have been held for better prices. Mr. Mungan must bear in mind that he is working for the manufacturer, and not the jobber.[182]

This was about as sharp a letter as Horatio Colony wrote to his agents. There were no such acrimonious exchanges as, for example, those between Michael Craven and his agents, and on the whole relations between the Cheshire Mills and Faulkner, Page, and Company seem to have been mutually satisfactory.

Thanks mainly to the excellent payroll records, which are

173 *Ibid.*, pp. 827–828.

174 *Cheshire Republican*, Special Trade Edition, Aug. 18, 1899; CMR, Copybook "B," pp. 838–839, 880–881.

175 CMR, "Consignments to Faulkner, Kimball, and Co.," 1861–1871.

176 CMR, General Journal, 1872–1882.

177 *Ibid.*

178 "J. B. Lorge & Company" appears to have acted as a selling agent for the Cheshire Mills. CMR, Copybook "B," pp. 701–703, 849–850.

179 CMR, Copybook "B," pp. 838–839, June 12, 1900.

180 *Ibid.*, pp. 305–306.

181 *Ibid.*, pp. 796–797, 932–933.

182 *Ibid.*, p. 926. This letter contains some illegible words.

complete from 1862 down to the present, a more systematic study of the Cheshire Mills' workers, wages, and conditions of employment is possible than has been the case for the other Harrisville mills.

The average number of workers employed at the Cheshire Mills during these years was in the vicinity of ninety, but the actual number varied considerably. During depression years or in a bad season for the industry, the number was sharply reduced, and occasionally the mill shut down completely for a short while. Aside from these periods the trend was toward a steadily lengthening payroll.

Thus, after the boom years of the Civil War the number of employees fell off slightly in the late sixties. Better prepared for it than Milan Harris, the Colonys survived the Panic of 1873, though belatedly they did feel the effects of the depression. The Cheshire Mills shut down for five months in 1875, in November of that year there were only fifty hands at work, and the payroll remained short until the following August. Like Craven and Willard, the Cheshire Mills were hampered by the low water during the winter of 1880–1881, but for the rest of the decade the company enjoyed full employment. Again, in 1890, the "mill was closed first two weeks in June on account of poor auction sale and dull market."[183]

The severe depression that began in 1893, like that of the seventies, affected production and employment only belatedly. Eighteen ninety-six was the worst year of the depression for the woolen industry in general,[184] and so it was also for the Cheshire Mills. The number of workers fell to fifty-three in February, and by August the company was only working three days a week.[185] Conditions improved the next year, and by September 1897, the employment rolls reached a new high of 120. It stayed close to that figure for the rest of the century.[186] Full employment was not then universal in the industry, as this note in the *Sentinel* indicates: "Although many other mills are either suspending business or 'running on half time' we note with satisfaction that Cheshire Mills 'pursues the even tenor of its way' undisturbed by Southern or other competition."[187]

In this period the labor supply available to the Colonys was probably never very short. The influx of the French Canadians and the disasters that overtook the Harris mills must have guaranteed an easy labor market, and the good reputation of the Cheshire Mills did the rest. Furthermore, the Colonys did not have the reservations about hiring newly arrived immigrants that those who ran the Harris mills seem to have harbored.

183 CMR, Payroll, June, 1890; *NHS*, June 4, 25, 1890.

184 Clark, *op. cit.*, III, 200.

185 *NHS*, Aug. 19, 1896.

186 CMR, Payroll, 1862–1900.

187 *NHS*, Aug. 31, 1898. Cf. *NHS*, Aug. 23, 1899; Clark, *op. cit.*, III, 200–201.

This last fact is apparent in the make-up of the Cheshire Mills' labor force. Among the different nationalities the Irish had been hired as soon as they began to arrive in the early 1850s, and thereafter Irish names were always in evidence on the payroll. These Irish mill hands sometimes gave their naturalization papers to the Colonys for safekeeping, and a few of these musty documents are still lying in the company's vault. It was the French Canadians, however, who made up the largest immigrant group among the Cheshire Mills' workers. They began to appear in numbers about 1867 and increased rapidly thereafter.[188]

From the information in the census schedules some quantitative analysis can be made of the national origins of the mill workers. Since in the census schedules it is impossible to tell by which company a woolen mill worker was employed, exact information on the birthplace of, say, those employed by the Cheshire Mills is limited to those living in the company boardinghouse. In 1870, there were thirty-six persons living there of whom over three fourths were foreign-born: eighteen Canadian, nine Irish, and one English (in contrast to the Milan Harris boardinghouse where less than half of the boarders were foreign-born).[189] In 1880, there were twenty-two adults and seven children living in the Colonys' boardinghouse. Of these, all the adults were Canadian-born, and the majority of them were women (compared with 38 percent foreign-born, mostly Irish, and mostly men, in the Craven and Willard boardinghouse).[190] Since it was the newly arrived immigrants who were most likely to live in the boardinghouses, these figures may not give an accurate picture of the whole work force. The total number of mill workers living in the town in 1880 was 184. Of these, 100, or about 55 percent, were foreign-born. Their national distribution was naturally much the same as the foreign-born population of the town generally: sixty-one Canadian, twenty-five Irish, twelve English and Scotch, two German.[191]

The pattern of nationalities in the Harrisville mills was somewhat at variance with that in the American woolen industry as a whole. Cole points out that the industry was affected more than most American industries by the infiltration of foreign-born workers. There is not necessarily any contradiction between this and his earlier statement that some woolen manufacturers avoided hiring the newly arrived immigrants; that reluctance probably applied more to earlier years and to mills in small communities, not to the growing number of large woolen mills in the industrial cities of the Northeast. The industry first experienced the influx of the English, the Germans, the Irish, and

188 See Appendix 7.

189 Census, 1870, Vo. XVI, Dublin, p. 1; Vol. XV, Nelson, pp. 10–12.

190 Census, 1880, Harrisville, pp. 16, 18.

191 Census, 1880, Harrisville.

the French Canadians. These nationalities continued to enter the industry after 1870, but in reduced numbers. Also, in the latter part of the century they were increasingly found in the positions of supervisors and skilled workers. Representatives of the New Immigration, from Southern and Eastern Europe began to enter the woolen industry in numbers about 1890, and thereafter the greater part of the labor supply, especially for the less skilled positions, was furnished by this group. By the early years of the twentieth century the representation of the New Immigration in the woolen mills of America was described as "peculiarly great."[192]

In Harrisville the representatives of the "Old Immigration" appeared on the Cheshire Mills' payroll later than they did in the larger textile centers, and only a very few Germans ever came there. Neither did the members of the "New Immigration" ever appear in any numbers. Just after the century ended, Harrisville saw the beginning of the last great wave of foreign immigration in its history. These were the Finns, who first began to arrive in numbers in the summer of 1902.

In the context of the times, working conditions in the Cheshire Mills must have been good. The buildings were undoubtedly more healthful than the average woolen mill.[193] There were occupational hazards, to be sure. There were accidents, instances of blood poisoning from the dyes, and the like. When such a mishap occurred, the victim could probably count on little more from the company than payment of the initial medical expenses. His only recourse was to bring suit against the company, as occasionally happened.[194] Still, when one thinks of the disastrous fires, accidents, and occupational hazards in other American industries during this same period, the Cheshire Mills would appear to have been a relatively safe place in which to work.

There is little indication in the mill records of the length of the workday. It appears to have been eleven and a half hours in 1882, about average for the industry, and ten hours in 1900, by which time it was a matter of state law.[195] When there was work, the mill ran six days a week. Actually, as the payroll shows, even in good times the mill hands, at least those on piecework, had a good many short weeks and part days. How many of these were because of lack of work, and how many because the operatives wanted time off is problematical, but there certainly were times when the operatives wanted, and took, time off for other things. They could also rearrange their work schedule. Thus in the summer of 1900 the *Sentinel* reported that the

192 Cole, *op. cit.*, II, 112–114.

193 *NHS*, Oct. 13, 1870.

194 *NHS*, Dec. 2, 1880, CMR, Copybook "B," pp. 957–959.

195 CMR, Payroll, Sept. 1882, Pat Whalen in Finishing Room; Payroll, Nov. 1900, Finishing Room; cf. Cole, *op. cit.*, I, 373, II, 116.

employees were able to have their Saturday afternoons off "by beginning work at an earlier hour."[196] Finally, an occasional holiday—without pay—might also break the monotony. In 1899, the Cheshire Mills did not work on Thanksgiving or New Year's Day, but apparently there was no stoppage for Christmas Day.[197]

In housing the Cheshire Mills' workers fared well. The boardinghouse and tenements have already been described. In the early sixties, the company charged their male workers $7.50 a month for board, the same rate charged in the fifties. By 1867, the rate had been doubled for male workers but remained the same for women. This rate amounted to nearly half the pay earned by a spinner or weaver, and there were complaints.[198] These rates do not seem exorbitant. Protests notwithstanding, these rates with some variations were maintained for the rest of the nineteenth century.[199]

The worker with a family stood a better chance of securing a tenement from the Colonys than he did from the Harris mills. For most of their tenements the Colonys charged five dollars a month rent. This was the rate in 1862, and it was the same rate more than half a century later. Elsewhere in the country rents rose steeply during this period. In the cities especially, rents rose as much as 85 percent between 1862 and 1869.[200] Also, it has been the policy of the Cheshire Mills for many years to do the major items of maintenance and repair on its tenements[201] and to allow the tenants to deduct from their rent any expenditures they might make in improving their homes. It is also true that the company very often did not collect rents in bad times when its tenants were not working.[202]

All this is not to say there were not difficulties between the company and its workers. An unusual letter, which has lain in the company's vault for nearly a century, reveals one example.[203] It was written in 1869 to Henry Colony by one of his employees, who signed himself "Spy." The writer revealed to his boss that some of the spinners were systematically defrauding the company. The writer did not make it entirely clear just how it was being done, but one method current then was known as "kicking the clock." A spinner, by a well-timed motion, could make the clock on his spinning jack or mule register twice instead of once each time the carriage returned.[204] It all added up when one was working on a piece rate. This letter, obviously unsolicited, has the ring of truth about it, and what it says about the spinners is consistent with the bad reputation enjoyed by the spinners of that day.[205] The letter is also a valuable document for showing

196 *NHS*, May 23, June 20, 1900.

197 CMR, Copybook "B," pp. 778, 782.

198 See Appendix 8.

199 See Appendix 7.

200 *Historical Statistics of the United States*, p. 129, Series E186.

201 *NHS*, June 14, 1899.

202 Note from John J. Colony, Jr., Aug. 26, 1963.

203 See Appendix 8.

204 Interview with John J. Colony, Jr., Feb. 16, 1961.

205 Cole, *op. cit.*, I, 360.

that even in those days the mill worker was not necessarily on the side of the angels.

The labor force in the woolen industry never organized effectively during the nineteenth century, but occasionally small groups of skilled workers acted together for their own interest.[206] During the last half of the nineteenth century there is record of only three strikes or walkouts at the Cheshire Mills. These all took place within a space of less than three years. The first was a brief walkout in 1887 and was occasioned by the fact that both Christmas and New Year's Day fell on Sunday. As the *Sentinel* said, the workers felt aggrieved: "It appears that they asked to have the mills shut down so as to give them a holiday, but the employes [sic] declined to grant their request, whereupon all went out except three. Probably Monday morning will find them all ready for work."[207]

This walkout may have expressed broader discontent than the *Sentinel* suggests, for the next week the weavers at the mill went on strike. Their complaint was over the company's increasing the length of the cuts of cloth, by which their wage rates were determined. The strike lasted only a week and was ended by a compromise settlement.[208]

The strike that occurred in 1890, and again described in the *Sentinel*, ended differently:

A strike took place last week at the Cheshire Mills. The spinners went out on account of the discharge of one of the operatives. Mr. Faulkner, superintendent, went to Lowell and soon had their places filled. It seems to us the strikers have made a mistake, especially at this season of the year, and when making so good pay.[209]

In neither of these strikes is there any suggestion that the rest of the operatives took any action. Inevitably, they were feeble protests. The company showed that though it might tolerate such an action as the New Year's Day walkout in 1887 and be willing to adjust differences, as it did with the weavers, it nonetheless could and would take prompt and effective action to end strikes when it saw fit. Yet when viewed against the background of industrial warfare that was rocking the country in these years, these disturbances in Harrisville seem insignificant.

There were other company policies to cause profanity on payday, such as fines levied on the weavers for poor work and on those in the card room for damaging the card clothing. These fines ranged from $.25 to $1.00, and occasionally were numerous. In two months in 1897, thirty-two workers suffered such fines.[210]

206 *Ibid.*, II, 123–126.

207 *NHS*, Jan. 4, 1888, "O.C."

208 CMR, Payroll, Jan. 1888; *NHS*, Jan. 11, 25, 1888, "O.C."

209 *NHS*, Oct. 1, 1890, "O.C."

210 CMR, Payroll, March, May, 1897.

Despite these examples of disharmony between company and workers, there was contrasting evidence testifying to a happier relationship, such as this item from the *Sentinel* in 1886: "We hear that the operatives at the Cheshire Mills have subscribed sufficient funds to purchase a silver tea service to present to their late super, W. F. Turnbull, as a token of respect."[211]

Finally, there is the matter of the wages paid by the Cheshire Mills. These are shown in detail elsewhere.[212] The payrolls for these forty years show that the company paid its workers monthly, frequently made advances to them before payday, and made deductions, not only for rent and board but also for a variety of commodities that the company apparently sold its employees from time to time. Flannel, kerosene, and cordwood were frequently listed items, but occasionally there were more unusual ones: one pig for $5.00, a barrel of apples for $4.50, suits for $6.00, and "three feather beds" for $3.00.[213] The payroll also showed instances of workers receiving extra pay for overtime work, especially those in the finishing room.[214]

It is difficult to trace the trend of wages at the Cheshire Mills over the years. Not only did the workers, bookkeeping methods, and the value of money change, but sometimes the job itself did, spinning for example. In the woolen industry generally, according to Cole, the trend of wages was more persistently upward than that of such other items as the price of raw wool or finished woolens. In good times wages shared in the general upswing, while in bad times they receded less considerably. For the Cheshire Mills the figures are lacking to correlate the trend of wages with the prices of commodities, but the upward trend of wages was true enough. Between 1862 and 1897, the number of workers rose 35 percent, but the gross pay-roll rose 95 percent. Similarly, during the same period the average daily cash wage rose 43 percent, from $.95 to $1.36. Most of this gain came in the sixties. After that, except during the bad years of the seventies, the average daily cash wage remained essentially unchanged for the rest of the century.[215]

It is difficult, too, to compare the wages paid by the Cheshire Mills with those paid by other woolen mills. Information of this sort was not given wide circulation. From some fragmentary evidence, which relates to the years from 1890 to 1900, it would appear that the rates in Harrisville lagged behind the median rates paid by the woolen mills of New England.[216] However, in contradiction to this, the *Sentinel* claimed in 1900 that wages at the Cheshire Mills "are still, as a rule, in excess of those paid at manufactories of the kind."[217]

211 *NHS*, June 9, 1886, "O.C."

212 See Appendix 7.

213 CMR, Payroll, 1862–1900, *passim.*

214 CMR, Payroll, Sept. 1877, Sept. 1882.

215 Cole, *op. cit.*, II, 131. See Appendixes 7, 7a.

216 U.S. Bureau of the Census, *Twelfth Census, Special Reports, Employees and Wages, 1903*, pp. xlii, xliii, p. 638.

217 *NHS*, Jan. 31, 1900.

More important than the exact wage rate was the certainty of the pay envelope, especially during the difficult decades of the seventies and the nineties. In this respect the Cheshire Mills undoubtedly did better than the average company. In 1893, the woolen industry instituted a general wage cut of 10 percent that was not restored until 1899.[218] The Colonys made no such wage cut at their mill during these years.[219] Though the company did not entirely escape from the depressions of the seventies and nineties, it seems to have felt their impact later and recovered sooner than the industry in general. One can only conclude that, though the picture concerning relative wage rates is unclear, in the steadiness of its wages and employment the Cheshire Mills must have compared favorably with the average woolen mill in this period.

In conclusion it can be said that the years from 1860 to 1900 constituted a period of widely fluctuating conditions for the Harrisville woolen mills, calling for the utmost in skillful management. The Cheshire Mills met the test of the times; the Milan Harris Woolen Company did not. In the later years the Cheshire Mills were limited in expansion by the difficulties of the woolen industry, but at the end of the century the company was in a sound and prosperous condition and was still the mainstay of Harrisville's economy.

218 Clark, *op. cit.*, III, 201.
219 *NHS*, Jan. 31, 1900.

7 THE NEW TOWN AND
ITS SOCIAL LIFE, 1870–1900

By an act effective July 2, 1870, the New Hampshire Legislature created the new town of Harrisville.[1] It was formed by taking a strip of land from the southern part of Nelson and a somewhat larger strip from the northern part of Dublin. Joined together, these lands formed an irregularly shaped area about eight miles east to west and three miles north to south. This was a rather small area for a New Hampshire town, but it was more than ample for a railroad right-of-way.[2]

This legislative surgery dealt a severe blow to the parent towns. Both Dublin and Nelson were considerably reduced in size, but they were even more reduced in population. Nelson lost 271 inhabitants or 36 percent of its total. Dublin lost more, about 435 inhabitants or nearly 47 percent.[3] As prescribed by the act, Dublin and Nelson divided their assets and debts with the new town on the basis of how much population they lost.[4] In October the selectmen of the three towns perambulated their common boundaries.[5] There is nothing to suggest that there was any ill feeling to hamper the carrying out of these arrangements between the towns. Fittingly, the agent chosen by the new town for the settlement of affairs with Dublin and Nelson was Milan Harris.[6]

On August 13, 1870, Harrisville held its first town meeting. "It was a bright, sunny day of the latter part of the summer, when nearly every voter in this new town assembled to take part. . . ."[7] The town chose for its selectmen Darius Farwell, Samuel D. Bemis, and George Wood. Darius was the grandson of early settler John Farwell and lived on the farm cleared by his grandfather midway between Harrisville Pond and Breed Pond.[8] Bemis, who lived in Pottersville until his death in 1918, was for twenty years chairman of the town's board of selectmen.[9] George Wood was born in Dublin, lived in various cities, and was a selectman in Dublin in 1870.[10] Somewhat remarkably, it would seem, this mill town's first three selectmen were all farmers.[11] Perhaps the agricultural element in the new town was stronger than the prominence of its manufactures would lead one to suppose.

1 *Laws of New Hampshire*, 1870, XLIV.

2 More precisely, Harrisville was formed by taking from Dublin that town's three northern ranges and from Nelson the three southernmost ranges on the east side of Breed Pond and the two southernmost ranges on the west side of that Pond. The new town's total area amounted to about 9,625 acres. The average size of New Hampshire towns at the present time is 24,387 acres. Cf. Hamilton Child, *Gazetteer of Cheshire County, New Hampshire*, p. 175; Fisk and Wadsworth, *Map of Dublin, N.H.*, 1853; C. H. Rockwood, *Atlas of Cheshire County, N.H.*, 1877; *New Hampshire Manual of the General Court*, 1963.

3 Michael G. Hall, "Nelson, New Hampshire, 1780–1870," p. 53, says that Nelson lost "almost half" of her population of 744 people, but that is a little high. The *Nelson Clarion*, May 1871, puts the number at 271 people. The 1920 edition of the *History of Dublin, N.H.*, p. 464, says only that when that town was divided its "population was materially diminished." However, from an

The years from 1870 to 1900 brought only modest changes in the new town's appearance. In the rural area the principal modifications were the deterioration of some of the farms and the erection of summer cottages along the lake shores. The mill village expanded, but its appearance was not greatly altered. As nearly as can be determined, the number of dwelling houses there rose from 23 in 1858 to 58 in 1870, to 70 in 1880.[12] After the fire that destroyed the New Mill in 1882, construction slackened. A map of 1906 indicates that the village was hardly larger then than it had been twenty-five years earlier.[13]

Most of the houses built during and after the Civil War were frame and not the architectural equals of the handsome dwellings built by the Harris family. Many of them were probably occupied by tradesmen, mill overseers, and workmen. At least until 1883, housing was much in demand.[14] If the new construction did not add to the elegance of the village, it at least provided decent housing for the mill workers and is good evidence of their rising standard of living. For the forty years after 1862, the records of the Cheshire Mills show a steady increase in the number of houses that the company rented to its employees. In 1902, the company was collecting rents on twenty-eight dwellings, the most in its history. And, until the late nineties, the mill records show a corresponding decrease in the number of workers living in the company's boardinghouse.[15]

The expansion of the village added other new buildings.[16] Harrisville never has gotten around to building a town hall, but "Eagle Hall" was built just behind Bradley's store and was in use for public gatherings and social affairs by 1871.[17] It was privately owned, eventually by the Winn family, and the owners charged the town an annual rent of thirty, later twenty, dollars for using the hall to conduct town meetings.[18] (Plate XXVIII) A small frame building intended to house the town library and town office was placed next to the same store in 1880.[19] And there were additions to the village of a different sort, as a local *Sentinel* correspondent wryly noted in 1882: "Mr. F. Pike has opened a pool room . . . in the rear part of his building, which, it is said, is largely patronized. . . . We do not know whether the influence of such a resort is 'good, bad, or indifferent,' but it shows we are becoming more city-like."[20]

In addition to new construction, there were changes in communication and means of travel that brought the new town into closer contact with the world beyond the highlands of Cheshire County. Short of ideal though it was, the long-awaited railroad was the principal improvement. Operating in close

examination of maps and the census schedules it appears that Dublin lost 435 persons. Many of these lived in Pottersville. The fraction used for dividing Dublin's public property and debts with the new town of Harrisville was 47/100 (*Laws of New Hampshire*, 1870, XLIV). This fraction applied to Dublin's population gives a figure of 437 as the number of residents lost to the new town, close enough to the calculation of 435 to indicate that this number is approximately correct.

4 Dublin Town Records, Vol. V, p. 395; Nelson Town Records, Vol. XIII, 1870, 1871.

5 Dublin Town Records, Vol. V, p. 394; Nelson Town Records, Vol. XIII, Oct. 1870.

6 D. H. Hurd, *History of Cheshire and Sullivan Counties, New Hampshire*, pp. 210–211.

7 *Ibid.*

8 S. G. Griffin *et al.*, *Celebration of the One Hundred and Fiftieth Anniversary of the First Settlement of Nelson, New Hampshire, 1767–1917*, pp. 66–67; C. H. Rockwood, *Atlas of Cheshire County, N.H.*, p. 41.

9 Leonard, 1920, p. 718; *NHS*, Aug. 21, 1918.

10 *Ibid.*, pp. 480, 948.

11 *Ibid.*, pp. 718, 948; Census, 1880, Harrisville, p. 22.

12 *Map of Cheshire County, N.H.* (1858); Census, 1870, Vol. XV, Nelson, Vol.

cooperation with the railroad was the Dublin Stage Company, owned by Francis Stratton of Harrisville. The horse-drawn vehicles of this company carried passengers, mail, and freight to and from the railroad station in Harrisville.[21] The stage line served as Dublin's main connection with the railroad and allowed the fastidious residents of that town to keep "in touch with the outside world without the unpleasant accompaniment of a railroad station nearer than three miles."[22] Then, within a half-dozen years after the railroad was completed, the first telephone line was run into Harrisville.[23] The first telephone in town was probably in the postoffice, the store, or the Cheshire Mills.[24]

The railroad and telephone brought improvement in communication, but the roads in Harrisville changed little in this period. More were discontinued than were built.[25] A new road was built from Dublin to Harrisville,[26] and the level of the road past Harrisville Pond was raised after forty years of annual flooding and disputation between the mill owners and townspeople.[27] Most of the few new roads were built for the benefit of summer people with cottages on the lake fronts.[28] The selectmen recommended this construction for they thought it a good policy for the town to provide these new taxpayers, increasing annually, "with all reasonable facilities. . . ."[29] For the village the main improvements were a dozen street lamps, some sidewalks, and a new iron bridge over the canal.[30]

The upkeep of the roads continued to be an important and frequently troublesome part of town affairs. Between 1880 and 1900, the annual cost of repairing roads and bridges, together with the cost of breaking roads in winter, averaged about one sixth of the town's total expenditures. In 1880, this work cost just over a thousand dollars and, in 1900, thirteen hundred dollars. Until the very end of the century Harrisville spent more money on its roads than it did on its schools.[31]

The explanation for this high cost lay in the various difficulties in keeping the roads passable. The town's system of taking care of the roads was not always the most efficient. In 1880, the town employed five road agents, assigning to each a certain amount of highway.[32] The principal advantage of this system was the political one of having not one but five jobs open for appointment each March. Indeed, the opinion was publicly voiced that the appropriations for road repairs were "too much like a bundle of boodle."[33] By the end of the century the town had switched to employing a single road agent. At that time it was Charles Blake, the hotelkeeper. He hired men to work on the roads and worked himself perhaps ten full days out of a month. For his labor he

XVI, Dublin; Census, 1880, Harrisville, Houses 105–175. Child, *op. cit.*, p. 176, says that the village contained "about one hundred dwellings" in 1885, but this was an exaggeration.

13 Fisk and Wadsworth, *Map of Dublin*, N.H. (1907).

14 *NHS*, Feb. 26, 1880, "F. C. Pike . . . intends to finish off two tenements, which are much needed, as none are to be had."

15 See Appendix 7. Of course, many of the houses owned by the Cheshire Mills were not built by the Colonys but simply bought by them as they came on the market. Real estate values were low after 1883. In 1889, the Cheshire Mills bought the estate of C. C. P. Harris at auction for $1,250. Cf. *NHS*, June 12, 1889.

16 For the various shops and stores that appeared in the village during these years, see ch. 5.

17 *NHS*, Feb. 16, 1871.

18 *HAR*, 1874.

19 *NHS*, March 18, 1880.

20 *NHS*, April 5, 1882, "O.C."

21 Leonard, 1920, p. 500.

22 H. H. Piper, "A Sketch of Dublin," *The Granite Monthly*, XXI, No. 2 (Aug., 1896), p. 94.

23 *NHS*, Nov. 12, 1884.

24 The Cheshire Mills were paying regular monthly bills to the New England Telephone Company by 1886. See CMR, Cashbook "C," 1882–1889, *passim*.

billed the town at the rate of twenty cents an hour.[34] Until 1893, residents of New Hampshire were permitted to "work out" their highway tax,[35] a system long recognized as inefficient.[36] And there was always the problem of road equipment. Even in the eighties there were complaints about the "old fashioned road scrapers"[37] and the shortage of available oxen for work on the roads.[38]

All these difficulties were minor compared to those caused by the elements. Harrisville residents took in stride the clouds of dust that rose from their roads in summer and the washouts of spring and fall, but the deep snows of winter and the mud of early spring brought travel to a virtual standstill. A heavy snowstorm with high winds might block the main roads for days before teams pulling heavy snow rollers could pack it down.[39] At such times people with sleighs often found it easier to use the frozen ponds for travel.[40] The great Blizzard of 1888 hit Harrisville as hard as it did the rest of New England. The local correspondent of the *Sentinel* wrote that because of the "great storm of the past week" no mails were delivered from Monday noon to Saturday afternoon.[41] It cost the town over six hundred dollars to make the roads passable after that storm, more than double the usual amount annually appropriated for breaking the roads in winter.[42]

Then every March and April the spring thaw and rains could be expected to turn the roads into ribbons of mud, making travel even more difficult than in winter.[43] Under these conditions a new road agent or an improved road scraper was hardly the solution. Real improvement had to wait for the automobile and the hard-topped road. This innovation in the means of travel so, momentous for all the country, made its first appearance in Harrisville in 1900. The event was briefly chronicled in the *Sentinel*: "The first appearance of an automobile was Sunday evening. It moved swiftly, smoothly, and noiselessly."[44]

Concerning its general appearance, the village probably lost some of the charm and quaintness of the antebellum period. There is no reason to believe, however, that it suffered very greatly. The mills did not deteriorate as they did in so many other New England towns. The old Upper Mill standing in the center of the village was preserved by the Colonys,[45] and for many years its appearance was enhanced by the ivy that covered its walls.[46] When the Colonys bought the property of the Harris mills in 1887, they promptly cleaned up the ruins of the burned mill.[47] There were also occasional efforts to make the village more attractive. Arbor Day, 1886, was observed by planting a number of shade and ornamental trees.[48]

25 *HAR*, 1886.

26 *NHS*, Nov. 6, 1889.

27 *NHS*, June 1, 8, 1898.

28 *NHS*, April 3, 1890.

29 *HAR*, 1891, 1896.

30 *NHS*, July 3, 1889, Sept. 9, 1896, July 24, 1900.

31 *HAR*, 1881, 1886, 1891, 1896, 1901.

32 *HAR*, 1881.

33 *NHS*, March 18, 1896.

34 *HAR*, 1901.

35 M. H. Robinson, *A History of Taxation in New Hampshire*, p. 212.

36 *HAR*, 1886.

37 *HAR*, 1881.

38 *NHS*, May 26, 1886.

39 *NHS*, Feb. 25, 1885.

40 *NHS*, March 4, 1891, "O.C."

41 *NHS*, March 21, 1888.

42 *NHS*, March 6, 1889.

43 *NHS*, April 7, 1886.

44 *NHS*, Aug. 1, 1900.

45 *NHS*, Nov. 9, 1898.

46 Leonard, 1920, p. 573.

47 *NHS*, Nov. 9, 1887.

48 *NHS*, May 5, 1886.

Had Harrisville's appearance deteriorated very much, complaints would have been forthcoming in the *Sentinel*, for there were complaints about nearly everything else. The only one, however, came at the end of the century and concerned a familiar evil:

49 *NHS*, July 11, 1900.
50 See Appendix 4.
51 *NHS*, July 16, 1890.
52 Census, 1880, Harrisville. See Appendix 6.
53 See Appendix 6.

In driving about town, one cannot fail to lament the general disfiguring of sign-boards by advertising placards. In Massachusetts a fine of fifty dollars may be imposed on anyone convicted of thus decorating even a tree by the wayside, and it is certainly highly objectionable to have thus forced on one the very doubtful merits of some horse liniment or patent medicine.[49]

In summary, the new town during the last part of the nineteenth century experienced moderate growth and improvement of communications without suffering serious impairment of its charm and attractiveness.

Turning from the town to its people and their life, this period between 1870 and 1900 saw Harrisville's population reach its peak. It increased about one hundred and fifty in the seventies, fell by more than a hundred in the eighties, and recovered about half that number in the last decade of the century.[50] Had it not been for the shutting down of the Harris mills, there would probably not have been this cleft in the peak of Harrisville's population curve.[51] Despite these vicissitudes, at the end of the century the population was relatively stable.

The census schedules of 1880, the last ones now available for examination, provide the material for a quantitative analysis of Harrisville's population just before it began to decline. The native-born showed a 6 percent decline since 1870 but still constituted three quarters of the total. Though decreasing, the New Hampshire-born made up the bulk of the native-born and still constituted 55 percent of the town's total population. Nor were these people second-generation Americans; almost 80 percent of the New Hampshire-born were the children of native-born parents. In distribution, these native-born residents were to be found in almost equal numbers in the village and in the rural area surrounding it.[52]

The foreign-born had increased a corresponding 6 percent during the seventies and, in 1880, made up 25 percent of the total. French Canadians made up the bulk of the newcomers. Between 1870 and 1880, their numbers more than doubled, and in the latter year they constituted a majority of the immigrants in the town.[53] The "New Immigration" from southern and eastern Europe barely touched Harrisville in these years. A gang of twenty-five Italians, brought to Harrisville to work in a gravel bank in 1885, were enough of an oddity for the local *Sentinel* correspondent to write about them and describe their diet.[54] With few exceptions the foreign-born lived in the village, and the great majority of those employed worked in the woolen mills.[55] In fact, opportunity for working in the mills was the valve that controlled the influx of immigrants into Harrisville. Since it was in the early eighties that the two mills employed the maximum number of workers, or at least before 1900, it was probably then that the town had its greatest proportion of foreign-born residents. Thus, at their peak strength the foreign-born constituted a minority large enough (25 percent and concentrated in the village) to have a great effect on the life of the town but not large enough to dominate it.

Only scattered evidence is available to show the manner in which Harrisville lost population through emigration. Some loss of population was undoubtedly incurred after the failure of the Milan Harris Woolen Company in 1874,[56] but probably no great number left, as the mill was soon back in operation, and in that depression year opportunities for work were apt to be no better elsewhere. The loss of population after the burning of the New Mill must have been considerably greater. The *Sentinel's* local correspondent reported in April 1883 that "The destruction of the mill and the dispersion of the operatives in consequence has diminished the valuation of the town almost $50,000. There seems to be almost a stampede from the village. The diminution in the number of polls from last year is about sixty."[57] The flight of the operatives caused a decline in business which in turn forced others to leave, including two store-keepers, a livery stable keeper, and a doctor.[58]

Just where this emigration went is scarcely indicated. The operatives seem to have sought work in other mill towns in New Hampshire and Massachusetts. Others in the village who joined the "stampede" went in various directions. The shoe merchant went to nearby Gardner, a mill town in Massachusetts. The doctor went "West."[59] A few men, with old-stock names, went to such varied places as New York State, Texas, and Colorado.[60] Sketchy though the evidence is, these cases are probably representative of the pattern of migration from Harrisville.

Concerning the health of the town's population, many of the conditions in the earlier part of the century continued to apply in the period after 1870. Still, there were new aspects. Beginning

54 He wrote that they boarded themselves in a large shanty. "Their bill of fare is peculiar, soup with plenty of onions the staple dish, with hard bread and macaroni which they make themselves." *NHS*, May 6, 1885, "O.C."

55 Census, 1880, Harrisville.

56 Cf. Bemis MSS, Box 6, XXXII, p. 4, "The church in Harrisville." Bemis wrote, "The business depression of the 70's which closed the Harris mills so scattered the strength of the church that it could no longer support a minister."

57 *NHS*, April 25, 1883, "O.C."

58 *Ibid.*

59 *NHS*, March 28, 1883.

60 *NHS*, May 13, 1885, July 29, 1885, Oct. 26, 1887.

in the eighties the town kept a record of its births, deaths, and marriages. Without knowing the age distribution of the town's population, however, these statistics are of limited value. For example, the town's crude death rate in the nineties does not appear to be lower than that prevailing in Dublin during the period from 1820 to 1850,[61] but the average age in Harrisville near the end of the century was probably higher than in the earlier period.

Among the causes of death infectious diseases were declining but still took many lives. With a gradual improvement in living standards, tuberculosis had declined; among the causes of death in 1900, it ranked third nationally, behind cardiovascular disease and pneumonia and influenza.[62] This same order may have applied in Harrisville. Certainly there were a great many deaths due to pneumonia.[63]

Although tuberculosis was no longer mentioned in quite the same awed tones used by Levi Leonard in the middle of the century, nonetheless it still caused numerous deaths. In 1888, Zophar Willard lost a twenty-seven-year-old son to the disease a few months after he had belatedly taken him to a ranch in Texas to restore his health.[64] The town's most tragic case was that of Chauncey Barker's family. Every few years during the late sixties and seventies the *Sentinel* reported the death of one of his daughters, usually in their late teens or twenties. Finally, in 1880, this item appeared:

The wife of Henry Stevens of Nelson died at her home on Sunday. . . . The deceased was a child of Mr. and Mrs. Chauncey Barker of this town; and is the seventh daughter which they have been called upon to follow to the grave. Every one of the seven died of consumption.[65]

The white plague was still, as Leonard had said, the killer of youth.

Other infectious diseases, though they took fewer lives, were prevalent and troublesome. Among these were bronchitis, typhoid fever, "La Grippe," measles, whooping cough, mumps, and scarlet fever.[66] Occasionally, there was an outbreak of "that dreaded disease," diphtheria.[67]

Because of the adoption of vaccination, smallpox was no longer a major cause of death.[68] Harrisville occasionally authorized its resident physician to vaccinate at the expense of the town.[69] This may have been regular practice, but more likely it was only done on the rare occasions when an outbreak of smallpox threatened the town.

61 In the earlier period, Dublin's crude death rate averaged 13.9 per thousand population. In 1890, Harrisville's was 17.3, and in 1900 it was 11.4. See *HAR*, 1891, 1901.

62 Cf. *Historical Statistics of the United States*, p. 26, Series B114–128, p. 30, Series B155–162.

63 *NHS*, May 27, 1880, Feb. 24, 1881, May 30, 1888, March 13, 27, 1889, Feb. 5, 1890.

64 *NHS*, Nov. 30, 1887, Feb. 15, 1888.

65 *NHS*, Aug. 12, 1880, "Z." Cf. *NHS*, Aug. 1, 15, 1888.

66 *NHS*, Oct. 10, 1872, May 27, 1880, Feb. 24, 1881, April 22, 1885, Dec. 25, 1889.

67 *NHS*, Jan. 15, June 4, 1890.

68 Cf. *Historical Statistics of the United States*, p. 30, Series B155–162.

69 *NHS*, Feb. 15, 1882; *HAR*, 1886, 1901.

Accidental deaths seem to have been remarkably few. The woolen mills maintained the good safety record they had had from their beginning. On the railroad there were accidents, derailments, and minor collisions, and very occasionally one of these resulted in a fatality. In 1884, a French-Canadian laborer was killed in a collision of two gravel trains, leaving behind a wife and five children in "destitute circumstances."[70] The most frequent accidents, some of them fatal, involved mishaps with horses and teams, such as a thrown rider, an overturned buggy, or a man run over by the wheels of his wagon.[71]

Violent deaths were rare, especially considering the reputation that Harrisville enjoyed in those years. A young man named Roebuck, from Cornwall, Ontario, committed suicide one morning, after having quarreled with another man in Milan Harris's old brick boardinghouse. During the quarrel, Roebuck drew a revolver and shot his antagonist. The victim was only slightly wounded, but he yelled murder so loud he apparently convinced Roebuck that this was his crime, for the gun wielder then shot himself, with greater effect.[72] A few years later a maid at the "Nebaunsit House" attempted suicide by taking poison, with unknown results.[73] No other suicides, or even attempts, were reported in the *Sentinel*. Perhaps it was their inclination to externalize their aggressions that saved the town's inhabitants from self-destruction.[74]

If the people of Harrisville were enjoying better health and living longer at the end of the century than people did in 1850, it was probably due more to an improved standard of living than to any great improvement in medical care.[75] A number of doctors practiced in Harrisville in this period, but they appear to have stayed in town no longer than had their predecessors in earlier years. That they moved on for much the same sort of reason is indicated by a comment of a *Sentinel* correspondent in 1880: "Dr. Perry has left town. This is the sixth physician who has located here within the last ten years and soon departed in search of 'fresh fields and pastures new.' "[76]

The last resident physician in Harrisville was Dr. Arthur G. Byrnes, who came there from Madison, New York, before the end of the century and practiced there until about 1920.[77] Reportedly, the doctor first appeared in the region selling remedies for the Kickapoo Indians; if so, it would have been fitting.[78] Though he was skillful and popular at first, he was not in good health and drank too much. He left town several times for short periods but always returned to resume practice.[79] Probably his liabilities prevented him from succeeding in any

70 *NHS*, Nov. 12, 1884.

71 *NHS*, May 6, 1885.

72 *NHS*, Oct. 25, 1877.

73 *NHS*, Dec. 10, 1884. Note, name of hotel had apparently been changed from the "Nubanusit House," as it was called in 1881.

74 Harrisville's record was good. Nationally, in 1900, accidents and suicides accounted for 82.5 deaths per hundred thousand population, as many as caused by typhoid, diphtheria, and measles combined. See *Historical Statistics of the United States*, p. 26, Series 114–128.

75 Nationally, the average life expectancy at birth had risen from about forty-two years in 1850 to about fifty years in 1900. Cf. Richard H. Shryock, *The Development of Modern Medicine*, p. 333.

76 *NHS*, April 12, 1880, "Z."

77 *NHS*, March 28, 1900.

78 Interview with Guy Thayer, Aug. 30, 1966.

79 *NHS*, March 28, Sept. 4, 1900; *KES*, Aug. 6, 1901.

better practice; certainly in his later years in Harrisville he deteriorated badly.

There were no resident dentists in town. Doctors made the difficult extractions. One summer an itinerant "dentist" stopped briefly in town and attracted some patients by promising painless extractions. The injection he used was evidently poisonous, and the local doctors were kept busy for a time.[80] Little wonder that the people of Harrisville had limited confidence in doctors and dentists.

If the residents of Harrisville had doubts about their doctors, there is evidence they were aware of the responsibility of the community in matters of health. Vaccination of school children at the expense of the town was one example.[81] Also, in line with the developing national interest in public health, Harrisville had by 1900 set up a board of health. It consisted of the town moderator, the overseer of the poor, and the town's resident physician.[82]

Closely related to this matter of public health was the care of the poor. In the previous century a number of New Hampshire towns, including Dublin, had met the problem by taking advantage of a law that allowed them to "warn out of town" all newcomers likely to become public charges. What the towns did was simply to warn out *all* new arrivals, thus acquitting themselves of responsibility. Then, in the early years of the nineteenth century Dublin changed its policy to, as the expression was, "venduing the poor." Under this system the town held a special annual meeting at which the town's poor were farmed out for the year to the lowest bidder. This was shabby enough, but it was the extraordinary practice for the town to furnish free liquor for all those present at these meetings, "serving to make," according to one contemporary, "the most prompt bidders of a class of men the least fitted to have charge of the bartered victims."[83]

Certainly an improvement on that system was the town farm which Dublin established in 1837. This farm was maintained for about thirty years in the northeast corner of Dublin within the present boundaries of Harrisville. Thereafter, in accordance with a new state law the poor were sent to a new county farm, located in Westmoreland. At that place there was also a house of correction, and provision was made for the insane poor. The new system was criticized for heartlessly removing old people from their friends and familiar surroundings, but the financial advantage of the system was overriding.[84]

Like other towns, Harrisville had an Overseer of the Poor to handle these matters. The selectmen were Overseers of the Poor,

80 *NHS*, Aug. 23, 1899, "Hermes."

81 *HAR*, 1886, 1901.

82 *HAR*, 1901.

83 Leonard, 1855, pp. 26–27.

84 Leonard, 1920, pp. 498–499.

ex officio, but usually there was an individual appointed and paid a nominal salary. In 1880, it was George Davis, who also ran the Cheshire Mills' boardinghouse,[85] and at other times it was one of the storekeepers. The town tried to choose agents with practical experience in provisioning.

The assistance furnished to the poor was all itemized in the published annual town reports, including the names of the individuals and the amounts paid. The town paid the board of its pauper residents who were confined to institutions, such as the state asylum or the reform school,[86] and it paid for the support and medical expenses of its poor who had been moved to the county farm.[87] It also appears that under some circumstances the town supported poor families while they lived in town. This support was quite limited in amounts and numbers. In 1873, three people received assistance, in 1880 seven people or families received some support, and in 1900 one "dependent soldier" was the only resident receiving help. The expense for these cases, shared by the county and the town, amounted to $180 in 1873, $389 in 1880, and $60 in 1901.[88]

There was one other category of poor relief. Beginning in the 1870s, there appeared in Harrisville that product of hard times and freight trains, the tramp. On larger towns like Keene, tramps often descended in numbers, created quite a nuisance, and were roughly handled. Relatively few tramps reached Harrisville, and though the town's response was varied, it did assume responsibility for the problem. The Overseer of the Poor provided meals and lodgings for these "transient paupers." By 1880, a "tramp house" was set up at some distance from the village. When the town turned literary that year, it moved the tramp house into the village and made a library out of it, but another shelter was provided. These were undoubtedly shacks of the rudest sort. The one in use in 1900 was described as "so unfit that it would discount any of Dickens' most touching sketches . . . ."[89] It was probably the practice of the town to furnish any tramp who requested help with a night's lodging and provisions with which to make a meal on condition that he not linger in Harrisville.

The severe depression of the 1890s brought many more tramps to the town, enough to provoke this commentary in the *Sentinel*:

The tramp nuisance is becoming very annoying indeed. Some days they seem to be pickets of Coxey's army. The men seem to be well dressed, able-bodied, and claim to be discharged operatives from some shop in search of a job, and so far as we know they are gentlemanly in

85 *HAR*, 1874, 1881; *NHS*, Nov. 7, 1888.
86 *HAR*, 1881.
87 *HAR*, 1886.
88 *HAR*, 1874, 1881, 1901.
89 *NHS*, March 21, 1900.

conduct. Cleveland times have worked disastrously for the wage earner. This town not having a place for the accommodation of tramps, a town meeting has been called to act upon the subject.[90]

This sympathetic tone was confirmed by action, for the tramp house was built,[91] and employers offered work to those so inclined. In the spring of 1894, the Cheshire Mills hired twelve tramps as weavers.[92]

As the decade wore on, however, the town's attitude towards tramps began to harden. In the fall of 1895, a *Sentinel* correspondent noted, "The police have posted placards through town warning tramps to flee from wrath to come if found plying their vocation in the limits of the town, and none too soon, as they had become too numerous and not always of the best behavior."[93] The easing of the depression also served to diminish sympathy since, as one expressed it, "with prosperous business conditions, there is no excuse for idleness."[94] Nonetheless, limited numbers of tramps continued to appear in town and to receive help. In the supposedly prosperous year of 1900, there was a demand for construction of a new tramp house, and the Overseer of the Poor furnished lodgings to forty-five tramps and provided forty-eight meals, at a cost of fifteen cents each.[95] It would seem that tramps were for the new town a perennial minor nuisance requiring action and always threatening to become a major problem during a business depression.

An examination of the other end of the social scale indicates that the social and political leadership in the town was broadening during these years. The two first families of earlier years were little in evidence. Henry Colony moved back to Keene about 1870, and it was to be seventy years before another Colony returned to live in his house in Harrisville. Milan Harris was Harrisville's first representative in the state legislature and served two terms in the senate,[96] but that about marked the end of his public life. He was now an old man, sick in body and grieved by his losses. He stayed on in Harrisville for the rest of the decade, occasionally selling off pieces of property.[97] He spoke at the "reunion of old residents" in Nelson in 1879.[98] The next year his second wife died,[99] after which his daughter came and took him to spend his last years with her and her husband (Milan's former overseer), J. K. Russell, in Massillon, Ohio.[100] Those last years, in exile as it were, must have been bitterly unhappy ones for old Milan. He died intestate in July 1884, and his body was returned to his old home for burial in the "island cemetery."[101] Milan's youngest brother, Charles C. P. Harris, lived

90 *NHS*, April 25, 1894, "O.C."

91 *NHS*, May 16, 1894, "O.C."

92 *NHS*, Dec. 20, 1893, May 2, 1894, "O.C."

93 *NHS*, Oct. 16, 1895, "O.C."

94 *NHS*, Jan. 31, 1900, Oct. 13, 1897.

95 *NHS*, March 21, 1900; *HAR*, 1901.

96 Child, *op. cit.*, p. 178.

97 *NHS*, Oct. 5, 1876, Oct. 25, 1877.

98 *NHS*, Sept. 4, 1879.

99 *NHS*, Dec. 30, 1880.

100 *NHS*, Feb. 3, 1881.

101 J. K. Russell was appointed administrator of Milan's estate, which had shrunk to $1,800 in personal property at the time of his death. Cf. Application for Letter of Administration, The State of Ohio, Probate Court, Stark County (Doc. G.—p. 10—# 2319), 18 August, 1884.

on in the town until his death in 1888.[102] The next generation of this family all either died or emigrated. By the beginning of the twentieth century there was not a single Harris living in the town.[103]

Those men who succeeded the Harrises and Colonys in the leadership of town affairs were still from the old-stock families. Samuel D. Bemis, a farmer, was the grandson of one of the early settlers in Pottersville.[104] George Davis, a carpenter by trade, was from nearby Hancock.[105] Darius Farwell, a farmer on the Nelson side of town, was the grandson of a Revolutionary veteran and early settler in Nelson.[106] Zophar Willard, owner of various enterprises in Harrisville, was the cousin of the famous temperance speaker, Frances Willard, and grandson of the Reverend Elijah Willard, Revolutionary veteran and the first Baptist minister in Pottersville.[107] Luther P. Eaton was the son of Moses Eaton, the celebrated wall-stencil designer who lived in the eastern part of Harrisville.[108] Francis Stratton, owner of the Dublin Stage Company, had come to Harrisville from Massachusetts.[109] These were the men, among others of the same sort, who served their town as selectmen, representatives in the legislature, and other town officers. On the average they may not have been men of the same stature as Levi Leonard, Henry Melville, and Milan Harris, but there is nothing to suggest that they did not do their jobs as capably and as conscientiously as most of their predecessors in the parent towns.

As a result of the increase of immigrants and mill workers in town, the balance between the political parties was relatively even during these years. In the eight presidential elections between 1870 and 1900, Harrisville went Republican four times and Democratic four times. Bryan, in 1896, made the poorest showing of any major party candidate in the town's history. That "Wild Jackass of the Prairies" had no appeal to the mill workers in Harrisville.[110] In sharp contrast to Dublin and Nelson, which maintained their unbroken record of opposition to all Democrats, the new town of Harrisville showed itself to be of a decidedly independent turn of mind in politics.[111]

In its early years, however, the town seems to have had difficulty in learning how to conduct orderly elections. Perhaps the Democrats, so long suppressed under the control of Dublin and Nelson, were a little too anxious for victory. Only occasional comments on the March and November elections for state and town officers appeared in the Sentinel, but those were candid and revealing, as one account written by an indignant Republican in 1876 illustrates:

102 NHS, Oct. 3, 1888.

103 The Invoice and Taxes of the Town of Harrisville, N.H., Taken April 1, 1906.

104 Leonard, 1920, pp. 716–718; Child, op. cit., p. 180.

105 Child, op. cit., pp. 180–181.

106 Griffin, op. cit., pp. 66–67.

107 Leonard, 1920, pp. 942–944.

108 Leonard, 1920, p. 743; J. Waring, Early American Stencils on Walls and Furniture, p. 23.

109 HAR, 1906.

110 The strength of this feeling is suggested by this commentary in the Sentinel: "We learn that the delegates from this town to the Democratic state convention refused to sanction the doings of the Populist majority, and, to their honor, left the convention." NHS, Sept. 16, 1896, "O.C."

111 Cf. Manual of the General Court of the State of New Hampshire, 1889–1960.

Our opponents commenced desperate and unscrupulous warfare by intimidations, falsehoods, lavish use of money in the purchase of votes, and inducing aliens to procure papers free of expense. . . . The conduct of the majority of the board of selectmen in making the check list was simply outrageous. . . . When the town clerk administered the oath to the selectmen at the opening of the meeting, and had proceeded as far as that part relating to the correctness of the check list, one of the selectmen lowered his arm and took his seat, thus exhibiting a *tableau* which we hope never to see repeated.[112]

112 *NHS*, March 30, 1876, "Exposition."

113 *NHS*, April 6, 1876.

114 *NHS*, Nov. 22, 1882, Nov. 10, 1886.

115 *NHS*, Nov. 22, 1882.

116 *NHS*, Nov. 10, 1886.

117 *NHS*, Nov. 12, 1890, "O.C."

Some of the accusations made by this writer were temperately and convincingly answered by Samuel D. Bemis, the Democratic moderator of the meeting, in a return letter to the *Sentinel*.[113] Whatever the truth of the matter, the bitterness of the Republicans is obvious. Also, the case of Bemis shows that party alignment in Harrisville was not entirely a matter of national origins. There *were* Democrats among the old-stock families.

In the next decade the *Sentinel's* news items from Harrisville continued to feature complaints about the "outrageous rulings of the supervisors" and the Democratic candidates' unscrupulous bids for the labor vote.[114] The electorate may have been gaining some sophistication about conducting political affairs, but not so much that a quiet election did not cause surprised comment. Thus, in 1882, "the election . . . was the most quiet and orderly that has been held for years. . . ."[115] And in 1886, "the election passed without the usual excitement. . . ."[116] But disorderly town meetings were not yet a thing of the past, as shown by this description of one in 1890:

The town meeting was fully attended; very few of the voters were absent and the Democrats had everything their own way. The Republicans polled every vote claimed in their canvass. The naturalization of voters and the addition of names to the check-list by the supervisors was too much to overcome. We are pained to say the meeting was very disorderly from the opening to the close, and the conduct of some was not becoming to good citizens.[117]

Whatever the justice of these Republican complaints, and there must have been grounds for them, it is clear that elections in Harrisville were free-swinging affairs. At least there were no complaints about voter apathy.

A closely related matter is the political role and consciousness of the foreign-born in Harrisville. There seems no doubt that the great majority of the immigrants who voted were apt to vote Democratic. The frequent complaints of the Republicans on this score[118] and the fact that the new town frequently did go Democratic, even in its first gubernatorial election,[119] are sufficient proof.

Just how many of the foreign-born took an active interest in politics is a more difficult question. Naturalization would be one indication. Although only the 1870 census schedules give this information, those figures are helpful. The following list shows the number of foreign-born male residents, twenty-one or over, by country of origin, and their citizenship status in 1870:[120]

Birthplace	Unnaturalized	Naturalized
England	5	1
Ireland	13	19
Canada	14	4
Germany	1	1
Italy	1	0
	34	25

Whatever the explanation, in 1870 the Irish were the only nationality of whom a majority were naturalized, and only 42 percent of all the new town's foreign-born male adults were citizens.

The immigrants probably took a more active part in political matters as time went on. The complaints about town elections would so indicate. Another indication of increasing political consciousness on the part of the immigrants in particular and the townspeople in general is to be seen in a comparison of Harrisville's voting record with that of its neighbors. In the census and presidential year of 1880 when its recent immigrants were probably most numerous, Harrisville's vote in proportion to its total population lagged behind those of Dublin and Nelson. In the census and presidential year of 1900, Harrisville's turn-out for voting exceeded that of both the other towns.[121]

National and international affairs touched the inhabitants of Harrisville but little. In 1889, "O.C." might comment in the *Sentinel* that "a new administration . . . will show to the world that we are an American nation and will submit to no *snubbings*, either by Canada or Hayti, and be felt as a power in the Samoan muddle."[122] But though the writer might be in tune with the rising tide of nationalism in the country at large, he hardly spoke for Harrisville. So singular was this comment that one can almost feel the embarrassed silence of the town.

Even the Spanish–American War, most popular of all our foreign wars, was viewed with ambivalence and misgiving.[123]

118 *NHS*, March 30, 1876, Nov. 10, 1886.

119 *NHS*, March 23, 1871.

120 Census, 1870, Vol. XV, Nelson, Vol. XVI, Dublin.

121 The figures are as follows:

Year	Dublin
1880	126/456 or 27%
1900	89/620 or 14%
Year	Nelson
1880	109/328 or 33%
1900	64/295 or 21%
Year	Harrisville
1880	171/870 or 20%
1900	179/791 or 23%

Obviously, a better basis of comparison than the total population would be the total number of eligible voters in each town in 1880 and 1900, but these figures are not available for the earlier year. For the elections of 1900, 1904, 1908, the voter checklist figures are available for Harrisville and Dublin. In all three elections Harrisville turned out a higher percentage of its registered voters than did Dublin. See Appendix 14.

122 *NHS*, Feb. 6, 1889.

123 *NHS*, April 13, May 25, 1898, "Hermes."

One resident, traveling to Boston in May 1898, found it "surprising to see the amount of patriotic sentiment displayed. Flags and bunting were everywhere."[124] The Grange discussed the war at its meetings. In one debate the opponent of our involvement spoke of the horrors of war and closed with the unimpeachable dictum that "our country's greatest necessity in time of war is peace."[125] And when our "glorious little war" was so successfully concluded, the local *Sentinel* correspondent commented with restraint that "all our townspeople are glad that peace has come, but believe its problems will be as complex as those of the war."[126]

During these years election days in Harrisville were not the only occasion for disorder. The fifties and sixties had seen enough drunkenness, lawlessness, and violence to indicate a declining morality. In the period from 1870 to 1900, this trend continued. Although there was nothing to match the drama of the Buchanan "celebration" and though the newspaper comments are understandably more suggestive than descriptive, the evidence indicates a further decline in the town's moral level. Brief notices in the *Sentinel*, like one in 1873 which said that "special efforts for improvement in the moral, educational, and social influences of the community are heartily supported,"[127] really testified more to the need than the achievement. Just how bad conditions actually were is more difficult to determine. Some evidence has already been described, such as the dubious atmosphere of the hotel, the defrauding of the Colonys by some of their spinners, the strong suspicion of arson in the fire that destroyed the New Mill, and the disorderly and fraudulent elections. But there is plenty more.

The Protestant churches of the town were in a low condition, and the closely related temperance movement as well. The Good Templars had disbanded in 1873,[128] even before the bankruptcy of patron Milan Harris. Two other societies were soon formed, the Reform Club in 1876[129] and the "G. W. P. Sons of Temperance, Silver Lake Division No. 65," instituted in Harrisville in 1879,[130] but it is doubtful that there was any sustained support for either of them. One of the *Sentinel's* local correspondents wrote at the beginning of 1880, "This village boasts of two temperance organizations but . . . we do not hear of any results of either club."[131] Later the same year the Reform Club was so near being defunct that it could not find anyone to take the office of president, and the same correspondent asked in dismay, "Is it possible that there is no man in town who has either sufficient interest in temperance, the time to spend, or the qualifi-

124 *NHS*, May 11, 1898.

125 *NHS*, April 27, 1898.

126 *NHS*, Aug. 17, 1898, "Hermes."

127 *NHS*, March 6, 1873, "H."

128 *Ibid.*

129 *NHS*, Oct. 24, 1877.

130 *NHS*, Jan. 30, 1879.

131 *NHS*, Feb. 12, 1880, "Z."

cations to fill the office?"[132] The answer was evidently "no," for the *Sentinel* made no report of temperance activity in Harrisville for the next twenty years. Then, at the end of the century a solitary notice to the effect that "a temperance meeting was held in the church . . . under the auspices of the W.C.T.U. which has recently started here and in Nelson," suggested new interest in a cause long moribund.[133]

The other side of the coin was the town's liquor trade. At the time the state laws provided for local option in each town, to be decided by the selectmen.[134] This made every election for selectmen also a referendum on allowing the sale of spirits in town. On this subject the town frequently changed its mind. Just how close the vote frequently was on matters concerning the sale of liquor can be seen in this account of a vote on a resolution in an 1890 town meeting, when a shift of two votes would have changed the result. A resolution was

presented by Chauncey Barker in substance instructing our representative to oppose the repeal of the prohibitory law. It was met with a motion to indefinitely postpone, upon which a vote of the check-list was called for. The result was fifty-one in favor and forty-eight against. We noticed some blue ribbon men voted for postponement.[135]

The town reports show that in some years there was a town liquor agent. His job, it appears, was to buy liquors for the town and sell them to the several dealers, such as hotelkeeper C. A. Blake. In some years at least the agent also sold small amounts directly to individuals.[136] The few statistics provided show that the sale of spirits considerably increased. In 1873, the town paid out $114 for their purchase; in 1890, the town paid out $416. The agent's report also shows that the town made a profit of $120 in this latter year.[137] The details are unclear, but such, in brief, seems to have been the state of the legal liquor trade in Harrisville during these years.

It is hardly to be supposed that those with a taste for hard liquor would have been obliging enough to "go on the wagon" every time the town went dry. There would have been no town agent in those years, but as far as the dealers in town were concerned it almost seems that the only difference was that in the dry years the bottle was kept under the counter instead of on top of it. The authorities seem to have been willing to look the other way unless there were complaints. Then there would be a temporary suppression of the illegal traffic.

132 *NHS*, April 15, 1880, "Z."

133 *NHS*, Aug. 29, 1900. Miss Frances Willard, president of the World's W.C.T.U., was a descendant of the Rev. Elijah Willard of Pottersville and occasionally visited relatives in the vicinity. Her hand may be seen at work here in the revival of the temperance movement in Harrisville. Cf. *NHS*, July 28, 1897, "C.S."

134 Leonard, 1920, p. 548.

135 *NHS*, Nov. 19, 1890, "O.C."

136 *NHS*, March 26, 1890.

137 *HAR*, 1874, 1891.

Among those most apt to make complaints, with good reason, were the mill operators. Thus Craven and Willard wrote succinctly to one of the liquor dealers in 1881:

Mr. J. L. Burbank   Feby 24th 1881
Dear Sir
    We understand you are selling liquor to our help
    We do not want you to sell any liquor to our help and if we find out you do will prosecute you to full extent of the law
                                                    Yours Respt
                                                    Craven and Willard[138]

The effect of this warning was apparently limited, for Burbank was arrested the following fall.[139]

Others also protested. In 1887, an item in the *Sentinel* announced that a petition had been presented to the selectmen to suppress the sale of intoxicating liquors in town and concluded, "The dealers have been duly notified, and it is the intention of the authorities to enforce the law."[140] A year later, if not before, new and more strenuous efforts were necessary. The circumstances were described in the *Sentinel*:

A week ago Sunday a party of young men from abroad came to town and filled themselves with tanglefoot and created much disturbance by their howling and fighting in the street. Believing that something should be done to prevent such occurrences, a petition to the county solicitor was drawn to apply the statute to abate such nuisances, and it promptly received the names of leading men. On Friday last a sheriff's posse made a raid on the places of C. A. Blake, W. Halpine and D. Barry who were cited to appear at Keene the next day. It is to be regretted that the reputation of our town should suffer for the benefit of unprincipled men.[141]

The delicate reference to "young men from abroad" suggests what is known to have been the case, that thirsty parties in the bone-dry towns around Harrisville came there to do their carousing. Nonetheless, most of the liquor sold in Harrisville was sold to the town's own residents.[142]

Also noteworthy in this period was the number of assaults that occurred. The shooting and suicide in the boardinghouse have already been described. A few years earlier the *Sentinel* carried another account of the free use of firearms in town. A young man in Pottersville was returning after escorting home from a sewing circle the daughter of the village minister. Two "young scamps" threw some apples at him, and the swain responded by drawing a revolver and firing at them. In the dark, no one was hit apparently, but the local correspondent concluded

138 CMR, C. & W. Copybook "D," p. 377.

139 CMR, C. & W. Copybook "D," p. 916.

140 *NHS*, March 16, 1887.

141 *NHS*, Feb. 29, 1888, "O.C."

142 In 1889, the town Liquor Agent made 481 sales, of which 348, or 72 percent, were to Harrisville residents. Cf. *NHS*, March 26, 1890. The next year about two thirds of the sales were to town residents. Cf. *NHS*, March 4, 1891, "O.C."

in dismay, "I cannot help thinking that it was not the practice, formerly in this quiet village, for young men to carry side arms when going home with girls from sewing circles and prayer meetings. . . ."[143]

That incident had its amusing aspect, but others did not. In the summer of 1890, two brothers named Dinsmoor were mowing when they got into a quarrel. One drew a knife and badly slashed the other in a "murderous assault."[144] One spring night in 1892, young Miss Clara Wood returned to her father's house after an evening at singing school and was putting up her horse when her lantern went out and she was seized and thrown down by a man. "After a hard struggle she managed to get clear of the miscreant but she was severely injured and frightened. The fellow was tracked to the village," the newspaper account concluded.[145]

Three years later there appeared in the *Sentinel* a front-page item entitled "Stabbed Six Times: A Harrisville Spinner Savagely Assaulted." The trouble had started at a Saturday night dance in Eagle Hall. There had been "more or less drinking," when spinner Edwin Davis and Walter Anderson, "a Scotchman employed as a painter," got into a dispute over one of the local belles. A fight ensued in which Anderson was worsted. About midnight, having secured a "knife or chisel," Anderson attacked Davis and stabbed him six times before he could be rescued.

The newspaper account continued that the Harrisville authorities took no steps whatever to secure Anderson's arrest, and he left town at his leisure on Monday morning. By that time the county solicitor in Keene had heard of the affair and sent word to arrest Anderson at once, but it was too late. A friend of Davis said that no one in Harrisville knew how to make out a warrant and that the police were old men who would do nothing. The editor speculated, "What sort of an assault the Harrisville authorities would deem of sufficient consequence to warrant an arrest is an open question."[146] Stung to retort, the selectmen lamely replied that neither they nor the policemen knew of the assault before Monday morning when Anderson had already left town.[147]

Occasionally the town suffered from vandalism or, as correspondent "O.C." described it, "pure cussedness." Sometimes it was damage to the village school, described in 1896 as "quite too common in these latter days."[148] Another time it was damage to the public watering trough and the theft of a number of minor and valueless items of property.[149] Still another time it was the removal of road signs.[150] The value of the property damaged or stolen was not great and as isolated incidents would

143 *NHS*, Oct. 29, 1874, "Veritas."

144 *NHS*, July 30, 1890, "O.C."

145 *NHS*, May 11, 1892.

146 *NHS*, Nov. 20, 1895.

147 *NHS*, Nov. 27, 1895.

148 *NHS*, Feb. 22, 1888, April 29, 1896.

149 *NHS*, Aug. 1, 1888, "O.C."

150 *NHS*, July 9, 1890, "O.C."

be unimportant. But when the vandalism was troublesome enough to be commented upon in the newspaper and when it is seen in relation to the other examples of lawlessness occurring during the same years, it takes on greater significance.

What does it all add up to? Was the amount of drunkenness, lawlessness, violence, and vandalism no more than "normal," no more than would be found in other towns of a similar size? It would seem that there *was* more, and there certainly was more than the town had previously known. Harrisville's record was not one to be proud of, as the town's thoughtful citizens well knew. During the struggle to have a decent town library, one of them pleaded for it in the pages of the *Sentinel* on the grounds that it would be a salutary influence on the youth of the town. He concluded by posing a remarkably frank question, one that goes far in revealing contemporary opinion with regard to the moral climate of Harrisville:

Shall we have good schools and a library to direct their tastes, or shall we continue to produce representatives of the 'Harrisville Roughs' which have so long been flung in our faces, and keep up our reputation of figuring more conspicuously in and costing the county more for court and trials than any other town?[151]

The schools of Harrisville, in which so much hope was placed for the improvement of the citizenry, have a well-documented history. Every year, a "report of the Superintending School Committee" was included in the published town reports. Despite variations from year to year, these reports contain a great deal of useful information. Even in those days committees were prone to provide a redundancy of statistics, but there was no jargon, and for the most part the reports are succinct, candid, and informative.

In 1870, the town supported five district schools of varying size: one in the village, one in Chesham, one in the small settlement known as East Harrisville, and two others in the rural sections near Dublin and Nelson.[152] (Plate XXVI) All of these were common or ungraded schools, and, except in the village where there was an assistant, each had but a single teacher. The village school was the largest, with an average of fifty-eight pupils for the two terms. Chesham's was the next largest. The rural district near Dublin was the smallest, having only twenty-one scholars. In the years that followed, the number of pupils kept pace with the trend of the town's population.[153] By 1873, it had been necessary to divide the village school into a primary and a

151 *NHS*, March 4, 1880, "P.A."
152 Leonard, 1855, Map.
153 See Appendix 15.

grammar School, though usually both were held in the two-story building erected in 1858. Even after the division, these two were the largest in town. In 1895, the former had fifty pupils and the latter thirty-eight. By that year, the school in Chesham had declined to twenty-six scholars, the one in East Harrisville to a mere twelve, and the two in the rural districts had completely disappeared.[154]

The school year was gradually lengthened during this same period. In 1870, and for some years before, the average had been two terms of nine weeks each. In the first town report, the school committee complained that this was too short and hoped that the town would provide for at least twenty-four weeks.[155] This was finally done, but it took time. Occasionally one of the districts would provide an extra term, paid for by subscription.[156] Chesham was the first to have regularly three terms of eight to ten weeks each. Marked improvement came in the last years of the century, and in 1900 the school year was a uniform thirty-two weeks.

Concerning the teachers, some were men, especially in the village grammar school, but the majority were women. Many of them were from Harrisville or neighboring towns. Members of the school committee periodically visited their classrooms, perhaps as often as several times each term,[157] and then sometimes gave their opinions in the annual report. These appraisals minced no words, as the following examples will illustrate:

> For the first term of the grammar school, we engaged Mr. Seth E. Pope of Gardiner, Me., a recent graduate of Bowdoin College, and doubtless a very capable young man, but we fear that school teaching is not his forte. During our visits at the commencement of the term there was nothing very striking, either good or bad, in the appearance of this teacher. We made all the suggestions to him which we thought were demanded and felt quite confident of his success. We cannot quite pronounce him a failure. He resigned at the close of the ninth week. . . .
>
> The next teacher [at the Chesham School] was Miss Keefe, of Danvers, Mass., a graduate of Salem Normal School. She was a total failure; there was no order whatever in the school during our visits, neither did she make the least effort to have any. We discharged her at the end of the first week. . . .[158]

Whatever their competence, one cannot help feeling that the teachers in Harrisville earned their salaries. In 1873, the average salary was $7.67 a week; in 1895, it was $9.45. Boarding in the nineties cost a standard three dollars a week.[159] Just how under-paid the teachers were in this period is shown by the comment

154 *HAR*, 1871, 1874, 1896.
155 *HAR*, 1871.
156 *HAR*, 1871.
157 *HAR*, 1896.
158 *HAR*, 1896.
159 *HAR*, 1874, 1891, 1896.

of the school committee in 1871 that "the compensation is frequently less than a person not able to read and write can earn in the factory or at housework."[160] It all has a familiar ring.

For these wages the teachers in Harrisville taught six hours a day, frequently under difficult conditions. Maintaining order and discipline was the first requirement for successful teaching, and in this many failed. Even if discipline was maintained, the classes were apt to be too large for proper instruction. Said the school committee of the Chesham School, when it had thirty-seven scholars, "The best results cannot be reached when the teacher has two or three times as many recitations as can be profitably heard in six hours."[161] For many years the failure of a number of students to possess textbooks was "a constant cause of disorder."[162] In the ungraded schools, where thirty or forty scholars ranging in age from five to sixteen were all working in the same room, problems of discipline and instruction might be insoluble.[163]

There were special difficulties in the village. In 1895, the primary school had fifty pupils from five to fifteen years old. There was also a language problem, for these were the children of the mill workers. They were of several nationalities, and a large percentage of them "were not accustomed to the use of the English language."[164] The large number of absences was complained of in all the districts, but nowhere more than in the grammar school. At least some of the pupils there were over fifteen, and, when the woolen mills became busy, they were likely to drop out and go to work.[165] These older students could prove unmanageable too. One report referred to "the demoralized state of this school," assaults on teachers were not unknown, and some years the building "suffered shamefully" from vandalism.[166]

All things considered, it is a wonder that anything got taught. Yet the records show that, in 1896 for example, all but one of the schools in town taught reading and spelling, penmanship, arithmetic, geography, and grammar. The East Harrisville and Chesham schools also taught such extra subjects as history, algebra, bookkeeping, and physiology.[167] Except for reading, this was much the same curriculum that was in use in the 1870s.[168]

Efforts were continually made to overcome the great handicaps under which the teachers labored. Discipline was supported by classroom visits from the school committee and parents. Corporal punishment was resorted to on occasion.[169] Perfect attendance was rewarded by an honor roll, with the names published

160 *HAR*, 1871.

161 *HAR*, 1871.

162 *HAR*, 1891.

163 *HAR*, 1906.

164 *HAR*, 1891, 1896; *NHS*, July 9, 1890.

165 *HAR*, 1881.

166 *HAR*, 1881; *NHS*, June 22, 1892; *NHS*, Feb. 22, 1888, "O.C."

167 *HAR*, 1896.

168 *HAR*, 1874. In 1885, the following textbooks were used in Harrisville's schools: Arithmetic—Greenleaf's; Algebra—Robinson's; Bookkeeping—Mayhew's; Geographies—Harper's; Grammar—Kerl's; History—Swinton's; Physiology—Hutchinson's; Readers—Lippincott's; Spellers—Swinton's; Writing Books—Spencerian Series. Cf. *HAR*, 1886.

169 *HAR*, 1874.

in the town reports. After 1890, as a result of a new state law, the town provided every student with free textbooks.[170] Teachers had the opportunity to attend free training classes in teachers' institutes. Usually two sessions of a week each were held annually.[171] Many of the parents, including those of foreign birth, visited the schools and took an active interest in their affairs.[172] Despite the shortcomings of the schools, attributable to both the general state of public education and to the special difficulties of Harrisville, there seems no question but that the children of Harrisville were receiving a better education in 1900 than they had in 1870 or in 1850.

Soon after the town was formed, there was a movement to establish a public library. There appears to have been some sort of a library in existence by 1873, but it probably possessed only a few hundred volumes and was not supported by the town.[173] At its March meeting in 1877, the town voted to create a public library and appropriated $250 for the purchase of books. Another $200 was contributed by individuals. For the next three years the collection was stored in the house behind the Congregational Church.[174] There the town residents could repair on a Saturday afternoon or evening and take out novels by T. S. Arthur, Cooper, Dickens, E. P. Roe, Defoe, Marryat, or Scott. If they wanted more substantial fare, there were biographies by Bancroft and James Parton or the historical works of Macaulay, Napier, and Lossing, as well as books on travel, poetry, and science.[175]

In 1880, the town acted to provide a building to house the library. Henry Colony offered to lease to the town, either free of charge or for a nominal sum, a plot of ground next to the brick store.[176] At the March town meeting, "a trifle tempestuous,"[177] the voters authorized the library committee to remove the town's tramp house to this central location and fit it up for a library. This was done, the rear part of the building being reserved for a "lock-up," for a total outlay of $165.25.[178]

It was a modest enterprise, but the new library had only been brought into being over strenuous opposition. In the weeks before the March meeting, the local column in the *Sentinel* contained numerous references to the issue. A few weeks after the victory, there appeared a long and interesting letter that told of the library's usefulness and also shed some light on the nature of the opposition. All the local correspondents for the *Sentinel* signed their contributions with either a pseudonym or one or two initials, so this writer is unidentified, but it is possible that he was the prominent entrepreneur and town officer, Zophar Willard.

170 *HAR*, 1891.

171 *HAR*, 1871.

172 *HAR*, 1881.

173 *NHS*, March 6, 1873, "H."

174 Home of John T. Farwell. Cf. Hurd, *op. cit.*, p. 216.

175 *Catalogue of the Public Library of Harrisville, N.H.* (1878).

176 *HAR*, 1881; *NHS*, March 18, 1880, "O.C."

177 *NHS*, March 18, 1880, "P.A."

178 *HAR*, 1881.

[The library] has been in operation while the M. & K. Rr. was in building, and all who wished of the 'tramps' were allowed books to read. The mill help is also of an emphatically 'floating' character, yet no one ever stole a book. . . . One man *has* carried off a book, but he pretended to be a gentleman, boarded at the hotel with a lady (?); when they went away they had a young infant with them, but somehow that got spilled out, and was found on a doorstep. . . . That book might have got spilled out too, for we can't find it. That is all we have lost. The mill help read and call for the best class of works, mainly history, biography, and Sir Walter Scott's works, and they display a knowledge of literature that probably few Americans here possess, if one may judge by the nature of the opposition to the library. The enemies of it denounce it as being 'infidel,'—call Wilkie Collins' and George Eliot's works unfit to read,—say that 'reading injures the brain,' yet those same men look on the 'mill help' with contempt. I wish those opponents could know how in three instances I was applied to for a library book on Saturday, but after reading the rules and regulations the parties said 'I see you do not let out books to transients,' (they were every one of them foreigners) and I *took* the responsibility of letting them have a book, and they scrupulously returned it on Monday. They might have been carousing round all day Sunday but for that. . . . During the winter of 1878 many railroad hands were here without work and no money. They had all the books to read which they chose to take, and on going away in the Spring one of them said, 'I don't know what we should have done if it had not been for this library. We wanted something to read, but had no money to buy a newspaper.'[179]

After 1880, the town regularly voted sums of from fifty to one hundred dollars to purchase books, and the library's collection steadily grew. In 1880, there were 823 volumes; in 1900, there were 1802.[180] The catalogue of 1896 showed twenty-seven pages of fiction, including works by such worthy writers as T. B. Aldrich, Jane Austen, Edward Eggleston, George Eliot, W. D. Howells, and Sarah O. Jewett, along with some ephemeral novels. The catalogue also listed ten pages of juveniles, including books by Louisa May Alcott, Hans Christian Anderson, Oliver Optic, and, inevitably, Horatio Alger. These two categories made up the bulk of the library's holdings, but there were also a number of works of history, poetry, religion, and travel, not to mention bound periodicals and public records.[181]

For those who wished to read, there was plenty of good reading matter in the library. Somewhat strangely, however, just as steadily as the number of volumes owned grew, the number of volumes issued shrank. In 1880, 3,176 volumes were issued; in 1900, only 1,870 volumes were issued.[182] Of course the population also declined during this period, but not nearly so much as

179 *NHS*, April 8, 1880, "Z."

180 *HAR*, 1881, 1901.

181 *Catalogue of the Harrisville, N.H. Town Library, 1896.*

182 See Appendix 16.

did the use of the library. In fact, the trend of the population would seem to have little to do with it, for the circulation of books rose again after 1905. The explanation for the decline is not immediately clear, but the fact remains that after the enthusiasm of 1880 the use of the library declined until it reached an all-time low about the turn of the century. Nonetheless, the evidence indicates that the library, like the schools, had a meliorating influence on the population of Harrisville, and both signified progress from the conditions of mid-century.

Whether there was any corresponding improvement in the state of Harrisville's churches is more doubtful. The Congregational Church in the village was in no high estate in 1870, and the ensuing decade brought further decline. The church had been in the "spiritual doldrums" since the 1850s, little more than a decade after its founding. Sporadic efforts to improve conditions had been to little avail.[183] In its whole history there has been only one extensive revival, and that one was of dubious benefit to the church. It came in the winter of 1873. There is no mention of it in the *Sentinel*; the only record is in a manuscript account of "The Church at Harrisville," written by Charles A. Bemis in 1913. Though he wrote forty years after the event, there is no doubt that Bemis had witnessed this revival firsthand when he was a young man living in Pottersville.[184]

Reminiscing, Bemis wrote that "looking back from the distance of forty years one may think he can discern some of the factors that finally produced the revival." He then cited three: the coming to the village of one Washington Phillips, "a plain farmer" who worked devotedly for the church and especially the Sunday school; secondly, the arrival of a new pastor, Rev. Ames Holbrook, a "warm-hearted, earnest, evangelistic pastor"; and thirdly and most important, "the spirit of God came upon the community."[185]

To let Bemis continue the account of the revival in his own words,

Just how it first manifest itself no one seems to know. It was not of human device. No committees had been appointed, no special meetings had been planned. In the winter of 1873–1874 people became serious. There were spiritual awakenings and inquiries for spiritual help. The presence of God was in the midst of the people. When this was discovered special meetings were announced and largely attended, neighboring pastors were called in to assist, the claims of the gospel were presented and the response was marvelous. The pastor gathered the young people at his home and organized the Covenant Band. . . . This revival touched

183 Bemis MSS, Box 6, XXXII, pp. 5–7.

184 C. H. Rockwood, *Atlas of Cheshire County, N.H.*, p. 41.

185 Bemis MSS, Box 6, XXXII, pp. 5–7.

all classes. . . . As a result of this revival 32 united with the church that year, all but one on profession of faith.[186]

Bemis's account and explanation of the revival has the virtue of being firsthand if retrospective, but it needs amending. He does not exclude the possibility of there being other causes, and it would seem that there was an important one he overlooked or forgot. The winter of 1873 saw the country in the grip of the worst depression it had experienced up to that time. There is no doubt that it hit Harrisville; the Milan Harris Woolen Company, with a local reputation almost as high as that of Jay Cooke himself, was then on the verge of bankruptcy. The good times of the sixties were over, and, similar to what would happen after 1929, the long, hard winter of 1873–1874 was a time of reexamination. Economic hardship fostered just such a turning to religion as Bemis described.

As to its extent, Bemis's recollection that "this revival touched all classes" just was not so. His account includes a list of the names of those thirty-three persons who united with the church in 1874. Without exception, they are the names of old-stock families in the area. There is not an obviously Irish or French-Canadian name among them. Though there was still no Roman Catholic Church in town, this revival won no converts from Catholicism. Probably no attempt was made to proselytize among the immigrant mill workers. Furthermore, most of those who united with the church lived in the rural areas, and two thirds of them were women.[187] This revival won neither the numbers nor the classes that were needed to reinvigorate the church, and the opportunity was not to be repeated.

Though the depression may have helped to bring about the revival of 1873, its ultimate effect on the church was deleterious. Bemis himself wrote that "the business depression of the '70's which closed the Harris mills so scattered the strength of the church that it could no longer support a minister."[188] That strength was the Harris family, and their loss was not offset by the thirty-three new members. By 1880, the resident membership had declined to about forty. At the time the explanation was that so large a portion of the population were foreigners and Catholics,[189] but the Census of 1880 shows that *all* the foreign-born, plus all those of foreign parentage, amounted to only 43 percent of the town's population.[190] That still left several hundred people in town, more than enough to fill the church on Sunday mornings.

The next year, the Harrisville church arranged with the church

186 *Ibid.*, pp. 7–8.
187 Bemis MSS, Box 6, XXXII, p. 8.
188 *Ibid.*, p. 4.
189 *NHS*, Feb. 12, 1880.
190 See Appendix 6.

in Nelson (also in a state of decline) to share together in the ministry and support of one pastor, and this arrangement lasted for the rest of the century.[191] In this period the two congregations had the services of six ministers, which meant they had a new one on the average of every three years.[192] Thus, in adversity, the Congregationalists in Harrisville again made common effort with the church in Nelson, from which they had separated forty years earlier.[193]

Although far from flourishing, the Baptist Church in Pottersville survived this period somewhat better. (Plate XXVII) It did not depend on the leadership of one family, and the mill failures in the village of Harrisville had little effect on it. It was said of the Baptists in 1880, "They have a neat commodious church, located in a community not occupied by any other denomination, and with a minister possessed with a candid, catholic spirit, a respectable congregation can be assembled."[194] That "candid, catholic spirit" was, alas, hypothetical, for it appears that the church did not have a minister at the time. Indeed, during the last thirty years of the century, the church had at least nine different ministers.[195] The tenure of most of them was too short for an effective ministry. The size of the membership varied, but in 1885 the church claimed seventy members, plus sixty scholars in the Sabbath School.[196] If this number was representative of the period, then the church was considerably larger than the Congregational Church in Harrisville.

To stimulate interest, the Baptists continued to hold social gatherings, but they met with indifferent success. They tried bean suppers on Saturday nights at ten cents per meal.[197] After a similar affair one member pointed out the trouble:

> The 'pound party' held at the vestry of the Baptist church on Tuesday evening of last week was poorly attended, owing partly no doubt to the unfavorable weather but largely to other causes. The success of such gatherings, and in fact the energy and vitality of any society or community, depends mostly upon its younger members. . . .[198]

Through the efforts of its older members, the church managed to maintain its independent existence for the rest of the century,[199] but they could not do much more. The Sabbath School might be crowded to overflowing, but most of the children there would soon be leaving this declining rural community. The church, like the community, suffered from this continual exodus.[200]

Undoubtedly a contributing cause in the decline of the Protestant churches in Harrisville was the fact that many of the new

191 Bemis MSS, Box 6, XXXII, p. 4.

192 Leonard, 1920, p. 341, contains a listing of the pastors down to 1916.

193 Of course, the decline of the Congregational Church in Harrisville was not exceptional. One historian has written that the Congregational Church in New Hampshire reached its maximum membership in 1845 and that there then followed a "fluctuating decline" until 1915, the time of his writing. Cf. Everett S. Stackpole, *History of New Hampshire*, IV, 231.

194 *NHS*, March 11, 1880, "O.C."

195 Leonard, 1920, p. 337.

196 Child, *op. cit.*, p. 181; Hurd, *op. cit.*, p. 216.

197 *NHS*, Feb. 19, 1880.

198 *NHS*, April 4, 1883, "Star."

199 Book of Records for the Baptist Society in Dublin, 1815–1903.

200 The number of persons in Pottersville between the ages of twenty and forty fell from 50 in 1860 to 11 in 1870. Cf. Census, 1860, Dublin, pp. 87–91; Census, 1870, Vol. XVI, Dublin, pp. 19–20.

arrivals in town were Irish or French-Canadian Roman Catholics. These Catholics reached sufficient numbers for a mission church, St. Denis's, to be organized in Harrisville in 1874. Its early records are few and scattered, for it was run as a mission of churches in Keene, Wilton, and Marlborough, successively, before it was organized as a parish in 1906. The first services may have been held in private homes, but by 1880, when there were about two hundred Irish and Canadian immigrants in town, plus their families, church services were regularly held in Eagle Hall.[201]

Occasional items in the *Sentinel* suggest that the mission church was well supported by the Catholics in town. Unfortunately for the record, the Catholic immigrants were not yet enough at home in the community for them to send their own notices to the newspaper, and these items were written by outsiders. The Catholics held fairs in Eagle Hall, they decorated the hall appropriately for such occasions as Palm Sunday, and sometimes they held special services, such as this one described in 1890: "The Catholics had a large gathering at Eagle Hall on Saturday last, and it is said to have been a sort of revival meeting. A priest from Montreal was present."[202]

Twenty years after the mission was organized, the Catholics erected a church in Harrisville. (Plate XXVIII) A lot for this purpose, located next to the library building, was donated by the Cheshire Mills.[203] The parishioners contributed the necessary money,[204] which could not have come easily in the depression years of the nineties. In the fall of 1894, the foundations were laid,[205] and the following July the *Sentinel* announced the church's completion and consecration: "The Catholic church was consecrated Sunday by Bishop Bradley of Manchester, assisted by priests. The church is a neat tasty edifice. There was a large audience and the services were impressive."[206]

The growth of the Roman Catholics in Harrisville contrasts sharply with the record of the Protestant churches, but their strength should not be exaggerated. Far more than the Congregational or Baptist Church, St. Denis's depended on the "floating population" employed in the mills, and its poorest element at that. The Winns were probably the only Catholics of means in the town. At the end of the century the church was still being run as a mission of the Catholic Church in Marlboro.

In the local news column of the *Sentinel* after 1870, the items mentioning religious activities are far exceeded by the number describing Harrisville's secular social life and recreation. Not that the social life was especially active. "Our town is very quiet— almost impossible to find an item worth reporting,"[207] was a

201 *NHS*, Feb. 12, 1880, "Z."

202 *NHS*, March 10, 1885; March 28, 1888; July 30, 1890, "O.C."

203 CMR, "Records of the Cheshire Mills," July 12, 1893; *NHS*, July 26, 1893, says "It will be located . . . on the spot where stood the large Clement stable."

204 Interview with Charles M. Bergeron, Nov. 10, 1962.

205 *NHS*, Oct. 3, 1894.

206 *NHS*, July 10, 1895, "O.C."

207 *NHS*, March 7, 1883, "O.C."

frequent comment of the local *Sentinel* correspondent. "We seem very stupid when compared with other towns and their amusements," complained another writer.[208]

Nonetheless, there were efforts to promote what can broadly be denominated as cultural activities. There were occasional lectures on current subjects, such as this one in 1872: "Miss M. M. Parker of Keene gave a lecture . . . in Jones' Hall; subject, 'Anti-Women Suffrage.' The attendance was good and the lecture was well received by the audience which was composed of the best citizens of the place. Her arguments were well chosen and unanswerable. . . ."[209] A few years later an attempt was made to interest the towns-people in the "Chautauqua Literary Scientific Circle."[210] The title may have scared them off; at any rate nothing more was heard of it. There was for a time a singing school in the village[211] and a dramatic club which had occasional periods of activity.[212] One Saturday in February 1882, it presented a drama entitled "Better Than Gold," which one may assume was at least as elevating as it was entertaining.[213]

More popular were social activities not so consciously bent upon "improvement." At any season of the year there were large house parties and family gatherings. Space was no great problem in farmhouses and farm kitchens. Thus, when Mr. and Mrs. Howe celebrated their twenty-fifth wedding anniversary, about one hundred of their friends came to join in the festivities.[214] When Darius Farwell held a family gathering, there was no trouble in providing an "ample dinner" for half a hundred people. For diversion at that reunion one could go inspect the new barn, join the younger members in a game of baseball, or, like the great-grandmother of ninety-two, just sit on the front porch and watch the people.[215]

Even with skating, sleigh rides, and ice boating, the long winter months went slowly. It was then that the townspeople might resort to a dramatic presentation, give a party for the minister,[216] or attend a "social festival" at Eagle Hall put on by the "Ladies of Harrisville."[217] (The civilizing hand of the women-folk is to be noted in these affairs.) The Grange in Chesham and a new "Nubaunsit" local in Harrisville[218] gave suppers and held meetings to discuss topical questions such as free mail delivery to farmers or "Should the Grange favor or discourage the reading of the scripture in the public schools?"[219] And one evening in the autumn of 1900 there was held in Eagle Hall a "moving picture and stereopticon exhibition."[220] More exclusively masculine ways of passing the time would have included

208 *NHS*, March 4, 1880, "P.A."

209 *NHS*, Feb. 1, 1872.

210 *NHS*, Oct. 7, 1880, "X."

211 *NHS*, March 28, 1883.

212 *NHS*, March 4, 1880, "P.A."

213 *NHS*, Feb. 22, 1882, "O.C."

214 *NHS*, Oct. 7, 1880.

215 *NHS*, Sept. 23, 1885, "H."

216 *NHS*, Jan. 23, 1889.

217 *NHS*, Feb. 16, 1871.

218 *NHS*, Dec. 1, 1897.

219 *NHS*, Jan. 5, 1898, Feb. 10, 1892, "O.C."

220 *NHS*, Nov. 14, 1900.

excursions to Keene or relaxed get-togethers at the blacksmith shop, the hotel, or the livery stable.

The approach of spring and summer brought more, and more varied, activities. Young lovers might spend a Sunday afternoon at "The Falls," a romantic spot near Chesham.[221] There was always fishing and swimming. In June, Barnum's Circus came to Keene, and, despite the amount of drunkenness and disorder attending this occasion, many people went down to see the street parade, the trained animals, and the "Oriental Scene."[222] There was apt to be some celebration of Independence Day, though instead of the traditional liturgy it might be an informal celebration at Silver Lake, featuring the genteel amusements of boating and croquet.[223] The baseball craze hit Harrisville in the seventies, and a team, the "Crescent Club," was organized to play against teams in neighboring towns. Its offensive game was apt to be better than its defensive, if one may judge by the club's 25-to-22 victory over Marlborough in 1874.[224] The Grange in Chesham continued to be active socially and held picnics and outings at Silver Lake.[225] In the late nineties the bicycle craze swept the country, and for the robust there were "century runs" of a hundred miles over the rough, hilly roads.[226]

Indeed, the wonderful New England summer, always too brief, was not to be wasted indoors. Those people in Harrisville for a vacation had the edge on those who had to work for a living, but both shared in the enjoyment of nature, as items in the *Sentinel* show: "Excursions and picnics are now in order and almost everybody is taking an outing either in camp or on the shore of some pond, climbing to the summit of old Monadnock, or going on some of the numerous advertised railroad excursions."[227]

One of the highlights of the summer season was the Nelson town picnic. This was first held in 1878 and developed into an annual reunion of the town's current and former residents. In 1883, about a thousand people attended this picnic, including most of the citizens of Harrisville.[228] There was music provided by a band and a quartette, a dinner, exercises, and a great many speeches.[229] Among special events the summer and autumn of 1893 saw a number of Harrisville people travel to Chicago to see the Columbian Exposition, though not so many as might have gone, it was allowed, had it not been for the hard times then prevailing.[230]

The coming of autumn and winter brought hunting season and the glorious foliage of October. People with the leisure and inclination took carriage rides through other parts of the state or into Vermont.[231] They were not apt to find more spectacular

221 *NHS*, Jan. 30, 1895.

222 *NHS*, June 12, 1889.

223 *NHS*, July 9, 1884.

224 *NHS*, July 16, 30, 1874.

225 *NHS*, Aug. 11, 1886.

226 *NHS*, June 14, Aug. 9, 9, 1899.

227 *NHS*, Aug. 11, 1886. Cf. *NHS*, Aug. 7, 1889.

228 *NHS*, Aug. 15, 1883, "O.C."

229 *NHS*, Aug. 22, 1883, "Recorder."

230 *NHS*, Aug. 16, Nov. 15, 1893.

231 *NHS*, Oct. 15, 1890.

scenery than they could see in the vicinity of Harrisville: the early scarlets of the swamp maples, the yellows of the birches and ashes, the russets of the oaks and the shades of green provided by the pines, spruces, and hemlocks, all set off by the stretches of clear blue water of Harrisville Pond, North Pond, and Silver Lake. The thrilling sight of such color on a bright, clear October morning could compensate for many other aspects of life in Harrisville.

Then, when the colors had become more subdued, there was Thanksgiving, always a major holiday in New England. The next month, there was apt to be a community observance of Christmas, sometimes in both Chesham and the village of Harrisville.[232] One held in Eagle Hall in 1882 was replete with songs, ice cream, presents, Christmas tree, and Santa Claus.[233]

Such were the opportunities for social activity and recreation in Harrisville at the end of the nineteenth century. Many of them, or the ones approved by society at any rate, may seem tame if not dull to later generations. In fact, they probably seemed tame to many of the young people of the time and increased their restiveness to go west or to the city. Still, the diversions available in 1900 reflected both a considerable increase in leisure time and a broadening of interests and outlook from the conditions of mid-century.

232 *NHS*, Dec. 25, 1889.
233 *NHS*, Jan. 3, 1883.

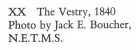

XX The Vestry, 1840
Photo by Jack E. Boucher,
N.E.T.M.S.

XXI Union District No. 8
Schoolhouse, ca. 1857
Photo by Jack E. Boucher,
N.E.T.M.S.

XXII Evangelical
Congregational Church,
1842
Photo by Jack E. Boucher,
N.E.T.M.S.

XXIII Blake's Hotel,
"Nubanusit House," ca. 1869
Photo by author in 1960

XXV Cheshire Mills in 1888. Sketch by Otto Barker for Century Co., October 1888. Courtesy of Horatio Colony, Keene, N.H.

183

XXVI District No. 3 Schoolhouse, ca. 1883
Courtesy of Mr. and Mrs. John Clark, Harrisville, N.H.

Teacher: Ada Farwell
Back row: Mary Rutherford, Cora Farwell, Edith Farwell, Nettie Hagan, Mabel Farwell, Fred Knoulton, Fred Rutherford, Fred Holmes, Frank Rutherford
Middle Row: Julia Hagan, Grace Farwell, Flora Holmes
Front Row: Henry Barry, George Knoulton, Curtis Farwell, Charles Rutherford, Harry Mason

XXVII Chesham Baptist
Church
Photo by Jack E. Boucher,
N.E.T.M.S.

XXVIII St. Denis's Roman
Catholic Church, Selectmen's
Office and Town Library,
Eagle Hall, Store
Photo by Jack E. Boucher,
N.E.T.M.S.

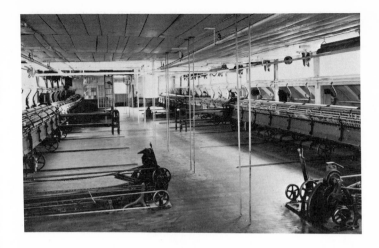

XXIX Self-operating Spinning Mules in Attic of Granite Mill, Installed ca. 1895, Now Unused. Picture shows clerestory lighting Photo by Jack E. Boucher, N.E.T.M.S.

XXX Cheshire Mills Work Force, ca. 1903, Including New Finnish Operatives.
Reprinted from *Amerikan Albumi*, 1904. Courtesy of Levy Luopa, West Swanzey, N.H.

XXXI Conference in President's Office. l. to r., Warren Thayer, Office
Manager, John J. Colony, Jr., President, Charles Colony, Treasurer
Photo by Jack E. Boucher, N.E.T.M.S.

XXXIV John Johnson
Doffing Spinning Mule in
Bldg. No. 2
Photo by Jack E. Boucher,
N.E.T.M.S.

XXXII Toivo Luopa
Preparing Stock Dye Kettle
in Dye House of Bldg. No. 4
Photo by Jack E. Boucher,
N.E.T.M.S.

XXXIII Matilda Winters
and Floyd Chamberlin
Carding Stock in Card Room
of Bldg. No. 6
Photo by Jack E. Boucher,
N.E.T.M.S.

XXXV Frances MacKenzie
Checking Filling Pattern on
Loom in Weave Room of
Bldg. No. 6
Photo by Jack E. Boucher,
N.E.T.M.S.

XXXVI Vestry and Church from Storehouse
Photo by Jack E. Boucher, N.E.T.M.S.

XXXVII Post Office
Photo by Jack E. Boucher, N.E.T.M.S.

XXXVIII   View to Southeast, showing Cheshire Mills' Bldg. No. 6,
Granite Mill, Storehouse
Photo by Jack E. Boucher, N.E.T.M.S.

XXXIX   View to South, showing Cheshire Mills' Boardinghouse, St.
Denis's Church, Library, Store
Photo by Jack E. Boucher, N.E.T.M.S.

XL   View to Northwest, showing Cheshire Mills' Storehouse, Granite
Mill, Brick Mill
Photo by Jack E. Boucher, N.E.T.M.S.

8 ECONOMIC AND POLITICAL LIFE OF HARRISVILLE, 1900–1940

The Economy

From the turn of the century down to the Second World War, the mainstay of Harrisville's economy continued to be the Cheshire Mills. Amid wildly fluctuating conditions of wartime boom and business depression, the survival of the Cheshire Mills was undoubtedly due in large part to the ability of its management. At the beginning of the century all the principal offices were still held by Horatio Colony. Although he shortly began to share control with his son John, he retained the presidency down to his death in 1917. This came, in very modern fashion, as a result of an automobile accident. So died the last of the sons of Josiah Colony and the president under whom the Cheshire Mills enjoyed its most prosperous decades.[1]

Horatio's successor was his eldest son, John Joslin Colony. He was born in Keene in 1864, graduated from Harvard College in 1885, and for two years thereafter studied at its Law School. His name first appeared in the Cheshire Mills payroll for 1887 when, like his younger brother Charles T. Colony, he received $41.67 a month. (Sixteen years later he was still receiving only $50 a month.) These wages were in line with what the more skilled workers were then earning. Exactly what he did for his wages is not entirely clear, but it can be safely assumed that he earned them and that he was learning all aspects of the business of manufacturing woolens. As some of his uncles before him had done, John Colony lived in the company boardinghouse for about a year and took his meals with the other boarders at the long tables in the dining hall.

All this was slow and careful preparation, for the elder Colony was certainly in no hurry to share control and responsibility. Only when he was sixty-nine did he turn over to his son, then forty, the important post of treasurer. John Colony served as treasurer from 1904 to 1917 and received for it a monthly salary of $83.33. In the latter year, with the war boom well under way, his monthly salary was substantially raised to $200. Then the next year he became president of the Cheshire Mills.[2]

1 *Granite Monthly*, XLIX, (Nov.–Dec. 1917), 231.

2 CMR, Payroll, "Records of the Cheshire Mills"; Letter from Horatio Colony (nephew), July 17, 1963; Hobart Pillsbury, *New Hampshire*, V. 304.

John Colony remained president of the company until his death in 1955, which meant that the company during its first century of operations had only three chief officers, each of whom headed it for approximately one third of that time. John Colony was more willing to share control than his father had been. His brother, Charles T. Colony, succeeded him as treasurer and retained this post until his death in 1944.[3] Charles T. Colony rarely came to Harrisville, however, for he lived in Keene and did his work in the company's office there.[4]

John Colony was also more willing than his father had been to supervise personally the running of the mill. Though like his father he lived in Keene, it was his regular practice to take the early morning train to Harrisville, arriving there about 5:30 or 6 a.m. On dark winter mornings he might catch a little extra sleep in his office while he waited for the power to come on. Until 1926, the plant generated its own power, and the president would not have the wheels started early just for his office light. Then, usually after a walk through the mill and conferences with the superintendent, he would take a late morning train back to Keene where he would see wool salesmen and the like at the company office.[5]

The personal qualities of the president of a small mill play a role in his success, and John Colony was undoubtedly well liked by both workers and townspeople in Harrisville. He was a familiar sight to both. Tall, athletic, always dressed in blue, he spoke little and presented an unassuming demeanor. On trips through the mill he never failed to greet the operatives with a smile. That the outward behavior indicated an attitude both kindly and considerate is well established.[6]

One of the most important officers of a woolen mill is the superintendent. He is the boss of the overseers, or manufacturing boss. He determines matters of design, layout, and production. He also helps the president to determine matters of company policy, salaries, selling costs, and so on. Since Horatio Colony had run the company from his Keene office and had seldom even come to Harrisville, the real boss of operations was the superintendent, a Scotsman named Alexander Taylor. He received a monthly salary of $125. Under John Colony's closer control, the position lost some of its importance. Succeeding Taylor was Horatio W. Colony, grandson of old Timothy Colony, and apparently inheritor of his temper and energy. He too commuted from Keene, was renowned for his acuteness in finding workers smoking or soldiering on the job, and received an annual salary of $3,000. After his death in 1935, his work was divided between

3 *KES*, Aug. 1, 1950.
4 Interview with John J. Colony, Jr., July 26, 1962.
5 *Ibid.*
6 Interview with Arvo Luoma, Nov. 10, 1962.

two of the overseers, and the superintendency lapsed for the rest of this period.[7]

During the time they were the officers, these three Colonys, John, Charles T., and Horatio W., not only ran the business but also did all the paperwork. As late as 1945, there were no stenographers or bookkeepers. There was not even a typewriter in either the Keene or the Harrisville office. Neither, as late as 1928, was there a telephone at the Cheshire Mills. Life was then indeed more leisurely.[8]

During the first forty years of this century the plant of the Cheshire Mills was enlarged and modernized. The original granite mill had been added to just before the Civil War by attaching a new brick mill at right angles to the front end of the original mill. Then a picker house had been attached to the rear end of the other side of the granite mill. Finally, in 1922, a two-story brick mill was adjoined to the picker house, so that the shape of the main plant from the air resembled one half a swastika. The new mill was built on an informal basis. The plans were adapted from those of another mill, and the construction was done largely by the Cheshire Mills' own employees with the aid of a gang of Italian laborers brought in from Boston.

Once finished, the new mill was furnished with looms and carding machines, some new and some taken from the granite mill. The new Davis and Furber cards were sixty inches wide and, for this reason, of somewhat controversial design. The Cheshire Mills were among the first to buy and prove the practicality of these cards. Also noteworthy was the power system in the second floor weave room. Hitherto, weave rooms had been cluttered with belts and shafts, posing hazards to safety and liabilities to efficiency. In the new weave room each loom had its own motor drive.

There were other changes at the Cheshire Mills reflecting those in the industry at large. There was continual replacement of machinery, especially finishing machinery, because of the requirements of new fancy fabrics. Also, the 1920s saw in the industry the introduction of synthetic dyestuffs, the result of Germany's scientific successes and military defeat. The development of a superior, independent wool-scouring industry and the need for space in the mill led to the removal of the scouring vats from the Cheshire Mills in 1940.

Like the machinery and manufacturing processes, the power plant also underwent change. At the beginning of the century the mill still ran on waterpower utilized through turbines and the famous rope drive. There continued to be the annual wrangles

7 CMR, Payroll; Interview with John J. Colony, Jr., July 26, 1962; Letter from Horatio Colony, July 17, 1963.

8 Letter from Horatio Colony, July 17, 1963; Interview with Arvo Luoma, Nov. 10, 1962.

and lawsuits with the town when the ponds used for storage overflowed in the spring, flooding roads and fields.[9] The proud tradition of a self-sufficient waterpower for the village mills came to an end in 1926. That year, for the first time, the power requirements of the new mill forced the Colonys to purchase electricity. Nonetheless, the waters of Goose Brook continued to furnish two thirds of the company's power requirements until 1947, when the rope drive stopped and the company began to purchase electricity for all its power needs.

Thus, the plant of the Cheshire Mills in this period continued to be maintained in first-rate condition. It was even something of a showplace. It is said to have had the reputation of being the cleanest mill in New England. When the new mill was opened, visitors came from distances to see the new power system in the weave room. And, of course, older features like the intricate rope drive and the architecture of the mill village continued to attract attention.

With all of these changes, it is easier to establish the quality of the Cheshire Mills than its relative size within the industry. Midway in this period, after the completion of the new mill, the Cheshire Mills plant probably came within the dimensions of what Cole calls the "representative woolen mill" in the country.[10]

This factual description conveys nothing of the very definite atmosphere of the mill, but a fragmentary impressionistic description written in reminiscence by the present Horatio Colony helps to fill this gap, even though it also adds a note of ambiguity to the view of the plant as being both immaculate and up-to-date:

The small room off the wet finishing department was against a little descending Niagara; the window panes of this room were always wet, and there was a constant chill and sound of watery tumult. The wet finishing room always gave the appearance of a room under water—more pleasant to my imaginative ego than to the persons who worked there. . . .

The plant's whole interior . . . gave the impression of some ancient house with uneven floor levels and many ratty staircases, and sudden steps up and down. It was full of nooks and cobwebs, and rooms high in the air. One of the spinning rooms was way up under the eaves—it was like spinning in the tree tops. When the mill was in full blast, the faint clash of the looms was pleasant and could be heard from . . . neighboring hillsides. Most of the windows in the old mill were small and never opened—the mill workers loved bad air and were susceptible to the slightest draughts.[11] (Plate XXIX)

In regard to operation, much of the company's purchasing, production, and marketing was, with the exceptions noted,

9 NHS, April 17, 1901, Sept. 24, 1902, May 4, 1904.

10 In 1922, the Cheshire Mills had 13 sets of small cards, and Cole says that 12 to 18 sets constituted a representative-sized woolen mill. Also, he says that the average number of wage earners per establishment was 112 in 1919, which the work force at the Cheshire Mills then would have closely approximated. Cf. Arthur H. Cole, The American Wool Manufacture, II, 209, 220–221.

11 Horatio Colony, Autobiography (ms), ch. XII, p. 131. (See Plate XXIX.)

carried on as in former years. Reflecting a trend in styles, the Cheshire Mills largely turned away from its once famous plain and mixed flannels[12] and made instead quantities of highly fancy fabrics, twist fabrics for suitings, and men's overcoatings. Perhaps its most important and most highly esteemed product was a double-faced fabric used for heavier coatings and known as Plaid Backs.[13] As had always been the case, the marketing of its fabrics was done through an agent. This continued to be the original firm of Faulkner, Page, and Company until about the time of World War I. Since that time there has been a succession of other firms. While the railroad ran, which was nearly till the end of this period, it was used for bringing in coal and wool and for carrying woolens to Boston. From there they were carried to New York by the Eastern Steamship Company. Of course, there were numerous changes in freight shipping to take advantage of the best rates.[14]

How did the Cheshire Mills fare in the wildly fluctuating business conditions between 1900 and 1940? It certainly was susceptible to the changes. The century started with very prosperous conditions for the industry at large, and also for the Cheshire Mills, which lasted until 1907.[15] The financial panic that hit the country that year was felt in the business community, and work was slow in Harrisville through the summer and fall. It was slow again the next year, from the fall of 1908 through the spring of 1909. Again in the summer of 1913, reflecting another national recession, the work force at the mill was reduced to about a third its normal size, and that number worked only about a third of the normal workweek. Thereafter, there was gradual improvement, and American entrance into the First World War again brought boom times and government contracts for uniforms and heavy overcoatings. The years, then, from the turn of the century through World War I were generally busy and prosperous, except for slack periods totaling less than a year and a half, or roughly 8 percent of the time.[16]

The widespread prosperity of the 1920s did not extend to the New England textile industry. The Cheshire Mills may have done better than other mills in the region, and it did have prosperous years, but it also had slack periods in every year except 1922, 1923, and 1926. There were also complete shutdowns, when all but a skeleton force of a dozen or so were laid off for a week or two. The summer months were most apt to be slow, but no season was immune to the curtailment of operations. At other times employment was maintained by working the mill only two or three days a week. These slack periods add up to

12 Cole, *op. cit.*, II, 164–180.

13 The Cheshire Mills has a sample library of all fabrics made there since the beginning of this century.

14 Except as noted, the foregoing material on the plant and operations of C.M. derives from interview with John J. Colony, Jr., July 26, 1962.

15 CMR, Payroll, "Records of Cheshire Mills"; *NHS*, Jan. 30, 1901.

16 CMR, Payroll; *NHS*, Oct. 16, 1907, Jan. 2, Jan. 23, 1918.

about a year and a half in the decade, or roughly 15 percent of the time.[17]

The decade of depression that followed the stock-market crash of 1929 was felt in Harrisville as it was in the country at large, though in some ways this community was better prepared than many others to weather it. It was not all bad. Occasionally the mill would get an order, rehire a full work force of 100 or 125, and work not only full time but overtime as well. Nonetheless, the mill can be described as operating fully in this decade only about 10 percent of the time. About 30 percent of the time the mill worked a drastically reduced schedule, and between 50 and 60 percent of the time the only ones on the payroll were perhaps a dozen supervisors and maintenance men. The mill lived from order to order, and bad times persisted throughout the thirties.[18]

Finally, what of the Cheshire Mills' employees and their conditions? As to the composition of the work force, it underwent a sudden change in the summer of 1902. The *Sentinel* took notice of it: "William Wilder, one of the employees of the Cheshire Mills, moved his family to Winchester last week. This makes the eighth family that has moved from town this summer. Their places are being filled by Finn families."[19]

There is a bit of mystery here in the flight of the older workers and the sudden influx of the Finnish immigrants. Neither the mill records, the *Sentinel*, nor any of the Finns still living in town have shed any light on it. Why did these workers, whose total number must have been considerably greater than that given by the *Sentinel*, decide to leave Harrisville for other New Hampshire mill towns at this time?[20] There is no evidence of any work shortage, strike, or other disturbance. Or was it the other way around, with the Finns appearing first and the native workers fleeing from competition with this mass of cheap immigrant labor? The Finns had been coming into New England to work in its granite quarries and textile mills for thirty years before this and locally had already arrived in Marlboro and Dublin. The Colonys had always been ready to hire immigrant labor and may well have sought out Finns to fill vacancies. Earlier, the Irish and French Canadians had appeared on the payroll in slowly increasing numbers over a period of years. In this instance, however, the payroll for 1902 shows no Finnish names through June, but then shows eleven in July, twenty-five in August, and forty in September.[21] However this replacement of the older workers by the Finns came about, it brought an important change in the company's work force and in the town's population. (Plate XXX)

17 CMR, Payroll.

18 *Ibid.*

19 *NHS*, Aug. 13, 1902.

20 *KES*, July–August 1902 *passim.* Other towns mentioned were Wilton and Hinsdale.

21 CMR, Payroll.

Because the census schedules are not open to examination, the composition of the Cheshire Mills work force by nationality cannot be determined as well for the first part of the twentieth century as it could for the latter part of the nineteenth. Nationally, we know that the woolen industry hired more immigrants than most other American industries and that in the first decade of this century the concentration of immigrants from Southern and Eastern Europe was, as Cole says, "peculiarly great." More specifically, they were in the majority in the many factory jobs of machine tending and those requiring more strength than skill.[22]

At the Cheshire Mills the only basis for calculating the number or percentage of immigrant workers are the Irish, French, and Finnish names on the payroll and the number of foreign-born living in the town. Obviously, judging a person's country of birth on the basis of his name is an imperfect system. It works better for the French Canadians and Finns than for the English and Irish, and the longer an immigrant group has been in the community, the more imperfect it becomes.

In 1902, about half the company's operatives had Irish, French, or Finnish names. This proportion declines to less than a third in 1917 and then rises to slightly over half in 1937. Assuming that, as time went on, an increasing number of these were actually second-generation Americans, it would appear that at the beginning of the century immigrants made up nearly half of the work force, that this proportion declined to less than a third at the time of the First World War, and that it rose only slightly during the twenties and thirties. If the ratio of foreign-born workers to foreign-born residents of the town in 1880 (the last year for which these statistics are available) is applied to population figures for 1940, then the percentage of foreign-born workers at the Cheshire Mills in the latter year would appear to have been about 35 percent, which figure closely coincides with that made on the basis of names.[23]

Among the immigrant groups in the work force there were changes in the balance of strength. Irish names made up more than a third of the total at the beginning of the century, but thereafter their number diminished to relative insignificance. The French Canadians, the most important group after 1870, were temporarily swamped by the Finnish innundation of 1902, but then regained first place and constituted about half the total number of immigrant names throughout this period. Allowing for the probably greater proportion of second-generation Americans among the French Canadians than among the Finns,

22 Cole, *op. cit.*, II, 112–114.

23 CMR, Payroll.

it can be safely said that these were the two important groups and that over these forty years there was a rough equilibrium between their numbers.

The conditions under which these operatives worked were, on the whole, good. The mill was clean and safe. There were no fatal accidents and few serious ones.[24] There were no fires of consequence. Throughout the period most of the operatives lived in town, many of them either in the company's boarding-house or its tenements. In good times thirty or forty operatives, nearly all men now, lived in the boardinghouse. They paid fifteen or sixteen dollars a month for board down to the First World War and seven dollars a week thereafter. The company rented on the average of eighteen houses for five dollars a month until the war and for an average of $6.25 during the twenties and thirties. Some prices, at least, rose in Harrisville during this period much less than they did nationally. Rents, for example, between 1912 and 1927 rose 66 percent nationally but only 13 percent in the company-owned houses in Harrisville.[25] During the Depression, when the Cheshire Mills were barely keeping their doors open and the payroll was short, the company simply wrote off the rents due from its unemployed tenants for, in some cases, as long as five years.[26] Fringe benefits like this were not easily matched elsewhere.

In the length of the workweek there were improvements coinciding with changes on the national scene. From the beginning of the century down to sometime during the First World War, the Cheshire Mills' operatives worked ten hours a day, six days a week. This was about the average in the woolen industry.[27] During this time the summer schedule was sometimes arranged to allow the workers to have Saturday afternoons off by beginning work at an earlier hour.[28]

By 1919, most employees were working a forty-eight hour week, and this prevailed during good times until 1933. During these years the mill bells in the cupolas atop the granite mill and the old Upper Mill rang each morning at 6:30, 6:55, and 7 a.m. There was an hour for lunch, and then the bells rang the end of the workday at 4 p.m. Of course, when the mill got very busy, overtime might extend the workweek to as long as sixty hours. A second shift, however, was not worked as a regular thing until the Second World War. Before that time the problem of lighting made it impractical.[29]

During the late twenties and throughout the thirties, the more important question was not how long was the workweek but how short. In August 1933, along with other mills in the vicinity,

24 Interview with Arvo Luoma, Nov. 10, 1962; Interview with John J. Colony, Jr., July 26, 1962.

25 *Historical Statistics of the United States*, Series E113–139, p. 126; Douglas, *Real Wages in the United States, 1890–1926*, p. 60; CMR, Payroll.

26 Interview with Arvo Luoma, Nov. 10, 1962; Interview with John J. Colony, Jr., July 26, 1962.

27 Cole, *op. cit.*, II, 116–117.

28 *NHS*, May 22, 1901.

29 Interview with John J. Colony, Jr., Nov. 11, 1962.

the Cheshire Mills adopted the National Recovery Administration's blanket code providing for a forty-hour week.[30] This standard was retained after the NRA was declared unconstitutional, but more as a goal than actuality. Occasionally, the workers might put in a full forty hours, but they were more likely to work thirty-two, or sixteen, or no hours at all.

Because the boilers had to be fired around the clock, one exception to the forty-hour week was the job of fireman. Two of them, a Finn named John Johnson and a French Canadian named Joe Bolio, each worked a twelve-hour shift for many years until 1938.[31] John Johnson, on the night shift, not only kept the boilers fired but also served as night watchman. It was a long shift, and the work was heavy. In the days before coal was used, Joe Bolio is reported to have handled as much as twenty-one cords of wood in a single day, which is roughly the equivalent of shoveling ten or eleven tons of coal.[32] It is also reported of the same fireman, however, that during his twelve-hour shift he regularly found time and energy to cut firewood for his own home, plant his vegetable garden, and even to make cherry wine up in the woods behind the mill.[33] A constitution like Bolio's complicates the problem of comparing working conditions then and now.

In the matter of wages and earnings the historiographical problem, in contrast to that of an earlier age when statistics were scarce, is to bring meaning out of a mass of records and statistics. The payroll continued on a monthly basis until 1919, after which it was weekly. At the beginning of this century the average daily cash wage in the Cheshire Mills was the same as it had been since the early 1880s. By 1907, however, it had risen nearly a third above the level of 1902. The average wage in 1912 was unchanged from that of 1907.[34]

The trend of real wages, in contradistinction to cash wages, is problematical. The several cost-of-living indices that have been devised for this period all show some increase, but they vary greatly. It might be expected that prices rose more slowly in the rural area of New Hampshire where the Cheshire Mills were located than they did nationally. On the contrary, the charges in the Cheshire Mills boardinghouse rose sharply in these years; in fact, they nearly doubled. Lacking more adequate information on local prices, it is impossible to make a conclusive statement on the trend of real wages, but in view of what is known, it seems doubtful that there was any significant increase in real wages at the Cheshire Mills down to the First World War.[35]

In relation to the woolen and worsted industry at large, it

30 *NHS*, Aug. 2, 1933.

31 Interview with Mr. and Mrs. John Saari (daughter), Sept. 6, 1960.

32 Interview with Newton Tolman, Feb. 15, 1961.

33 Interview with Arvo Luoma, Nov. 10, 1962.

34 See Appendices 7, 7a.

35 *Ibid.*

appears that annual earnings of employees at the Cheshire Mills had been in the 1890s about $15 less than the industrial average of $345. (This was, in turn, much less than the average of annual earnings in all industries, which was $423.) Then, from 1900 to 1914, cash earnings rose appreciably in the industry. They rose more slowly at the Cheshire Mills, and down to the First World War they remained about 20 percent below the average in the industry.[36]

With the coming of the war, wages began to rise faster than the cost of living. This was so in the industry, and it must have been true in Harrisville also. In 1917, the average daily wage in the Cheshire Mills was 53 percent higher than in 1912 and more than 100 percent higher than in 1902. Local prices must have gone up too, but the charges at the company boardinghouse were not much different from those of the preceding decade.[37]

After the war, wages continued to rise. In 1922, the company's weavers were earning nearly three times the cash wages they had earned in 1917, and the average daily cash wage of all the company's employees had risen about 75 percent in the same period. These wage levels were maintained through the twenties but were offset later in the decade by the dull business conditions and reduced workweeks already described.[38]

The stock-market crash of October 1929 and the ensuing business depression were felt at the Cheshire Mills by April 1930, and the wages and earnings picture for the following decade was dark. The company tried to maintain wage levels and to spread the work, but there were long periods when the mill was all but shut down. The overseers, traditionally well paid, had to take a 50 percent pay cut for several months at a time during the dark years of 1932 and 1933. In its turn, the company not only let the rents of the unemployed go uncollected but also regularly made pay advances to those who were working. Many recall how John Colony would go through the mill on a Monday morning with the petty cash book, asking each worker how much of an advance he wanted. In Harrisville, "welfare capitalism" was no slogan of a decade but the established order of things.

Average weekly earnings are difficult to determine when long periods of inactivity are broken by occasional busy periods. For the sharply reduced number on the payroll, a sampling shows an average wage of $20 for a fifty-three-hour week in September 1932[39] and an average wage of $15.91 for a workweek of twenty-seven hours in September 1937.

Wage rates, as distinguished from earnings, generally held up

36 CMR, Payroll, 1890–1914; Cf. *Historical Statistics of the United States*, Series D–603–617, pp. 91–92; Douglas, *op. cit.*, pp. 257–260.

37 See Appendices 7, 7a.

38 *Ibid.*

39 Forty-eight hours was the standard workweek at this time, so these figures include some overtime, paid at the regular hourly rate.

and compared favorably with those at nearby mills. In 1932, they had fallen off from the levels of the twenties, but by 1937 they were actually higher. Figures for the woolen industry are not available for the years after 1926. However, a local comparison can be made on the basis of a document, the like of which rarely comes to light in the woolen industry. In the summer of 1933, perhaps in connection with the drawing up of the industry codes of the National Industrial Recovery Act, the Cheshire Mills and four other Cheshire County textile companies exchanged information on their wage rates, usually a closely guarded secret in such a competitive industry. An overall comparison of these wage rates for specific jobs at the five mills, shown in a "Confidential" schedule that was drawn up and exchanged, reveals that the Cheshire Mills ranked first by a considerable margin.[40]

Given these working conditions, it is not surprising that the company's relations with its employees in this period were, on the whole, good. Relations had not been bad during the last part of the nineteenth century, but there had been difficulties. In this century, relations definitely improved. John J. Colony's presence at the mill, his friendliness, and his demonstrated concern for the workers were the main reasons for the improvement in labor relations. One indication of this was the absence of any interest among the workers in unionization. Even during the troubled 1930s, with flying squadrons ranging far and wide in the industry, no organizers came to Harrisville, and none were sought.[41]

In conclusion, this period between 1900 and 1940 was one of increasing stress for the Cheshire Mills, but it survived because of the special qualities it possessed to a high degree. The entire New England woolen industry suffered heavily after the First World War. For the most part, the mills that have gone under have been the large, remotely owned, and casually managed mills that produced great quantities of plain goods. Conversely, the mills that have survived have been the smaller mills owned by families who kept them up to date and able to produce specialty goods and who knew the industry, their plants, and their employees.[42] The Cheshire Mills fit the latter category of woolen mills in every particular.

Farming was an economic enterprise the condition of which cannot be described with the same precision or detail as the woolen industry. It helps little to say that farming was declining in this period, for it had been doing that for most of the town's history. Nonetheless, a few features and developments can be described with some exactness.

40 See Appendix 7a; CMR, "Schedule of Wage Rates," June 28, 1933.

41 Interview with John J. Colony, Jr., July 26, 1962; Interview with Arvo Luoma, Nov. 10, 1962.

42 Interview with Mr. Lightbody of the American Textile Reporter, Boston, July 2, 1964.

While the decline in farming in the years between 1900 and
1940 cannot be measured in terms of the people engaged, acreage,
or production, there is one set of statistics available that can
serve as an index. Each year the town reports listed in the town
invoice the number of oxen, horses, cows, neat stock, and sheep.
Between 1900 and 1940, every type of livestock declined by at
least 75 percent, and the total number of all types declined by
88 percent.[43] And, of course, the year 1900 itself represented no
golden age of agriculture in Harrisville.

Horses and oxen declined very greatly with the increase of
automobiles and tractors, but of all farm animals sheep declined
the most. In 1900, there were 210 sheep in town, roughly a third
of the 612 in the town back in 1873 when Harrisville was still
in the afterglow of the boom times of the Civil War.[44] At the
end of this period, the town invoice for 1940 listed no sheep at all,
and this for an area where a century before there may have been
ten thousand grazing on the open hillsides. Sheep raising declined
precipitously in the state at large in these same years, but not as
greatly as in Harrisville where in fact it disappeared.[45]

The town invoices indicate that cows and dairying fared better
than sheep raising but still declined greatly. In 1900, there were
229 cows listed, slightly more than half the 405 listed in 1873.[46]
In the forty years after 1900, the number of cows underwent a
further decline of 75 percent. Though they survived better than
all other types of livestock, there was in 1941 no herd containing
as many as ten cows, the minimum number estimated necessary
to show a profit in dairying.[47]

This local decline was in accord with regional developments.
As Vermont began to offer keener competition, the dairy industry
declined in New Hampshire, and particularly in the southern
part of the state.[48] Still, the farmers of Harrisville struggled to
stay in the dairy business, and when in the Boston Milk War of
1910, the New England dairy farmers withheld their milk from
sale until the Boston milk contractors would guarantee a fair
return,[49] the Silver Lake Grange met and passed a resolution
condemning the milk contractors for their ruinously low prices.[50]
The result of the war was an ephemeral victory for the farmers,
but in the long run there was no profit in dairying in towns like
Harrisville.

As dairying declined in the state, many farmers in the southern
part turned to poultry raising. In Harrisville, however, only a
few farmers were ever attracted to this industry, and none on
more than a modest scale.[51]

As for crops, the experience of farmers in the state is suggestive

43 See Appendix 11.
44 Harrisville Town Records, Invoice 1874; HAR, 1901.
45 H. F. Wilson, Hill Country of Northern New England, p. 224.
46 Harrisville Town Records, Invoice 1874; HAR, 1901.
47 HAR, 1942; Wilson, op. cit., p. 317, n. 46.
48 Wilson, op. cit., pp. 220–224, 316.
49 Ibid.
50 NHS, May 25, 1910.
51 Cf. Wilson, op. cit., pp. 220–221; Invoice and Taxes of the Town of Harrisville, 1906, 1914; HAR, 1934, 1942.

of local developments. The principal trend was the growth of intensive farming. Farmers increasingly tilled only their best lands. In 1910, New Hampshire farmers tilled only a little over a quarter of the lands they owned. The poorest farms reverted to forest. One study shows this trend of curtailment persisting from 1900 to 1930, and there is no reason to think that it did not continue after that date. "Only the most favorable (sic) situated farms and the acres most adaptable to cultivation were kept in use."[52] Consequently, the amount of improved land in the state diminished 51 percent in this period, and one half of the farms in the state were given up. The reasons for this curtailment were the economic competition from the western states, the isolation of the farm and the attractions of the city, and a frequently burdensome taxation. As usual, it was the young people who left the farms. Deserted by his sons, the farmer found it increasingly difficult to secure and keep dependable farm laborers.[53]

As a result of this intensive farming, yields on the good farmlands rose to exceed midwestern averages in several crops, especially potatoes, hay, and some grains.[54] New Hampshire farmers also found it profitable to grow market produce and quality apples and to make maple syrup. To some degree all these developments took place in Harrisville.[55] Nonetheless, the paucity of good farmland there is indicated by the fact that in 1930 the average value of farmland in Harrisville was only $24 per acre, less than half the national average value of farmland at that time.[56]

Probably of greater importance locally than any single crop or farm animal was the wood and lumber grown and harvested. New Hampshire had reached its lowest proportion of forest land in the early 1870s.[57] Judging from the number of sheep there, the area of Harrisville may have reached its lowest proportion of forest land somewhat earlier. Thereafter, as farmland and sheep pastures were given up, natural afforestation proceeded steadily. Early in this century, with encouragement from the state government, some farmers undertook reforestation programs. In 1914, the *Sentinel* described one in Harrisville: "Wellington Wells has cleared a tract of land on the Seaver farm for the purpose of setting out young pine. Already about fifty thousand pine saplings have been set out."[58]

Roughly in proportion to the years of growth, these new woodlands provided many varieties of a marketable product. Of the various trees having value, spruce was one of the most profitable of trees that sprang up naturally. Young ones could be cut after four or five years for Christmas trees and older ones for

52 Wilson, *op. cit.*, pp. 346–347.

53 *Ibid.*, pp. 349–358.

54 *Ibid.*, p. 214.

55 Cf. *NHS*, Oct. 30, 1907, Dec. 22, 1915.

56 Margery D. Howarth, *New Hampshire*, pp. 21–22; *Historical Statistics of the United States*, Series K 1–7, p. 278.

57 Wilson, *op. cit.*, p. 238.

58 *NHS*, May 13, 1914, B.F.B.

pulpwood or lumber.[59] In 1905, the Chesham correspondent for the *Sentinel* reported that "Parties have been here the past week procuring Christmas trees to be shipped to New York."[60] A more profitable market for both the larger softwoods and hardwoods were the local sawmills, box and chair factories, or, for cordwood, the Cheshire Mills.

In 1915, with the number of taxable livestock declining, the town began to invoice and tax "wood and lumber" as separate items, which suggests its increasing importance. By the mid-twenties, land that had been allowed to revert to forest since the Civil War was yielding an annual return of from three to five dollars an acre.[61] In Harrisville, wood and lumber produced reached a peak value for a year of $49,200 in 1924.[62]

What emerges here is a pattern of change and adaptation, as farmers struggled to survive in a declining way of life. Owners of farmland, some of them with forebears who had converted their fields from growing grain and produce to sheep raising and then to dairy farming, now converted their farms to raise poultry, trees, truck gardens, or perhaps a combination of all three. In addition, many kept a cow or two to supply themselves and their neighbors with fresh milk. To augment his meager income from these sources, the Harrisville farmer might also tap his maple trees in early spring, do occasional work on the roads, drive the school bus, or serve as town clerk or selectman.

What may seem surprising is the number of farmers who by these means managed to survive. In 1930, according to the federal census the "Rural farm population" in Harrisville numbered 127, roughly one fourth the town's total population. Ten years later the same element numbers 173, just over a third of the town's population.[63] The increase in these years was undoubtedly due to the Depression. For the whole period in the twentieth century the number of farms and farmers certainly declined. In 1880, near the peak in the town's growth, there were fifty-eight farms in town,[64] while in 1940 there could hardly have been more than twenty.[65] It would seem that though the farm population declined, it declined no more than the rest of the population and at the end of this period still constituted an important element in the town.

Of minor industry and commercial life there was no such diversity and activity as marked Harrisville in the previous century, and a number of enterprises that had once flourished declined and disappeared before 1940. Business failures were nothing new in Harrisville. They had been endemic there since the days of Jonas Clark. And, as Page Smith shows, what was

59 Wilson, *op. cit.*, pp. 238–239.

60 *NHS*, Nov. 29, 1905.

61 Wilson, *op. cit.*, p. 242.

62 *HAR*, 1925.

63 U.S. Bureau of the Census, *15th Census of the United States.*, *1930*, Pop. Vol. III, Pt. 2, p. 170, Table 21; U.S. Bureau of the Census, *16th Census of the United States.*, *1940*, Pop. Vol. II, Pt. 4, p. 796, Table 28.

64 *NHS*, Aug. 12, 1880.

65 Cf. *HAR*, 1942.

true here was true of towns across the country.[66] Twentieth-century Harrisville differed from the nineteenth-century town in the absence of new businesses to take the places of those that failed.

As one instance, at the beginning of the century the town still supported two blacksmith shops.[67] One of these, run by a Scot named MacColl, managed to survive into the 1930s, but this was the last one in town, and no auto mechanic could or did take his place.

Faring somewhat better was the wood and wood-products industry. Sawmills came and went according to the supply and the demand for lumber, and a box company or two survived for awhile in Chesham.[68] Most important was a chair factory run by the Winn Brothers. In 1914, Thomas Winn, Sr., and his brothers disposed of the general store they owned in the village and entered the business of manufacturing porch and veranda chairs and rockers. The enterprise proved highly successful for a time and, next to the Cheshire Mills, was the most important manufactory in town.[69] But like so many others this business, too, eventually failed.

The transportation industry felt the effects of economic decline while it was at the same time being revolutionized by the invention of the automobile. The Dublin Stage Company, that once vital link between the railroad and the outlying farms and estates, survived for a time hauling building materials brought from Boston on the railroad for house construction in the area. Fred Stratton, the owner, had from twenty to thirty horses in his livery stable and hauling business.[70] Nonetheless, the days of the Dublin Stage were numbered. The name survived the horses, for the company was taken over and run as a one-man taxi service in the 1920s. Then in 1928 the company came to an abrupt end when its Stanley Steamer exploded.[71] Similarly displaced by the motor car, the railroad, after several abortive attempts to give up its profitless service, finally stopped running through Harrisville in 1936.

Stores and shops declined in number and variety from the heyday of the eighties but held their own better than industry because of less competition. Mobility was improving but was not yet what it would be in the free-wheeling post-World War II days. Also, the increasing numbers of summer people helped the local stores, as indicated by such dispatches to the *Sentinel* as the following from Chesham: "Bemis Bros., our local grocery men, have done a rushing business the past week filling orders for summer residents who are about to open their houses for the

66 Page Smith, *As a City Upon a Hill*, pp. 84–109.

67 Interview with John Clark, Nov. 11, 1962.

68 *NHS*, June 4, 1902, Jan. 20, 1904; Interview with Ralph Bemis, Nov. 11, 1962.

69 Pillsbury, *op. cit.*, V, 18.

70 Interview with Charles M. Bergeron, Nov. 10, 1962.

71 Interview with Newton Tolman, Feb. 15, 1961.

summer season. An order amounting to $75 was received and delivered one day last week to a single family."[72] Then as such things came into use, up-to-date proprietors might add to their stores a soda fountain or gas pump to attract customers.[73] By such means it was possible even in the 1930s for the town to support three general stores in the village and at least one in Chesham.

The experience of one store owner, John Clark, was probably representative. He ran a general store in Harrisville from 1922 to 1950. At first he had a small room in one of the village houses and stored his goods in the barn. He sold hay and grain in addition to groceries and household supplies. He made deliveries all around the vicinity of Harrisville and Nelson, using a sleigh in winter to get through the deep snows. To supplement his income, he also worked as a carpenter.[74] By such means, without much margin and perhaps without much profit, the several local storekeepers could still stay in business.

In the whole picture of the town's economic life in these years, the brightest note was the increase in what were generically called the "summer people." These were, of course, well established in Harrisville at the turn of the century, with cottages and camps scattered along the wooded shores of the several ponds and numerous homes taking in summer boarders. Favored as the region was by climate, scenery, and opportunities for outdoor recreation, the industry based on these people continued to grow.

The town invoices show that the number of cottages, camps, and "summer residences" rose from 60 to 117 between 1906 and 1941.[75] The number of camps rose to twenty, the largest of which was still Marienfeld where up to 250 boys vacationed in the 1920s.[76] At the end of the period the valuation of nonresident cottages, camps, boat-houses, boats, and the lands these buildings occupied amounted to $143,875, which was more than the valuation of the Colonys' mill property and stock in trade.[77]

Some attempts were made to encourage the growth of winter sports in the area. Newton Tolman of Nelson offered skiing lessons in Chesham in the primeval days of that sport in the 1930s, and his ski reports were published in the *Boston Transcript*.[78] Winter sports contests were held during these same years on Harrisville Pond. Diverting though these may have been for the inhabitants, the Depression years were not the time to make a great financial success of such ventures, and the snow was not so dependable as it was farther north.

The summer visitors, however, continued to increase. Somewhat curiously, the number of cottages showed the greatest

72 *NHS*, June 25, 1902.

73 *NHS*, Aug. 30, 1905, June 25, 1924.

74 Interview with John Clark, Nov. 11, 1962.

75 *Invoice and Taxes of the Town of Harrisville, 1906*; *HAR*, 1942.

76 *NHS*, Jan. 14, 1925.

77 *HAR*, 1942. Of course, the Cheshire Mills also owned extensive real estate in Harrisville, so that the total valuation of its property was some $40,000 greater than that of the nonresident cottages, etc.

78 *NHS*, Nov. 15, 1933; Cf. Newton Tolman, *North of Monadnock*, Ch. 6.

increase in the Depression years. It rose from fifty-six to ninety-six between 1933 and 1941. Nineteen forty marked something of a turning point in the town's history. That year, for the first time, the town invoice showed the property valuation of the nonresidents exceeding that of the residents.[79] By this time too, the number of summer residents must have equalled the number of permanent residents.[80] Every index, in fact, suggests that at the end of the 1930s the summer residents or tourist industry was ready to become the largest element in the economic life of Harrisville.

Town Affairs

While retaining its essential features, the physical appearance of Harrisville in these years continued to undergo some moderate changes. The boundaries remained as they had been created, though there was at least one proposal to change them. In 1904, the *Sentinel* reported that "It is reported that some of our wealthy summer residents wish to have the existing town lines somewhat changed by taking the west part of Harrisville, the east part of Roxbury, and the east part of Marlboro, and making a new town under the present name, 'Chesham.' It has met with very little favor with our permanent residents. . . ."[81] Indeed it might—it is easy to imagine older readers in Dublin and Nelson smiling at the prospect of such poetic justice for the "railroading" they got in 1870, but no more was heard of this proposal. All these hill towns had and have some friction between their residents and their summer people, and Harrisville was no exception.

As might be expected from the decline in population, construction was very limited. In this span of forty years the number of houses probably did not increase by more than ten or fifteen.[82] The number of stores and shops increased not at all. The only significant construction was the expansion of the Cheshire Mills and the proliferation of summer cottages around the ponds.

Of the existing buildings many of the outlying farms were given up by their weary owners and either slowly reverted to woodland or were bought by summer people or immigrants from Canada or Finland. In the latter case they might still be run as farms on a reduced scale.[83] Samuel D. Bemis, speaking to a county Grange meeting in 1901, deplored the converting of "our best farms" to summer residences occupied only a few months of the year as harmful to the farmers' interests, but the trend continued and accelerated.[84] Since the working farm in New

79 *HAR*, 1941.

80 In 1932, there were 88 summer residents for every 100 permanent residents in the town, and, given the increase of cottages, the proportion of summer residents must have risen during the remainder of the decade. Cf. Howarth, *op. cit.*, pp. 21–22.

81 *NHS*, Oct. 19, 1904.

82 Cf. *Invoice and Taxes of the Town of Harrisville, 1906*; *HAR*, 1942.

83 Cf. J. W. Goldthwaite, *Town . . . Downhill*, p. 546.

84 *NHS*, Sept. 11, 1901.

Hampshire was not particularly a thing of beauty, the change often meant at least an improvement in its appearance.

Of the buildings in and around the mill village, most of those put up by the Harrises were maintained by the Colonys, who rented the dwellings and used the mills for storage. The sharp decline in population inevitably meant that some dwellings, like the great frame boardinghouse, remained unoccupied and steadily deteriorated.[85] A number of Canadian and Finnish immigrants, after several years of careful saving, bought houses. This was often a salutary development for the town's condition and appearance, for these immigrants moved into houses badly run down and vermin-ridden and promptly renovated them. This is not to say that they had modern conveniences. Even at the end of this period probably a large majority of the houses in the village did not have indoor plumbing.

It was at a special town meeting in the summer of 1914 that the town decided to have electric lights installed in the village.[86] (A few buildings, including the Cheshire Mills and the Winn brothers chair factory, had been generating their own electricity for a number of years.) The right of way was cleared and the lines and lights installed the next year.[87] The residents were, at first, more diffident about having electricity in their homes. Before the lines were strung, the power company took a house-to-house poll and found that many people did not want electricity.[88] This attitude, revealing as it was of the town's attitude toward "progress," gradually changed.

The overall appearance of the village was without doubt a varied one. The mill buildings and mill-owned houses were kept up better than others. With the increase in national groups there was greater diversity in matters of taste. More than one person, seeing the village for the first time, was struck by its drabness. It would probably be an exaggeration to describe it as run-down, but clearly the village in these years no longer struck the visitor as fresh, bright, and pristine.

Greater than the changes in its buildings were those in the town's means of transportation and communication. The railroad, though never profitable for the Boston and Maine, continued to provide dependable service to Boston and Keene until nearly the end of this period. The stations and properties were kept up, and the line even ran Christmas specials.[89] Nonetheless, the automobile and the truck gradually took away the riders and the freight, and in 1925 the railroad petitioned the Interstate Commerce Commission for permission to discontinue service on the branch line that ran through Harrisville. Cheshire County

85 In the boom days of the First World War there was something of a housing shortage in the village, but that condition was exceptional. Cf. *NHS*, March 27, 1918.

86 *NHS*, Aug. 5, 1914.

87 *NHS*, Feb. 3, May 19, 1915.

88 Interview with Charles M. Bergeron, Nov. 10, 1962.

89 *NHS*, May 26, 1906, Dec. 22, 1915.

promptly sent a committee to Washington to fight the case, and under such pressure the commission decided against the application.[90] The end, however, was now only a matter of time. The Great Depression brought a curtailment of service, and the spring floods of 1936 were the final disaster.[91] Little damage occurred in Harrisville itself, but in the valleys to the east and west it was great.[92] The Boston and Maine was forced to limit service to one train a day, and later that summer, by substituting a bus service for the area, it received official approval for a suspension of train service.[93] The public was satisfied with the bus service, and the next year the Interstate Commerce Commission allowed the final abandonment of the railroad.[94] The floods were, in fact, a mercy death for a railroad line that, like many others built in the ebullient years following the Civil War, was unprofitable even in the best of times and was impossible in adversity.

The automobile, which helped to put an end to the railroad's operations, required better roads than those the townspeople had contended with in the nineteenth century. Expenditures on highways, some part of which came from the state, increased greatly, especially after the First World War. During the 1930s the roads benefited from a number of work relief projects financed by the town, the state, and the federal government.[95] In 1940, highway expenditures accounted for 35 percent of the total town budget, a larger proportion than that devoted to any other item that year or in any decennial year in the period.

Some of this money went to replace the larger wooden bridges with iron or concrete structures, according to a planned program instituted in 1900, and toward installing electric street lights, which began in 1915.[96] A few new roads were built, most of them for access to the new summer cottages. Nonetheless, the amount of new road laid down did not nearly equal the mileage of road that fell into disuse and was abandoned.

Most of the highway money went toward widening and improving existing roads. During the greater part of this period the roads continued to be what the town called hard-surfaced, gravel roads, or what outsiders called dirt roads. Apparently, the first tarred road in town was a stretch built about 1920 at the instigation of one of the wealthy summer people, Mr. Arthur Childs. Many people came to see it, shook their heads, and expressed the opinion that it would not prove successful.[97] The doubts must have persisted, for the next record of a road being tarred was the "state aid road" in 1937.[98] Probably several more had been similarly resurfaced by 1940, but it seems clear that

90 *NHS*, Nov. 25, 1925.
91 *NHS*, Jan. 13, 1932.
92 *NHS*, April 1, 1936.
93 *NHS*, April 29, Aug. 5, 1936.
94 *NHS*, Oct. 27, Nov. 17, 1937.
95 *HAR*, 1934, 1936; See Appendix 13.
96 *HAR*, 1900; *NHS*, May 19, 1915; *HAR*, 1916–1926.
97 *HAR*, 1921; Interview with John Priest, Sept. 4, 1964.
98 *NHS*, June 16, 1937.

it took outside influences and the extensive public works programs of the 1930s to bring paved roads to Harrisville.

As the roads were improved, the automobiles that traveled them increased in number. In their early, uncertain years, cars were apt to be regarded with derision, as suggested by the following account in the *Sentinel* of an excursion in a "four-seated sightseeing automobile" run by the "Fitchburg Auto Transit Company" in the autumn of 1905:

> Several from this place went to the Brattleboro fair last Thursday making the 3 hour trip in an automobile. No doubt they enjoyed the ride, but could not have enjoyed the long wait of two and a half hours before starting in the morning and not reaching the fair grounds till 2 o'clock in the afternoon. It was near the small hours of the morning before the puffing of the machine told of the arrival of the auto on its return trip.[99]

The success of the automobile brought a change in attitude, and in 1924 another local correspondent proudly reported to the same newspaper that "there are 69 automobiles in the little village of Harrisville, probably more, according to its size, than any other place in Cheshire County."[100] In 1935, there were at least 106 motor vehicles in the town, and by 1940 the number had risen to at least 221.[101] Both the sudden increase during the years of the Great Depression and the total number are noteworthy. There were more motor vehicles in Harrisville in 1940 than there were horses in 1900, and the number of motor vehicles per capita in Harrisville in 1940 was distinctly above the national average.[102]

The great increase in the number of motor vehicles during the 1930s should not obscure the importance of the lowly bicycle as a means of travel during most of the period. Buying a bicycle was among the first objectives of the Finns who settled in Harrisville. One of them recalls a day when on his way to work in Dublin he counted fifty other Finns riding to work on bicycles, all with their lunch boxes behind the seats.[103]

Keeping pace with better transportation were improvements in communication. The telephone lines had first reached Harrisville in the 1880s, but there could not have been many subscribers. In 1902, a Harrisville exchange was established as a magneto office. At the time there were three subscribers. Thereafter, the number of telephones steadily rose, with the greatest increase coming in the decade after 1910. At the end of the period, in 1940, there were a total of ninety-three in the town. The increase in the number of telephones over the years compared

99 *NHS*, Oct. 4, 1905.

100 *NHS*, July 2, 1924.

101 *HAR*, 1936, 1941.

102 There were .434 per capita in Harrisville in 1940 compared with .261 nationally. Cf. *Historical Statistics of the United States*, Series Q 310–320, "Motor Vehicle Factory Sales and Registrations . . . 1900 to 1957," Item 314, p. 462.

103 Interview with Victor Santala, Oct. 17, 1960.

with the national average in a steadily improving proportion. In 1910, Harrisville's per capita distribution of telephones was only 61 percent of the national average. By 1940, it was 111 percent.[104]

It may be supposed that beginning at a later date Harrisville took to the radio as it did the telephone. Detailed statistics on the number of radios in Harrisville are lacking, but the *Sentinel* reported as early as 1925 that "there have been fifteen radios installed in the village up to this time."[105]

An authoritative study of the hill country of northern New England points out that the region in the early twentieth century not only began to achieve some economic stability but also felt some "quickening influences," developments that brought new life and vitality to the region. These included the increase in immigrant farmers, the inauguration of Old Home Week, the growth of the summer recreational industry, the introduction or expansion of rural mail delivery, the telephone, radio, and, most influential of all, the automobile. These means of contact with the outside world did much to diminish the feeling of isolation and to enliven the monotony of rural life.[106] As has been shown, Harrisville experienced all of these "quickening influences," especially during and after the First World War and even more so in the 1930s. By 1940, they had wrought considerable change in the life of Harrisville. By that date, and more so since then, they had helped to make life there as in other small towns far more pleasant and comfortable than it had ever been before. Part of the price, of course, was that as the small town was drawn into the national community and made a part of it, the town lost many of its traditional characteristics.[107]

Despite these winds of change, it is perhaps a typical example of cultural lag that the political leadership of Harrisville remained largely in the hands of those whose families had come to the region in the eighteenth century. The Harrises had vanished, and the Colonys, living in Keene, kept out of local affairs, but the names of Bemis, Clark, Farwell, Keniston, and Wright, among others, appear again and again in the lists of town officers. Bernard F. Bemis was selectman at the turn of the century, and he was selectman when the period ended in 1940.

This group did share political leadership with a few men whose fathers or grandfathers had immigrated from Ireland or Canada, but the newly arrived Finns held very few town offices during these years. Of the politically active nineteenth-century immigrant families, the most outstanding was the Winn family. Edward Winn had come from Ireland in 1852, in the aftermath of the Great Famine. Perhaps attracted by the name, he had

104 Information supplied by Bob G. Slosser, General Public Relations Supervisor, New England Telephone and Telegraph Co., letter of Dec. 16, 1966; *Historical Statistics of the United States*, Series R2, p. 480.

No. phones in Harrisville

1910	1920	1930	1940
31	64	85	93

No. phones in Harrisville per 1,000 population

1910	1920	1930	1940
49.8	114.6	166	182.5

No. phones in U.S. per 1,000 population

82	123.9	163.4	165.1

105 *NHS*, March 11, 1925.

106 Wilson, *op. cit.*, pp. 259–266.

107 Smith, *op. cit.*, pp. 300–301.

settled in Dublin, New Hampshire. He later worked for Milan Harris, lived in Harrisville, and, before his death in 1910, saw all of his six children graduated from high school.[108]

Edward Winn's eldest son, Thomas, went on to business college and, after gaining some practical experience, returned to Harrisville to help run the Winn brothers' business ventures there. Thomas Winn was highly respected by his fellow townsmen, and at various times he held most of the town offices.[109] His son, Thomas, Jr., succeeded him as town clerk and treasurer, and from time to time other members of the family also held various offices.[110]

The Winns owned not only the chair factory and the brick general store but also Eagle Hall, where town meetings were held, and Winn Grove, site of the annual Old Home Day picnic. Roman Catholics, ardent Democrats, and shrewd (some say artful) businessmen, the Winns were natural leaders for much of the mill town's immigrant population. Reportedly, in the early years of this century there was considerable rivalry between the Colonys and the Winns.[111] One can see the reasons for it, and they probably date back to Edward Winn's employment with Milan Harris, but the rivalry never assumed the epic proportions of the feuding between the Colonys and the Harrises in the nineteenth century.

The political leadership of the town was thus divided among a number of families, Yankee and Irish, with probably as much alternation of personnel as a town the size of Harrisville would allow. If the number was limited, it was not due to political exclusiveness but to the limited number of people, of any background, who could and would assume the responsibilities of office.

In New Hampshire towns the selectmen and town treasurer are responsible for the management of financial affairs and the budget. These aspects of a town's history can be traced in detail in that mine of information, the published annual reports. Harrisville's budgets are summarized elsewhere.[112] During these forty years there were changes in both the sources of income and the purposes for which it was expended, but amid all the complexities of the budget, increased by varying methods of bookkeeping, a close examination reveals certain features and trends.

Among the sources of income a minor one of long standing was aid from the state. In the forms of a savings bank tax and a library fund, it dated back at least to 1880. In the twentieth century, state aid increased steadily until the Great Depression, and most of the time that aid exceeded the taxes paid to the

108 Pillsbury, *op. cit.*, V, 18–19.

109 Cf. Obituary, *NHS*, July 2, 1937.

110 Pillsbury, *op. cit.*, V. 18–19.

111 Interview with Newton Tolman, Feb. 15, 1961.

112 See Appendix 13.

state by the town. Nonetheless, the amounts involved were relatively small throughout the period.

Poll taxes were paid only by adult males until woman suffrage in 1920. Until the First World War it seems that the poll tax was the amount of the annual tax rate on property, or usually between a dollar and a half and two dollars. After the war it was a fixed amount, at first five dollars and later two dollars.

The most important source of revenue for the town was then, as now, the local property tax. The holdings of every resident and nonresident property owner were periodically appraised, and every few years the town published a booklet, "Invoice and Taxes," describing each persons' property and listing its valuation and taxes. In recent years the invoice has been included in the annual reports. It can be safely assumed that no part of these annual reports is more closely studied by the citizenry than this listing of who pays what.

A table comparing the valuations of resident property owners at the beginning and end of this period can be found elsewhere.[113] It is noteworthy that, in 1906, 84 percent of the adult residents had valuations of less than $2,000 and that, even in 1941, 66 percent fell into this same category. At the other end of the scale there were in both these years only four valuations of $10,000 or over. These included the Cheshire Mills, the Winn Brothers, and two or three wealthy retired persons. Of course, these valuations may have been at a discount, but their public nature argues against too great a disparity. Considering the changes in the real value of the dollar, it would seem that there was remarkably little change in the wealth of the great majority of the town's residents between 1906 and 1941. And in terms of constant dollars the total valuation of the town's resident property owners probably declined in these same years by 25 percent.[114]

The tax rate was set annually by the selectmen on the basis of the annual appropriations in town meeting.[115] For the last decade of this period, and perhaps longer, the tax rate in Harrisville was lower than the average for the state.[116] Paralleling this development, and certainly related to it, was the increase in the number of summer people. As has been noted, nonresidents, and particularly owners of summer cottages, contributed an ever-increasing proportion of the revenues from the property tax until at the very end of the period they were paying a larger proportion than the residents. Furthermore, if the property valuation of the residents declined in terms of constant dollars by about 25 percent, nonetheless, because of the summer

113 See Appendix 12.

114 See Appendix 12. Cf. *Historical Statistics of the United States*, Series E113–139, Consumer Price Indexes, (BLS), Item 115 (Food) p. 125, Series F1–5, Gross National Product, Item 5, Implicit Price Index, p. 139.

115 For specific rates, see Appendix 11.

116 *HAR*, 1942.

residents, the total property valuation of the town in the same terms must have risen about 50 percent.

In the town's expenditures the three most important items were schools, highways, and payments on the town's indebtedness.[117] The relative size of these three occasionally changed, and over the years the trend was toward proportionally less money for the schools and more for the construction and maintenance of the highways. In 1940, the town spent 35 percent of its disbursements for this purpose, the largest proportional amount allotted for any purpose down to that time. The town's indebtedness rose and fell. Apparently, after the depression of the 1890s, it was usually necessary for the town to borrow money each year in anticipation of taxes.[118] After 1920, this item seems to have constituted the bulk of the town's indebtedness. An audit in 1942 by the state tax commission noted approvingly that "the problem of debt has not faced the town of Harrisville, at least in recent years, for it is to be noted that at the close of each fiscal year a surplus was indicated by the balance sheet."[119] Finally, the total budget steadily increased in these years until the Great Depression struck. Calculated in terms of constant dollars, the budget of 1930 was three times that of 1900, though the resident population had shrunk by a third during these years. Clearly, it was the summer residents who, despite the long economic decline climaxed by the years of the Great Depression, kept Harrisville's finances in the black.[120]

Concerning matters of police and fire protection, little needs to be said. Breaches of the peace there were, to be sure, but law enforcement in a rural small town like Harrisville, where everybody knows everybody and most of their affairs, tends to be individual and informal. In this century as in the last, the chief official was the town constable. He might be assisted when necessary by from one to three police officers. The police were needed mainly on special occasions such as election day or Old Home Day. Other situations, an act of vandalism, a disorderly drunk on a Saturday night, domestic discord of a violent nature, were handled by the constable if he could be found in time, but they were just as likely to be settled by some private citizen who might in fact command more authority than the constable.

Fire protection falls into something of the same pattern, but it also reveals the trend of decline and improvement that has been the story of Harrisville in the twentieth century. In the mid-nineteenth century the young town had, for the time, quite good fire protection, but by the end of the century this was no longer true. When a fire broke out at the "Monadnock Hotel" in

117 See Appendix 13.

118 *NHS*, March 7, 1900.

119 *HAR*, 1942.

120 See Appendix 13. Cf. *Historical Statistics of the United States*, (Index A) Series E157–160, Cost-of-Living Indexes, (Burgess) 158, p. 127, (Index B) Series E113–139, Consumer Price Indexes (Food) 115, p. 126.

Harrisville in January 1901, help was summoned from Marlboro by telephone. The Marlboro Fire Department made the eight-mile trip in an hour and ten minutes, but by that time the fire had been brought under control by a bucket brigade. The *Sentinel* remarked that "Harrisville has no fire apparatus and no water under pressure."[121] Soon after this, taking heed, the Cheshire Mills installed a new force pump and new hydrants, but (or perhaps because of this) the town seems to have done nothing. Yet the need remained, and in the next few years several houses were destroyed by fire.[122]

By the time of the First World War there was some improvement. In 1915, the town bought ten fire extinguishers and thereafter more equipment such as ladders, pails, sprayers, and more extinguishers.[123] By 1930, the activities of a "Fire Committee" were being included in the annual reports, and in 1935 the volunteer fire department acquired a suitable truck equipped with a stationary pump.[124] This improvement was probably due not only to the general revival of civic activity in the town but to the increased danger from forest fires as more and more farm-land reverted to its natural state. In 1940, the list of town officers included a Chief of Fire Department, four fire wards, and a Forest Fire Warden with an equal number of deputies. Characteristically, the Fire Chief and the Forest Fire Warden were John Clark and Bernard F. Bemis, both of whom had filled many other town offices for many years. Nonetheless, in 1940, in proportion to the danger Harrisville had reasonable fire protection.

Reflecting the economic changes in the life of the town and the "settling down" of its immigrant groups, political activity in these years continued to be a favorite pastime but with fewer rough edges than formerly. Though the electoral process still left something to be desired, the town seems to have outgrown its penchant for riots and near-riotous town meetings. It is said, without substantiation, that the Winn Brothers used to give voting orders to their employees and, according to a former town officer, candidates formerly printed their own ballots, big or small, so that they could spot their votes as they went into the ballot box. It may be assumed that votes were bought and sold and that some of the immigrants who voted were not citizens. But a *Sentinel* account of an election celebration in 1908 (not a very exciting election nationally) suggests a new spirit:

One of the most enthusiastic if not the largest election celebration and parade took place Saturday evening . . . marshaled by the newly elected

121 *KES*, Jan. 25, 1901; *NHS*, Jan. 30, 1901.

122 *NHS*, March 13, 1907, July 15, 1908.

123 *NHS*, May 5, 1915; *HAR*, 1916, 1926.

124 *HAR*, 1931, 1936.

representative of the town accompanied by one of the same from an adjoining town. The parade consisted of a drum corps, youthful torch-bearers and other interested individuals. That which caused the most merriment was the fulfillment of four election bets. Wheelbarrows containing four happy and contented faces (not all confined to the sterner sex either) were wheeled by puffing and steaming losers.[125]

Remembering the victory "celebration" that followed Buchanan's election in 1856, this parade of 1908 was surely a sign of progress!

With its balanced population of immigrant mill workers and native-born farmers, Harrisville in national elections continued to vote as a very independent-minded town in the middle of strongly Republican territory. In the eleven presidential elections from 1900 to 1940, Harrisville went Republican five times and Democratic six. It also voted for the winner ten times out of eleven, which suggests that most of the time the town was well in tune with national sentiment. Only in 1924 did it go against the current in preferring John W. Davis to President Coolidge. Without denigrating Davis, it would seem that the explanation lay in the economic condition of this mill town. The Cheshire Mills were shut down or running only part time from May to September of 1924, and not surprisingly the voters saw little advantage in "keeping cool with Coolidge."[126]

Important national events in this period were, of course, felt in Harrisville and generally more than in earlier years. But not all had the same impact. The Progressive Movement that swept the country in the years between the turn of the century and the First World War reformed many conditions of American life. But the movement was not strong in New England where many Progressive reforms had been effected years before, and the available evidence suggests that it was hardly felt in Harrisville. Unlike the reform impulse of the 1830s and 1840s, which had found a strong response in the hill towns of New Hampshire, that of the early twentieth century was city oriented. It confronted the problems created in part by the flight of the population to the cities from just such declining rural areas and small towns as Harrisville.

The First World War, on the other hand, had a considerable impact. Military orders for the mills again brought prosperity and vitality to the town, and from its beginning the tragedy elicited concern and sympathy. At special services held in the Congregational Church on October 4, 1914, the Baptist minister from Chesham wrestled with the question "Is War Ever

125 *NHS*, Nov. 11, 1908.

126 *NHS*, May 14, June 18, July 23, 1924; CMR, Payroll, 1924; *Manual of the General Court of New Hampshire*, 1901–1941. See also Appendix 14.

Justifiable?"[127] Benefits were run for the Belgium Relief Fund, and the Ladies Sewing Club made surgical dressings for the French Army.[128]

Once the United States entered the war, the tempo of activity increased. Business boomed, and the clatter of the looms was accompanied by myriad demonstrations of patriotism. An honor roll was unveiled at the Congregational Church, and a service flag was flown at the center of the village.[129] The ladies sent work to the Red Cross, and the young people sent eighty dollars which they raised by giving a benefit performance of the play "Alice in Wonderland" on the lawn at the residence of Dexter D. Dawes in Chesham.[130] The women in town heard lectures on "War Time Menus" and were quizzed by a Ladies' National Defense Committee on just how much sugar and flour they had on hand.[131] The new minister courageously asked himself "Where Was God When Belgium Was Being Murdered and Armenia Massacred?" and the town oversubscribed its allotment for the Third Liberty Loan five and a half times.[132] Leading citizens like Tom Winn and Arthur Childs headed committees, and Massachusetts State Senator Wellington Wells made a "stirring patriotic address" to a captive audience at a town meeting.[133] With the holocaust three thousand miles away and no local German population to persecute, the war effort in Harrisville had a brave, unclouded face.

The Armistice was occasion for special victory services at the Church, at which the parishioners considered "The Golden Age to Come."[134] The men who went to war from Harrisville, far fewer than those who went off to the Civil War from Dublin or Nelson, were all back in time to be feted at a celebration in August 1919, announced in the *Sentinel*:

The official welcome home for the Harrisville men who served in the war will be held Tuesday, August 19, in connection with Old Home day celebration at Winn's grove. . . . Bells will ring and whistles will blow at sunrise and noon. The honor roll, containing 40 names of the men who served in the late war will be dedicated at 11:30 a.m. . . . In the early evening the soldier boys will be given a banquet with an orchestra to furnish music.[135]

Somehow, the "Golden Age to Come" did not quite pan out. The 1920s brought short work weeks, diminished civic activity, and a further decline in population. The Great Depression meant a worsening of economic conditions but also resulted in some new life in the town. Politically, the town met the

127 *NHS*, Sept. 30, 1914.

128 *NHS*, Feb. 17, July 15, Sept. 29, 1915.

129 *NHS*, Jan. 9, Aug. 21, 1918.

130 *NHS*, Feb. 13, April 10, July 24, 1918.

131 *NHS*, April 24, May 1, 1918.

132 *NHS*, Sept. 4, 25, 1918.

133 *NHS*, March 20, 27, June 12, 1918.

134 *NHS*, Nov. 13, 1918.

135 *NHS*, Aug. 13, 20, 1919.

crisis by taking the necessary relief measures and by voting for Franklin D. Roosevelt.

It appears that Harrisville endured the years of the Great Depression better than did much of the country and better than the town had done in earlier economic crises. If this was so, it was because Harrisville had little fat to lose in 1929, because of the strength and paternalism of the Cheshire Mills, and because this community had built-in shock absorbers not found in larger towns and cities. As the period here examined drew to a close, Harrisville was far from finding any Golden Age, but it showed considerable evidence that it was as a town coming of age.

9 SOCIAL LIFE OF HARRISVILLE, 1900–1940

It will be recalled that Harrisville's population trends were not in harmony with those of northern New England in general. That region suffered its greatest decline during the last three decades of the nineteenth century and then began to regain some stability early in this century. Harrisville's growth had been erratic and had roughly corresponded to the growth and prosperity of its mills. At the end of the century the population still numbered close to eight hundred. Then, in the first decade of this century, Harrisville suddenly lost more than 20 percent of its people. This decline continued on a diminishing scale until the 1930s. The period ended with the population temporarily stabilized at slightly more than five hundred.[1]

In 1940, the town's population was composed entirely of white people, 84 percent of whom were native-born, 55 percent of whom were male, and nearly half of whom were thirty-five years of age or older.[2] Although the proportion of foreign-born in the town had declined since 1880, the proportion of second-generation Americans had risen. In 1930, 51 percent was either foreign-born or of foreign parentage, while in 1880 these totaled 43 percent.[3] For some reason the proportion of males was much greater among the native-born than among the immigrants. Finally, because of the growing number of summer people the town's population approximately doubled during that season of the year.[4]

The sharp decline during the first years of the new century coincided with the arrival of the last wave of immigrants to settle in Harrisville. These were the Finns. The relationship between the two events is not entirely clear, but the flight from Harrisville dates from 1902 when the Cheshire Mills hired its first Finnish workers. Though the decline began in a prosperous year, it was certainly accelerated by the Panic of 1907 and the slow business conditions of the years immediately following. The *Sentinel* correspondent from Harrisville reported in the spring of 1909 that "there are about fifty poll-tax payers less than last year."[5]

Perhaps some of the natives, already unhappy with small-town

1 See Appendix 4.

2 U.S. Bureau of the Census, *Sixteenth Census of the United States*, 1940, Population, Vol. III, Part 4, Table 28, p. 796. The local picture varied from the national in Harrisville's being obviously "whiter," and having about 5 percent more males and 7 percent more foreign-born. *Cf. Historical Statistics of the United States*, Series A22–23, A51–58, pp. 8–9.

3 *Cf.* Howarth, *New Hampshire*, pp. 21–22; Appendix 6.

4 *Cf.* Howarth, *op. cit.*, pp. 21–22. In 1932, there were 88 summer residents for every 100 permanent residents.

5 *NHS*, May 5, 1909.

life, saw the threat posed by the arrival of this new supply of cheap labor and decided to leave when they had the opportunity for work elsewhere. The payroll samples for 1902 and 1907 suggest a lowering of wage rates and, in addition, show a marked decline in gross payroll. The exodus continued for reasons social as well as economic, but the poor business conditions after 1907 prevented the Cheshire Mills from hiring enough immigrants to make up for the loss of native-born population.

The Finns who came to this country had been spurred to emigrate by "nature's vicissitudes, combined with Russia's arbitrary methods and the preoccupancy of the better regions of their country." Finnish sailors began to arrive in New England about 1870, but those who came in the early twentieth century were from the poorer rural districts. In general, the Finns settled in parts of the country that were geographically similar to those from which they emigrated, and, instead of crowding into the cities as did most immigrants, the majority of them settled in rural districts.[6]

Once here, the Finns found work in lumber camps, mines, quarries, and textile mills. Many had, and fulfilled, the ultimate objective of becoming independent farmers. Among Americans, the Finns became known as a persevering, courageous, tenacious people with an appreciation of education. Although often phlegmatic and stolid, when opposed the Finn could be quickly aroused. (And a habit of carrying long knives made an aroused Finn someone to be avoided.) Similarly, a reputation for hard drinking, extreme suspiciousness, and bad ventilation in their homes and places of work did nothing to improve their welcome in communities where they settled.[7]

Politically, the Finns had an affinity for socialism, which they managed to combine with their conservative Lutheran religion. Whether this political radicalism was a legacy of old world conditions or due to a disillusionment with what they found in America is a subject of some debate, but it is agreed that among the Finns, as time went on, there was a gradual change to conservatism.[8]

How do these generalizations apply to the Finns who settled in Harrisville?

Fortunately, there are a few of these early Finnish immigrants still living in Harrisville. It has been possible for the writer to interview four men personally and one woman indirectly through her daughters. Not surprisingly, their experience falls largely into a pattern familiar to all immigrants to this country.

All five of these surviving immigrants came to this country

6 E. Van Cleef, *Finland*, pp. 190–194.

7 E. Van Cleef, *Finland*, pp. 197–200; E. Van Cleef, "The Finn in America," *Geographical Review*, VI, 207.

8 E. Van Cleef, "The Finn in America," *Geographical Review*, VI, 208; Cf. E. Van Cleef, *Finland*, p. 202; J. I. Kolehnainen, *Sow the Golden Seed*, pp. 6, 83.

between 1899 and 1913, at ages ranging from twenty to twenty-four. Four came from rural, central Finland, and one came from the southern port city of Hango. All of them had relatives either then living in the United States or who had lived here at one time. Some of them had their passage tickets sent them by these relatives. Except for one who went first to Michigan, they came directly to Harrisville or the nearby towns of Dublin and Marlboro. One of them has lived in Harrisville since he arrived there, but the rest have moved about. None of them recalls receiving any ill treatment because they were immigrants.

On arriving in this country, none of these immigrants had a trade or more than an eighth-grade education. A striking fact is the number and variety of jobs that the men have done, either in Harrisville or elsewhere. These include not only the common immigrant jobs such as lumbering, road work, estate care, and factory positions but also quarrying, masonry, farming, shipbuilding, and sailing with the merchant marine. They had neither trained skills nor much education, but they were industrious, and they all prospered. All of them have married, raised families, and now own their homes. Some of their children are college graduates. Four of the five are naturalized citizens. Although they all came from large families, only the woman raised a large family herself; the four men all have either one or two children.[9]

The experiences of these five immigrants are not necessarily typical of the Finns who came to Harrisville, but they are probably suggestive. Many of the early arrivals were rough sorts, prone to brawling, drunkenness, and knife fights. Others were suspicious, taciturn, and generally unfriendly. Certainly not all Finns flourished and became such solid citizens as this group of survivors. Certainly there *was* bias and discrimination on the part of some of the older residents—in fact it is safe to say that all newcomers to the town throughout its history have met with some measure of hostility and suspicion.

Nonetheless, as a group the Finns had a rather different experience from that of the Irish or the French Canadians. That they were neither Roman Catholic nor as destitute as the earlier groups undoubtedly improved their reception and their chances of success. In contrast to the French Canadians, who concentrated in the village and worked in the mills, the Finns were more dispersed and more versatile. Whatever the reasons, the tidiness of their homes and their concern for education compared favorably with that of the earlier immigrants. Coincident with the arrival of the Finns, a few Socialist votes began to appear in the town's returns for presidential elections. However, perhaps

9 Interviews with Arthur Wikman, Sept. 1, 1960, John Saari, Sept. 6, 1960, Isaac Hakala, Oct. 12, 1960, Victor Santala, Oct. 17, 1960, Mrs. John Johnson, through her daughters, Mrs. John Saari and Mrs. Howard Main, August 1961.

because of better opportunities, they do not seem to have developed the political radicalism of Finns elsewhere. In short, the character of these Finns in Harrisville won from the community if not affection, at least respect.

What was the impact of this new immigration on the town? The Finns were no more able to dominate local affairs than the Irish and Canadians had been, but their influence was undoubtedly great. They prolonged the existence of a supply of cheap immigrant labor for the mills. They may have prevented the town's population from declining more precipitously than it did. And they added a new element to the cultural mix that was already a distinction of Harrisville.

In the area of vital statistics the limited data available allows only a few generalizations.[10] For one thing the number of children born to immigrant and native-born parents shows a distinct trend. In the 1890s, the total numbers of births to immigrants far outnumbered those to the native-born. Since the foreign-born in 1880 accounted for only a quarter of the town's population, and it was probably no higher in the 1890s, this trend in the birthrate would seem to reflect the greater youth and fecundity of the immigrant population. From 1900 to 1920, there was a rough equilibrium in the total number of births to immigrants and native-born. Thereafter, the number of births to native-born parents was consistently much larger than to immigrants. At the end of this period the foreign-born in the town had shrunk to less than 17 percent of the total population, and the number of children this group had was about the same percentage.

As for deaths, it is impossible to see any great difference in the number of stillbirths to the immigrants and native-born in this fifty-year period. This suggests that there was no great difference in the health and medical care of the two groups. The crude death rate shows little change from that in Dublin a century earlier, though obviously the age of Harrisville's population in the twentieth century was older, and so would have a higher crude death rate.

Beyond their limited statistical value, the annual entries of vital statistics in the town reports are a fascinating record of individual lives in Harrisville's history. In the reports for the early years of this century, one finds the deaths of a number of people who were grandchildren of the region's original settlers. Such a one was Aaron Smith, Jr., who died October 26, 1900, at the age of seventy-eight.[11] During his youth, he had worked with his father in one of the last potteries in Chesham. Later, he had been

10 Since 1887, state law has required the towns to include vital statistics in their annual reports. However, because the federal census records for the twentieth century are not open to inspection, it is impossible to collate the local record of births, marriages, and deaths with the census statistics to figure, for example, age-specific death rates. Nonetheless, a sampling of these vital statistics at five-year intervals does show some trends and developments. See Appendix 10.

11 *HAR*, 1901.

active in politics, local government, and temperance reforms.[12]

Smith was also for many years the local correspondent for the *New Hampshire Sentinel*, probably the one who signed his articles "O.C.," for "Old Cheshire."[13] After his death the *Sentinel* was no longer the rich source for local news that it long had been. This change was only partly due to "dull times" in the new century, for Smith himself had complained of the paucity of news. Rather it was due, at least in part, to the passing of men like Smith in Harrisville, who were not succeeded by others as able, interested, or as outspoken.

Another name found among the deaths early in the century was that of Alfred Romanzo Harris, who died in Haverhill in 1910 at the age of eighty. This youngest son of Milan Harris was brought back to his birthplace and buried in the "island cemetery" that his grandfather Bethuel had laid out for the family in the 1840s.[14]

The list of deaths for 1915 included that of eighty-four-year-old Irish-born John Hagen. He had been the night watchman at the New Mill in 1882 when it was destroyed by fire, and he may have taken with him to the grave the true story of that mysterious fire that proved to be such a turning point in the town's history.

Although the vital statistics do not specify causes of death, the *Sentinel* and the town reports cast occasional light on this subject. The new century began in Harrisville with a relic of the past, an outbreak of smallpox. There had been some cases in the adjoining town of Hancock, and the Harrisville Board of Health appointed the local physician, one Dr. Crediford, to vaccinate all the school children. It was too late. Several children came down with the dread disease, and the local doctor[15] made matters worse by diagnosing the cases as chickenpox. At length, someone from the state board of health in Concord came and made the correct diagnosis. Six houses were quarantined, with guards on duty from early morning until late at night. A smallpox hospital was also prepared, but the disease was arrested and it was not needed.[16] The *Sentinel's* headlines and coverage indicate that the incident caused more than a mild sensation in the county, though the paper assured its readers with what could be interpreted as satisfaction that "the cases have been confined chiefly to French-Canadian families."[17]

Other diseases one finds in the early years of this century were cholera infantum, scarlet fever, and diphtheria. Often a large family would be stricken with one of the last two, resulting in quarantines, school closings, and one or more deaths.[18]

The last great epidemic in the twentieth century was the

12 H. Child, *Gazetteer of Cheshire County, New Hampshire.* p. 179.

13 *NHS*, Jan. 3, 1900.

14 *HAR*, 1911.

15 Probably, but not certainly, Dr. Crediford. The wording in the town report is ambiguous.

16 *HAR*, 1901.

17 *NHS*, Feb. 6, 13, 26, 1901.

18 *NHS*, Oct. 8, 1902, Feb. 5, 12, March 25, 1908, Aug. 31, 1910.

influenza that swept over Europe and America at the close of World War I. Harrisville's isolation had spared it the cholera epidemics of the nineteenth century, but it no longer had this protection in 1918. Although public gatherings were canceled and the opening of school was postponed, the disease appeared in October 1918. Few deaths occurred, but quite a number of people were ill with influenza during that winter.[19]

What was true of Harrisville's doctors in the years before 1900 was true after that date: they were few and not very good. Little is known of Dr. Crediford, but the smallpox incident of 1901 hardly testifies to his professional skill, if it was he who made the wrong diagnosis. The notorious Dr. Byrnes continued to practice medicine for those brave or desperate enough to go to him. He came and went from time to time and does not seem to have been involved in the smallpox affair. In 1910, he was reduced to such straits that he was receiving poor relief from the county and coal from the town.[20] Dr. Byrnes gradually gave up practicing after the First World War, as his condition worsened, and he died in Marlboro sometime in the 1920s.[21]

Among those in the area who remember the doctor, opinion is divided. A few speak well of him. It is pointed out that he had ability, especially in dealing with pneumonia cases, and that in the 1918 flu epidemic he lost only one patient. Apparently, if he thought well of a patient, he would go to great trouble for him and make frequent calls, however inconvenient the time and distance.[22]

The majority of those who remember the doctor, however, have little if anything good to say about him. A number have described his "bedside manner" as being little short of terrifying. One man told the writer that Dr. Byrnes brought him into the world, though he was so drunk when he came to the house he had to go to bed himself and sober up somewhat before he could make the delivery. Another recalls that when, as a boy, he was vaccinated by the doctor, it was such a bad job he carried his arm in a sling for three months, and the infection was so deep he could see the bone. It is some measure of his slovenliness that to examine the school children's tonsils the doctor used a single spoon and that he extracted teeth with rusty instruments left lying about his desk. His office in Blake's hotel was, in fact, renowned for its squalor. Dr. Byrnes smoked endlessly, drank paregoric when he could not get whiskey, and regularly took some kind of mysterious little pills. Harrisville's last resident physician, repulsive, frightful, and tragic, was a far cry from the stereotyped image of the old-time country doctor.[23]

19 *NHS*, Oct. 9, 16, 23, Nov. 13, Dec. 25, 1918, Jan. 8, 15, 29, 1919.

20 *HAR*, 1911.

21 Interview with Guy Thayer, Aug. 30, 1966.

22 Interview with John Priest, Jan. 22, 1967; *Cf.* Newton Tolman, *North of Monadnock*, ch. 10.

23 Interviews with Charles M. Bergeron, Nov. 10, 1962, Arvo Luoma, Nov. 10, 1962, Newton Tolman, Feb. 15, 1961, Guy Thayer, Aug. 30, 1966, John Priest, Jan. 22, 1967.

Serving to make up somewhat for the lack of medical care in childbirth was "a wonderful, dynamic little woman named Mrs. George Desilets who acted as midwife for a generation" of Harrisville residents. She was born Julia Yandeau, in Michigan, of French-Canadian parents. She grew up and married there but then came to Harrisville with her husband in 1896. Here she began to practice midwifery. Though to some extent she worked with a doctor in Dublin, she was reportedly self-taught in her practice. She did this service mainly for "friends" in Harrisville, and compensation was on an informal basis. That she did it at all is noteworthy. In contrast to Europe, where it has remained an accepted profession in modern times, midwifery in America has been uncommon, if not rare, since the latter part of the eighteenth century. Harrisville, with its dubious doctors and its immigrant population, was a logical place for midwifery to reappear.[24]

Also serving to compensate somewhat for the poor quality of the local physicians was the town board of health and later the health officer, supported by the state board in Concord. The local officials acted to save the town from serious trouble in the small-pox outbreak. In 1916, they used the medium of the town reports to explain the state's new compulsory vaccination law and to call the public's attention to the dangers of moving a person with a contagious disease. This had happened the previous year and "practically cleaned out a whole family."[25]

Despite the efforts of these officials, the amount of money the town spent on public health was miniscule. In 1940, it amounted to $11.80 out of the total town expenditures of more than $28,000. Much remained to be done in this area, if we are to judge by one set of statistics that is available. Beginning about the time of the First World War, the schools gave annual physical examinations to their students. Then in the 1930s, they began to tabulate the results of these examinations in the town reports. Checking for such physical defects as under-weight, poor vision, hearing, teeth, breathing, and tonsils, the examinations of three sample years showed the following results:[26]

24 Letter from Mrs. John J. Colony, Jr., Harrisville, Dec. 2, 1966; Telephone interview with Mrs. John Silk, Sr. (daughter of Mrs. Desilets), Marlboro, New Hampshire, Dec. 29, 1966.

25 *HAR*, 1916.

26 *HAR*, 1931, 1936, 1941.

Year	1930	1935	1940
Students enrolled	85	75	80
Students examined	74	70	81 (*sic*)
Defective conditions	118	56	71
Defective conditions corrected in year	17	22	16

Paralleling these results were those obtained by a physical examination of children of preschool age ordered by the state board of health in 1930. This revealed that of twenty children examined, only "seven were without defects while thirteen were found to have one or more defects."[27]

This was certainly no health record for a modern society to boast of, but it was clear that improvement lay not only in increased expenditures but also in changing attitudes. In 1930, for example, a dental clinic was held in Keene, and of sixty-eight children with defective teeth in Harrisville, only seventeen went to the clinic. The same year, a tonsil and adenoid clinic was planned, but not one of the twenty-two children needing medical treatment responded.[28]

In the closely related matter of caring for the poor, much of the history has already been traced, and much of the earlier pattern continued to apply in the early twentieth century. Despite the town's loss of population and the vicissitudes of business conditions, caring for the resident poor was not a major problem until the Great Depression. In those years of hardship, the "Charities" item in the town reports greatly lengthened. In 1935, for example, the town provided twenty-one of its residents with such things as fuel, supplies, rent, and medical care. The names were all listed, along with the aid and the cost. There is no need to repeat them, but the lists included many names, both old stock and new, more usually found in the list of town officers. Many more were helped by the work relief projects undertaken by the town. These included not only work on the roads but such things as remodeling the town library and selectmen's office.[29]

A more constant problem, in the twentieth century as in the nineteenth, was that of tramps. At the turn of the century the disreputable tramp house was removed from its "undesirable location" (perhaps it was too prominent) to land behind the Henry Colony house.[30] There it remained through the 1930s. One of the overseers of the poor in the 1930s recalls that it was his practice to give to any tramp requesting help twenty-five or thirty cents worth of food supplies and allow him to put up in the tramp house. A policy of requiring the recipient of this bounty to chop wood in return was only partly successful. The tramp might chop what wood he needed to keep warm, but he was just as apt to break the axe handle. Unless the weather was bad, tramps put up for the night were expected to move on the next day.[31]

It is not easy to determine just how many tramps passed through

27 *HAR*, 1931, 1936, 1941.
28 *HAR*, 1931.
29 *HAR*, 1936.
30 *NHS*, June 14, 1905.
31 Interview with John Clark, Nov. 11, 1962.

Harrisville. The same overseer of the poor recalls that sometimes there would be three or four tramps in town and then none for a week. He estimates that there were fifty to sixty a year during the 1920s and no more than that during the 1930s.[32] The totals did not often appear in the town reports. However, in 1937 the *Sentinel* reported that "from July to November 12 there were 63 transients at the police station" in Harrisville, which suggests a considerably higher number of tramps.[33] Be that as it may, the evidence suggests that this problem, though constant, was not overly troublesome and that the town met it with a policy containing roughly equal measures of humanity and practicality.

As the town lost population and economic vitality in the early part of this century, it also lost some of the rough edges that it had shown for so long. Vice and lawlessness continued, but crime, like virtue, has its degrees, and it undoubtedly declined in both quantity and violence. There were still occasional reports of vandalism and theft: a house broken into, a fence torn down for an Independence Day bonfire, firewood stolen from the church shed.[34] Enforcement of the fishing laws sometimes met with resistance, and occasionally a trout brook was dynamited.[35] The Harrisville event that made the biggest headlines in the *Sentinel* occurred in 1925, when the authorities discovered a shortage in the tax collector's accounts of just over a thousand dollars.[36] And to bring the criminal record down to modern times, in 1936 another resident of Harrisville was "sent up" for sixty days after pleading guilty to a charge of drunken driving.[37] This is not, of course, the town's complete blotter, but these instances are probably representative. If so, they indicate a definite change from the days of the "Harrisville Roughs" in the latter part of the nineteenth century. Harrisville had not abjured crime, but those it now committed were the quieter crimes of age.

As the earlier history of Harrisville has shown, one index of morality in a community is the state of its liquor traffic. In this matter, too, the town seems to have experienced a sea change. Early in this century there were still four saloons in the village, and the residents voting in town meeting still changed their minds on the matter of prohibition with bewildering frequency. In 1903, by a margin of ten votes the town again decided for "license," as the sale of alcoholic beverages was somewhat unfairly dubbed. But this was the last time it did so before the national prohibition law went into effect in 1920.[38]

Several factors must have helped bring this change. The militant temperance (i.e., abstinence) movement that was part of

32 *Ibid.*

33 *NHS*, Nov. 17, 1937.

34 *NHS*, April 2, July 9, 1902, Jan. 30, 1918.

35 *NHS*, Aug. 14, 1918, May 12, 1937.

36 *NHS*, March 4, 1925.

37 *NHS*, July 8, 1936.

38 See *N.H. Manual of the General Court*. From 1904 on, the license vote was taken in alternate years.

the Progressive Era struck a responsive chord in this area where the temperance cause had deep roots. Perhaps, as Page Smith suggests, the cause of temperance in this period was closely linked to the cause of women's rights and the battle of the sexes.[39]

Admittedly, one finds fewer notices of temperance meetings then than in the nineteenth century, but there were fewer meetings of any sort. Occasionally there was a well-attended temperance rally held in one of the churches under the auspices of the New Hampshire Anti-Saloon League and featuring "out of town speakers."[40] The simultaneous sharp decline in the population was almost certainly related, though without the census schedules it is impossible to tell just what elements were leaving town. Finally, improving transportation may have helped ease the way for local prohibition, as drinkers with automobiles or trainfare could easily get what they wanted in places like Keene.

Whatever the reasons, prohibition sentiment grew and reached a peak in the years 1908–1910. In the town election the latter year, the *Sentinel* reported, "The sentiment as to license was nearly three to one against, or 28 yes to 79 no. The Democrats carried all the town officers."[41] It is also worth noting that prohibition was not a party issue. Democrats could be "drys."

Violation of the liquor laws also had a long tradition in Harrisville, and by no means all of the inhabitants were converted to prohibition by the Anti-Saloon League. A long-time resident says that at one time there were at least two illegal stills in town. One was run by Joe Bolio, the Cheshire Mills' fireman, in the woods behind the mills, and the other was in the vacant frame boardinghouse. Both were small-scale operations, designed more for private consumption than for commercial gain. Nonetheless, in wet times or dry they were illegal. The long arm of the law eventually caught up with Bolio in 1933; of the other still, there is at least no scent or sign at the present.[42]

A dozen years under the Volstead Act changed the thinking of many people about the wisdom of national prohibition, including the residents of Harrisville. Repeal was debated at a meeting held in the Congregational Church in June 1933. "One of the largest crowds ever assembled in town" attended, heard the speakers, and asked questions, "all of which tended to impress even the unprejudiced listeners that the 18th amendment was not practical."[43] At a special election later that month Harrisville residents voted 81 to 29 for repeal.[44] Though decisive, the smallness of the vote is striking, especially after the mass meeting at the church. The total vote was less than half the size of the presidential vote

39 *Cf.* Page Smith, *As a City Upon a Hill*, pp. 154–156.

40 *NHS*, Nov. 26, 1902, Oct. 13, 1915.

41 *NHS*, Nov. 16, 1910.

42 Interview with John Clark, Nov. 11, 1962; *NHS*, March 22, 1933.

43 *NHS*, May 31, June 7, 1933.

44 *NHS*, June 28, 1933.

the preceding November. Probably those voters who stayed home still felt, as did Hoover's Wickersham Commission, that the law was unenforceable but should not be repealed.

With the return to local option, Harrisville still chose to remain dry most of the time. At the end of the decade it voted against allowing the sale of either beer or liquor in the town.[45] But, wet or dry, it is clear that twentieth-century Harrisville took the matter of temperance much more in its stride than did the town a century before, without the polarization of opinion between rural old-stock and immigrant mill hand and without the Victorian intensity of moral indignation.

Also managing to get along without the clashes of Milan Harris's time was the town's school system. In the twentieth century a greatly increased school budget, a larger role played by the state, and changes in the teaching profession all combined to bring about considerable improvement in the quality of education.

The financial aspect of the town's educational history is complicated, but several key factors explain the important developments. Although the schools' share in the town budget actually declined in uneven fashion during the first four decades of this century, in terms of constant dollars it would appear that the town budget more than tripled and school expenditures nearly tripled in that period.[46] Or if, as may have been the case, prices did not rise as much in Harrisville as they did nationally, then the increase in the town budget and school expenditures was even greater. Also, during these same years the school population declined by more than a third.[47] The result of these trends appears to have been that in terms of constant dollars the town spent in 1940 nearly five times the amount per student that it spent in the 1890s.[48] In large part this increase was paid for by the summer residents. They contributed a steadily increasing proportion of the taxes paid (indeed, by 1940, the greater part), but they added nothing to the school population.

Also helpful was the larger role played by the state. This came principally in the form of financial aid and higher educational requirements. The financial aid was most important in the decade 1910–1920 when the town's population was still rapidly declining. In the earlier year, close to 20 percent of the school budget was paid by the state.[49] Among the helpful state laws one of the most important must have been the one passed at the end of World War I which gave the State Board of Education authority to "make such rules and regulations as may seem desirable to secure the efficient administration of the public schools," and

45 *NHS*, March 6, 1940. The vote on beer was 66 to 87, and on a State Liquor Store, 45 to 75.

46 See Appendices 13, 15.

47 See Appendix 15.

48 *Ibid.*

49 *Ibid.*

50 In *HAR*, 1921.

51 NHS, April 3, 1907.

52 *HAR*, 1926, 1931, 1941.

required the local boards and officials to obey these rules. Among other things the regulations adopted by the state board prescribed the duties of the local school boards and superintendent, provided a program of study for the schools to follow, required annual physical examinations of the students, specified the length of the school year, and established minimum standards for school-houses.[50] Such state laws did not solve all the problems of the schools, and there was always the question of how these laws were administered, but undoubtedly they removed many educational questions from the arena of town meeting and helped to bring long-needed reforms.

In the administration of the schools a new and important figure was the district superintendent. It was in 1907 that Harris-ville voted to unite with several other towns to form a supervisory district and employ a superintendent of schools, which the *Sentinel* drily acknowledged to be "a step in the right direction."[51] Though the town occasionally seems to have been without a superintendent in the years that followed and though, of course, the superintendent had to divide his time and attention among the several towns in his district, it is clear that, from 1907 on, this official took over much of the responsibility of the school board and became the central figure in the administration of the schools.

The reports of the superintendents, published in the town reports, are an excellent source of information on the schools. As full-time professional educators, they tended to report in much greater detail than had the school boards. The report for 1910 ran to six pages. It explained to the citizenry the duties of the superintendent, the work being done in the schools, and the problems to be faced.

One Frederick T. Johnson probably served Harrisville longer than any other superintendent. In 1940, he completed his eigh-teenth year in the office, and he continued to serve for several years more. His reports customarily began with a quotation from Horace Mann or Herbert Spencer and then went on to examine the objectives and procedures of the schools. Frequently, his reports reflected current concerns and trends in education. Thus, in the 1920s he stressed the concept of "service." In the second year of the Great Depression he pointed out "the need for training for leisure." And in the threatening atmosphere of 1941, he saw the students of our public schools as "our first and last line of defense," and demanded that "all isms and philosophies that are poisonous to democracy and the American Way of life must be banished from the public schools and colleges."[52] There was, comfortingly, nothing in the rest of his report to suggest that such "poisons" were present in Harrisville's schools, at least.

In the presence of the superintendent and the new state requirements, the school board's role quickly diminished. In 1910, its authority extended to "matters relative to the appropriation of revenue, establishment of schools, election and dismissal of teachers and employees, control of the program, and the adoption of rules, regulations, etc. . . ."[53] By 1921, it allowed meekly that its duties were "simply to 'comply with the rules and regulations of the State Board.' "[54] After World War I, the annual reports of the school board were largely limited to statistical summaries of attendance, revenues, expenditures, and the like.

Membership on the school board was not greatly sought after. Those who accepted the post often had other public duties as well and no doubt were not reluctant to have the professional superintendent take over much of the work. Among the members of the board who worked long and hard for the schools were Samuel D. Bemis, Frank P. Symonds, Helen B. Thayer, and Bernard F. Bemis. The last-named particularly deserves mention. Together with his wife, who was treasurer, he served as chairman of the school board for about forty years. One who served with him recalls how she "appreciated the dignity and dedication he brought to the office—especially his elegant little addresses at graduation before handing out the diplomas."[55] Other members were not well qualified for the overseeing of the schools. It would be difficult to compare the calibre of the membership with the board in earlier years. It seems fair to say that the quality of the members on the board probably did not decline but appeared to as the standards of education rose.

As for the schools themselves the town continued to have four of them until the mid-1920s: Chesham (grades one through eight); Harrisville Primary (grades one through four); Harrisville Grammar (grades five through eight); and Eastview School in the eastern end of town (grades one through six). By 1930, the dwindling number of students had closed the last-named school, but the Chesham and Harrisville schools, occasionally repaired or "remodeled," continued to serve the town not only until 1940 but for a decade thereafter. Often much of the superintendent's annual report had to be devoted to such topics as what to do about the "inadequate stove" or the "unsanitary outhouse." And the reports make clear, too, that it was only under the prodding of the state board that such conditions were corrected.

Before the First World War, at least, few young people in

53 *HAR*, 1911.
54 *HAR*, 1921.
55 Letter from Mrs. John J. Colony, Jr., Harrisville, Dec. 2, 1966.

Harrisville went on to high school. Those who did went to Keene or Wilton, and the town paid their tuition. To continue one's education required resolution. Mr. Guy Thayer, who graduated from Wilton High School in 1912, recalls that to make the twenty-five mile journey he had to leave home at 6:30 in the morning, and he did not get back to Harrisville until 7:30 or later in the evening.[56]

With regard to the actual work of the schools, the annual reports always stressed the importance of good teachers. Undoubtedly, the town had better teachers than it deserved. An examination of the real weekly earnings of Harrisville's school teachers indicates that they may have declined as much as 25 or 30 percent between 1895 and 1920.[57] During approximately the same period and measured by the same index, the average real weekly earnings in the Cheshire Mills rose about 70 percent.[58] Clearly, Harrisville's teachers were paying for a considerable part of the increased expenditures for education during this period down to 1920.[59] Helping to obscure this loss, and helping to make up for it as well, was the lengthening school year. This increased from twenty-two weeks in 1895 to thirty-six weeks by 1915. Therefore, to judge from sample years and the available indices, the average real annual earnings of Harrisville's teachers rose in the two decades from 1895 to 1915 perhaps 45 percent, and though the subsequent inflation brought a reduction, they remained above the levels of the 1890s.[60]

The prosperity and inflation surrounding World War I brought something of crisis in this matter. In his annual report for the school year 1920–1921, the superintendent explained that there was a national teacher shortage, particularly in rural communities, and that Harrisville was lucky to hold good teachers. The situation had eased somewhat by the time the superintendent made his report because of increased salaries and a lower demand for women in other occupations. "We may expect to be able to secure and hold good teachers, but we cannot anticipate any reduction in salaries."[61]

In view of the way that the teachers' real wages had been declining, this warning was ironic, but it was also apt. By 1925, the teachers' real weekly wages had risen at least to where they were in the 1890s, and they kept on rising. By the end of the period, in 1940, they were, insofar as can be judged over such a long timespan, at least 50 percent above the level of 1895.[62] On the same basis, because of the longer school year, average annual earnings in the same period seem to have risen at least 150 percent.[63]

56 Interview with Guy Thayer, Aug. 30, 1966.

57 See Appendix 15. Index A indicates a decline of 33 percent, and Index B a decline of 32 percent. Because prices probably rose less in Harrisville than they did nationally, the decline in the teachers' real earnings would have been less than these figures.

58 See Appendix 7. Cf. *Historical Statistics of the United States*, Series E157–160, Cost-of-Living Indexes, (Douglas) 159, p. 127; Series E113–139, Consumer Price Indexes, (Food) 115, p. 126. The first index indicates an increase of 68 percent between 1892 and 1922, and the second an increase of 74 percent.

59 Teachers' salaries also constituted a steadily declining proportion of the total school budget. In 1910, salaries accounted for 71 percent of the total expenditures. By 1940, they had declined to 37 percent of the total. In the latter year, almost as much was spent on high school tuition.

60 See Appendix 15. Again, if prices rose less locally than they did nationally in these years, the annual earnings of Harrisville's teachers may have risen much more than 45 percent.

61 *HAR*, 1921.

62 See Appendix 15.

63 *Ibid*.

64 See Appendix 15.

Despite this improvement, throughout these years the annual earnings of Harrisville's teachers lagged behind the national average. If Harrisville's average rose 150 percent between 1895 and 1940, the national average must have risen 175 percent. Furthermore, Harrisville's average was all along about 25 percent lower than the national average, and the disparity was rather greater at the end of the period than at the beginning.[64]

Statistics, however indispensable, cannot give a complete picture of the schools in Harrisville. It has been possible to fill part of this gap with the personal recollections of one who taught in Harrisville in the 1920s and lives there still. Louise Baldasaro, as she was then, came from North Walpole, New Hampshire. Her father had come there from Naples only to be killed in a railroad accident a few months after she was born. Life was not easy in a large, fatherless family, but Louise studied at the local high school for two years and then began teaching school in nearby Surry in 1917. Her district superintendent, whom she greatly respected, brought her to Harrisville to teach in 1923.

When she first saw Harrisville late that fall, young Louise was very depressed by the town's appearance. It had, she thought, a "worn look." It might well have had, and the season was the least flattering, without benefit of either the greens of summer or the manteling snows of winter.

Nor was she encouraged by what she had heard of the school where she was to teach. The superintendent had warned her that the discipline in the grammar school had been bad for some time. The schoolhouse itself was the barnlike structure built in 1857. It housed the village primary school on the first floor and the grammar school on the second. In 1923, the school still lacked electricity, and each of the two rooms was heated by a large wood stove which the older boys kept fired.

Despite her initial misgivings, Louise Baldasaro taught in this school for three years before her marriage to Charles M. Bergeron of Harrisville. She recalls that her salary the last year was one thousand dollars. Concerning the atmosphere of the 1920s in Harrisville, she recalls no political pressure on her teaching. At the same time she remembers that teachers like herself were liable to considerable scrutiny on personal matters such as dress and hair styles. Harrisville had a rough equilibrium between Protestants and Catholics that inhibited the discrimination and prejudice so rampant elsewhere. The school board steered a safe middle course by an unstated policy of hiring one Protestant and one Catholic teacher for the village schools. Overall, Miss

Baldasaro's teaching experience in Harrisville was a happy one, and she recalls it with the fondness of a good teacher.[65]

The number of students declined along with the general population, especially in the village and East Harrisville schools. Chesham did not change much. But this trend did not mean small classes. Except that the village school was divided into a primary school of grades one through four and a grammar school of grades five through eight, the town's schools were all ungraded. Louise Baldasaro, on her first day of teaching, faced forty-two students, some of them six feet tall and weighing one hundred and seventy pounds.

Little wonder that discipline, in the twentieth century as in the nineteenth, was still the first order of business. It was also the first test for the teacher, and some failed it ingloriously. Louise Baldasaro was among those who succeeded in maintaining discipline. The arduous years of her own upbringing stood her in good stead. She laid about freely with her heavy ruler and kept order.

After her marriage, the now Mrs. Bergeron retired from teaching and ran the town telephone office in her home until the system was changed in 1957. Then she returned to teaching in Nelson and Harrisville. When asked to compare the school-children of the twenties and fifties, she expressed the considered opinion that the children of the earlier period did more for themselves, worked harder, and were more responsive and self-reliant. Children in more recent times, through no fault of their own, were less attentive and less fearful of the teacher, more restless and more talkative.[66]

On being asked the greatest difference between teaching in Harrisville in the twenties and more recent years, Mrs. Bergeron thought it was in the materials of teaching. They were now "more concrete," generally better, and of much greater variety. And, indeed, her view is confirmed by the annual reports of the school board and the superintendent. Year after year they pointed out the need for better materials. The superintendent's report for 1910 listed the textbooks in use and added that "we are below the minimum requirements, in all grades for reading."[67]

In their annual reports the superintendents were wont to give more attention to the condition of the plumbing than to the curriculum, so there is limited information on the latter subject. In 1911, the superintendent emphasized the importance of learning to read in the first three grades and also mentioned the teaching of arithmetic, geography, grammar, writing, history, and physiology. In 1926, the superintendent reported of the

65 Interview with Mrs. Charles M. Bergeron, Feb. 19, 1966.

66 *Ibid.*

67 *Ibid.*; *HAR*, 1911.

Eastview School that "not only is efficient work being done in the fundamentals viz. reading, spelling, numbers, geography, history and English, but also commendable work in music and drawing. Through the efforts of teacher and scholars a Victrola has been obtained for the school."[68] In the thirties the superintendent said little about the actual work of the schools; he dwelt instead on their goals and objectives, usually in grand but empty phrases. "The boys and girls of today are looking not only onward but upward." Nonetheless, despite such pomposities, the evidence is overwhelming that the children of Harrisville were receiving a far better education by the end of the 1930s than they ever had before in the town's history.

In contrast to the almost steady improvement of the schools, the closely related history of the town library shows some interesting variations. It will be remembered that the library had been established at the end of the 1870s over strenuous opposition from the rural, old-stock residents. As if to prove them right, not much more than ten years later the use of the library began to decline. It is possible to check the number of books issued per capita in the decennial years. In each of those years from 1880 to 1930 the library issued fewer than four books per capita. The figures also indicate that the nadir of the library's usefulness came in the last decade of the nineteenth century and the first decade of the twentieth. The nineties are a puzzle. The mid-nineties were depression years, when one might expect people to have read more rather than less. Yet in 1895 the use of the library seems to have been as low as it has ever been. Fewer than a third of the townspeople patronized the library, and they on the average took out fewer than seven books each during the year. The poor use in the first decade of this century is more explicable. Then it was that Harrisville suffered its sharpest loss of population, while those coming into the town were principally the Finnish immigrants, not inclined by education or temperament to spend their time browsing in the library. In 1905, the library issued a mere 798 books.[69]

The dull picture indicated by the circulation figures is confirmed by the amount of money the town spent on books. In 1895, the state passed a law entitled "An Act to Aid Public Libraries," which directed selectmen to make an annual assessment for the purchase of books, the amount to be based on the apportionment of taxes. For Harrisville, in 1895, it meant the sum of fifty dollars. In the years that followed, down to the 1930s, it appears that the town spent on the library just what the state law required and no more. The amount remained roughly the same

68 *HAR*, 1926.
69 See Appendix 16.

down through the 1920s, but the inflation of the World War I period meant that in terms of real money the per capita expenditure reached low ebb about 1920.[70]

It was the decade of the 1930s that saw the first real improvement in the library. To some extent this was due to the enforced leisure of the depression years, the greater role of the state in local affairs, and the generally greater "maturity" of the town. But probably the most important single factor was the better education that Harrisville's children had been receiving. Statistics are lacking, but one resident long interested in the library recalls that it was the young people, those under twenty, who were responsible for the increased use of the library.[71] Whatever the reasons, there *was* improvement. In that decade both the per capita and total circulation of books more than doubled. Comparatively, a 1934 report, based on the findings of the New Hampshire Library Commission, came up with figures that, though higher than this writer can corroborate, show that the circulation of library books in Harrisville was distinctly above the average in the state, including those for neighboring Dublin and Nelson.[72]

The next year, 1935, saw developments to delight all those in Harrisville who read books. Through the use of W.P.A. funds, the library was remodeled and its size doubled.[73] The state's Public Library Commission extended invaluable assistance. One of its librarians came to Harrisville for several days, worked through the collection, and discarded all the library's "worthless" books. This weeding out reduced the collection by nearly half! The remainder were classified and catalogued, and at its summer school the commission instructed the Harrisville librarian in the classifying and cataloguing of new books. Finally, the commission also helped to fill the empty shelves by sending to the town selected lots of books on a rotating basis.[74]

The spirit of change and progress in the 1930s brought other improvements in the use of the library. Deposit stations were established in various parts of the town. The library's hours were extended. During the winter there was a story-telling program for the children, which resulted in their reading more books. All these improvements caused a far greater use of the library than it had enjoyed in its fifty-five-year history.[75] Both the per capita use and expenditure continued to rise in the years that followed. For all its ups and downs, the town library ended this period in a flourishing state.[76]

While the public records permit a relatively detailed history of such institutions as the schools and library, the records of

70 See Appendix 16.
71 Interview with Guy Thayer, Aug. 30, 1966.
72 Howarth, *op. cit.*, pp. 21–22.
73 *NHS*, Feb. 5, 1936.
74 *HAR*, 1936.
75 *HAR*, 1936.
76 See Appendix 16.

private organizations, like the churches, are apt to be few and far between. Such evidence as does exist of the town's religious life holds few surprises. Apparently, the fate of the churches conformed to the general trends of Harrisville's recent history; that is, they reached low ebb in the early decades of the century and showed some improvement by 1940.

The Baptist Church in Chesham, the oldest one in the town, showed the effects of continuing decline and secularization. Early in the century the *Sentinel* noted, as evidence of "the growing liberal tendency in religion," that the Baptist minister in Chesham had umpired a baseball game.[77] The next year the *Sentinel* carried an article about an evangelist who in a series of meetings in Chesham made an attack upon secret societies, "being especially bitter towards the Grange." However, continued the paper, "We hardly think the effect of the fulmination of this fellow will be visible to the naked eye."[78] Considering the local popularity of the Grange, the verdict of the newspaper was probably that of the people of Chesham. Then, in 1909, after one of its periodic renovations and rededications, the church moved to broaden its attractions further. The "free circulating library" of Chesham was removed from the home of Mrs. George F. Bemis to the church building. There it was housed in the "church parlor," which was set aside as a "reading and recreation room for the young men of the village."[79] These liberal and secular trends were not new but continuations of tendencies that dated back well into the nineteenth century. Whether such measures did anything to check the declining membership or simply reflected the declining vitality of a "covenanted community" is an interesting but unanswerable question.[80]

The Chesham church did have certain advantages that kept its decline from being more precipitous than it was. It was located in a rural area containing mostly Protestant families long settled there. There were two leading families who alternately (and with considerable rivalry) ran things. The state Baptist Convention lent assistance.[81] Nonetheless, the church found it could no longer maintain its traditional independence. By the end of the 1930s, it was sharing its minister with the Harrisville Congregational Church with services alternating between the two places.[82]

Faring much worse was Harrisville's Congregational Church. When the century began, it was suffering from a decline that dated back to the revival of 1873. It was, in fact, moribund. Mr. Guy Thayer, whose family came to Harrisville in 1906, recalls that often on a Sunday his was the only family in church. Even

77 *KES*, June 29, 1903; *NHS*, July 1, 1903.

78 *NHS*, March 2, 1904.

79 *NHS*, Nov. 3, Dec. 29, 1909.

80 Cf. Smith, *op. cit.*, ch. 4.

81 Interview with Guy Thayer, Aug. 30, 1966.

82 *NHS*, June 18, 1941.

in the 1920s, he does not believe the formal membership exceeded six or eight people.[83]

The arrangement with the "mother church" in Nelson, whereby the two churches shared in the ministry and support of one pastor, continued until 1909.[84] Thereafter, the ministers at Dublin and Marlboro were engaged to "supply." Altogether, between 1901 and 1916, Harrisville had the occasional services of about ten different ministers.[85] By the beginning of the 1930s, the only services were those held by student ministers in the summer months. Toward the end of that decade the community church idea was begun throughout New England to help just such foundering churches as the one in Harrisville. Here, at least, it did not work. Dublin eventually went its own way, and Nelson, which had some special funds, refused to join the common treasury. That left only the union with the Chesham church already mentioned.[86]

There were from time to time some organizations within the church. A Sunday School existed and had about a dozen members in 1912. In 1925, there was a Missionary Society and a Ladies Aid Club.[87] But it may be assumed that over the years these organizations had no more vitality than the church itself.

Money was always a problem. There were no large funds to carry the church over lean times. The state conference paid for some of the supply ministers. A few of the summer people, particularly Mrs. Arthur Childs, made occasional contributions. Maintenance was done by the members. By the early thirties the condition of the church was so bad that the members gave the state conference a "revisionary deed" to the property. A state law provided that under certain conditions of disuse church property could be confiscated by the state, and this revisionary deed was designed to prevent that from happening. It was, then, an action taken in anticipation of the end of the Harrisville church, a kind of legal extreme unction.[88]

It was close, but the worst did not happen. The population decline leveled off in the 1930s, and some new families came to town. Perhaps the Crash and the lean years that followed led to some soul searching. Perhaps it was the gathering war clouds or just the enforced leisure. Whatever the reason, as the thirties ended there was some slight improvement in the condition of the church. The financial crisis eased, and formal membership rose to about eighteen. That was about one third of the number who had founded the church just a century before.

Why this long period of stagnation and impotence? Most commonly, the difficulty has been attributed to the preponderance

83 Interview with Guy Thayer, Aug. 30, 1966.

84 Bemis MSS, Box 6, XXXII, p. 4, "The Church in Harrisville."

85 Leonard, 1920, p. 341.

86 Interview with Guy Thayer, Aug. 30, 1966.

87 *NHS*, Oct. 21, Nov. 18, 1925.

88 Interview with Guy Thayer, Aug. 30, 1966.

of Roman Catholics in town. Whether there were more Catholics than Protestants, the explanation is inadequate for the twentieth century as well as for the nineteenth.[89] The causes of the condition were more complex: the mixed, clashing cultures in town, the lack of strong social institutions in general, the absence of leading church-going families after the departure of the Harrises, and the decline of the town in the late nineteenth and early twentieth centuries. There just were very few who cared.

Saint Denis's Roman Catholic Church fared better than either of the Protestant churches. A mission church had been built in Harrisville in 1894, a permanent priest installed in 1903, and a parish established in 1906.[90] Its records appear to be little better than those of the languishing Protestant churches, so its history cannot be much more detailed. There may have been as many as three hundred Catholics in the town in 1900. At present the parish has between sixty and seventy families, many of whom live in Dublin. The parishioners have always been fairly well divided between Irish and French Canadians. This is hardly a statistical survey, but it is about all that can be said.[91]

In the seven decades since the church was built, there have been eight resident priests plus a few substitutes. For the most part the earlier ones were Irish and the more recent ones French.[92] There have never been any assistants, nor has there been much in the way of parish organizations.[93]

Undoubtedly, the better organization and discipline of the Catholic Church, the greater number of parishioners, and the longer tenure of the pastors all helped to strengthen Saint Denis's Church and give it greater success than that enjoyed by the Protestant churches. At the same time, it too had its handicaps. There was rivalry and antipathy between the Irish and the French Canadians, many of the parishioners were poor and transient mill workers, and the decline of the town could only have adversely affected the church. In short, the flourishing of the Catholic Church in Harrisville was more relative than absolute.

The latter part of the nineteenth century had seen the decline of religious activity in Harrisville and the increase of secular organizations and diversions. In its social activities and means of recreation, there were few differences between those just before and just after 1900, except that the latter years were quieter. There were few active groups or organizations, and most of those that have left any record of themselves were associated not with the mill village but with the older and more cohesive farming village of Chesham. A mothers' club and a baseball team had

89 There are no figures for the number of Roman Catholics and Protestants in town, but the number of foreign-born and those of foreign parentage totaled 43 percent of the population in 1880 and 51 percent in 1930. *Cf.* Appendix 6; Howarth, *op. cit.*, pp. 21–22.

90 *KES*, Jan. 14, 1903; *Historical Records Survey, Guide to Depositories of Manuscript Collections in U.S.—New Hampshire.*

91 Interview with Charles M. Bergeron, Nov. 10, 1962.

92 They have been Frs. Magill, Cotter, Belford, Driscoll, Vaccarest, Allison, Belanger, and Trottier.

93 Interview with Charles M. Bergeron, Nov. 10, 1962.

some kind of existence, and the Grange, long established among the farmers of the area, was very active and held many social affairs.[94] The Finns in Harrisville held some functions of their own, such as weddings, "socials," or, most commonly, church services.[95] But, as a group, the Finns were too new and too hard-working to contribute much to the town's social life.

Most notably lacking in the town's activities was anything even remotely cultural. Though the mill village had never been strong on self-improvement, throughout the nineteenth century a small minority had made efforts to raise the sights of their fellow townspeople. But in the twentieth century, and especially its first three decades, these efforts seem to have all but disappeared. Perhaps in extenuation it could be argued that because of the twentieth-century improvements in communication and trans-portation these local efforts were no longer as important as they had been.

From the turn of the century to the First World War, the town contented itself mainly with private parties, traveling troupes of entertainers, dances, and the like. A typical example of the first was a party for Mr. and Mrs. Michael Kennedy, who were returning to Ireland for the winter after twenty-five years,[96] or Chesham's celebration of Independence Day, 1914. Of this, the Sentinel reported, "All the people of the place were invited to the home of Dexter Dawes Saturday evening to witness a fine display of fireworks. After the fireworks, all were treated to refreshments of ice-cream and wafers."[97]

Occasionally, a traveling troupe of entertainers stopped in Harrisville. Shipman's Uncle Tom's Cabin Company played there in 1907. More frequently, there were traveling medicine shows, which included vaudeville acts and dances at Eagle Hall as part of the come-on.[98]

There were also dances provided by the Cheshire Mills in their own "Hall" or by local groups at Eagle Hall. Curiously, there was quite a spate of these in 1902. Coming in that year when the great exodus began, could they have been an attempt by the Cheshire Mills management to make life in Harrisville more attractive?[99] These dances, which were so popular, were rough affairs. There was always drinking, and there were usually fights. It was at such an affair that Edwin Davis was knifed in 1895, and knives were drawn at other dances that did not gain the attention of the newspapers.[100]

Some improvement in the possibilities for recreation came with the prosperous days of the First World War. Neighboring Nelson had commenced holding annual town picnics about 1878, and

94 *NHS*, July 27, Aug. 3, 1904, Jan. 29, 1908.

95 *KES*, March 24, 25, 1903, July 11, Aug. 22, 1906.

96 *KES*, Nov. 5, 1902.

97 *NHS*, July 8, 1914, B. F. B.

98 *NHS*, June 27, 1904, Feb. 26, 1908.

99 *NHS*, Feb. 19, May 7, 1902, Jan. 21, 1903; *KES*, July 29, Sept. 17, 1902.

100 Interview with John Clark, Nov. 11, 1962.

Old Home Week was officially established by the state in 1899, but, significantly, Harrisville did not officially observe the occasion until 1917.[101] Perhaps the town was finally spurred to activity by Nelson's gala sesquicentennial celebration that same summer. The *Sentinel* described the day:

> The town of Harrisville gave formal recognition to the Old Home Week season for the first time this year . . . August 22. The Marlborough Band gave a concert in the forenoon, followed by athletic events consisting of 100 yards dash, 75 yards dash, girls' race, three legged race, potato race, and a tug-of-war between the married and single men. . . . A basket picnic was enjoyed.
>
> A programme was carried out under the direction of Arthur E. Wright, chairman. Prayer was offered by Rev. Arthur E. Wetherbee, an address of welcome was given by P. W. Russell which was responded to by Rev. Ellsworth Phillips. Rev. Arthur E. Gregg gave an historical sketch, and the principal address was by Arthur L. Livermore. William Nye, the eminent Keene basso rendered several solos which were heartily applauded.
>
> At 3 p.m. a ball game between the Marienfeld Camp boys and a Harrisville team was the attraction. . . . (Score 4 to 3) Some of the residences in town were decorated with flags. . . . In the evening there was a . . . dance at Eagle Hall, the hall being filled to its capacity. . . . Fully 1,500 people from in and out of town were present during the day.[102]

After some debate in town on the subject, it was decided to celebrate the occasion again the next year.[103] The 1919 celebration included a welcome home to returning veterans and dedication of a tablet containing an honor roll of the names of Harrisville's forty sons who served in the Great War. Thereafter, the annual Old Home Day was usually, but not always, celebrated. It is a pleasant occasion, and the form of it has varied little in the half century since its inception.[104]

The flush times surrounding the war also brought several new organizations into being, including a Y.M.C.A., a veterans' post, and a "Cheery Chesham Canning Club."[105] But this burst of activity did not continue in the next decade. The twenties did not "roar" in Harrisville. Business was poor, and the population continued to decline. If the opportunities for recreation and cultural activities were better than before the war, and it would seem that they were, two factors were responsible: the better schools the town now enjoyed and the "quickening influences" felt throughout northern New England, such as the better communication and transportation.[106]

101 Cf. State Association, *Annual Reports of Old Home Week in New Hampshire, 1899–1915, passim.*

102 *Nineteenth Annual Old Home Week in New Hampshire, 1917. Report of the State Association,* pp. 45–46.

103 *NHS,* July 24, Aug. 7, 28, 1918.

104 Cf. State Association, *Annual Reports of Old Home Week in New Hampshire, 1899–1914, passim.*

105 *NHS,* March 12, July 23, Dec. 10, 1919.

106 In 1924, a feature motion picture was shown every Saturday night at Eagle Hall. The first was "Bulldog Drummond." *NHS,* Jan. 23, 1924.

It is an ill wind that blows no good. The 1930s saw a revitalization in Harrisville, and it had a certain connection with the circumstances of the Great Depression. With mass unemployment everywhere, the town may well have retained some talent that it would have lost in prosperous years. Also, there was a movement of the population from urban to rural areas, from which Harrisville may have benefited. Certainly, the population decline leveled off. The enforced leisure of people who had received a better education than, say, those living in Harrisville in the 1890s must also have had a bearing on this revitalization.

More fundamental, probably, was the effect of adversity itself. Martin Buber has written that the true community is "a community of tribulation, and only because of that is it a community of spirit. . . ."[107] For a considerable part of its sixty-year history before the Great Depression, Harrisville had known tribulation. The town had been born of the struggle between the farmers of Dublin and Nelson and the industrial population of the factory village. There followed the depression years of the 1870s and 1890s, the failure of the Harris mills, the clashes of ethnic groups, the disastrous fire that destroyed the New Mill, the disenchantment with the railroad, the exodus of the early twentieth century, and the hard times of the 1920s. But the town *had* endured and, by doing so, had gained the time so necessary for the development of social institutions and community spirit. By the 1930s, this endurance of adversity was beginning to yield some benefits.

Some examples of this new vitality have already been described. Old Home Day in 1933 was "one of Harrisville's most successful," with nearly a thousand people attending.[108] The schools made real progress, the library was rebuilt, and even the Harrisville church showed signs of shaking off its malaise.

New organizations sprouted from old stumps. Among them were a new dramatic club, a Y.M.C.A., and for the boys of Chesham a "Friendly Indian Club."[109] A boy scout troop was organized in the village, and later in the decade there were efforts to build a playground and a town beach.[110]

Sports, too, assumed a new importance. In 1932, the town voted to allow "Sunday Sports."[111] In winter, basketball was now popular, and in summer the town fielded a baseball team.[112] During the winter of 1932, there were horse races, that is to say, "a little friendly competition between local horses."[113] Utilizing the town's natural assets for recreation in an organized way, Dublin, Nelson, and Harrisville planned a winter carnival, which featured snowshoeing, skiing, and ice hockey, respectively, in the three towns on three successive days. The "Harrisville

107 Quoted in Smith, *op. cit.*, p. 8.

108 State Association, *Annual Report of Old Home Week in N.H., 1934.*

109 *NHS*, Oct. 19, Nov. 23, 1932, March 15, 1933.

110 *NHS*, March 23, April 6, June 1, 8, 29, 1932, July 14, 21, Aug. 4, 1937.

111 *NHS*, March 16, 1932.

112 *NHS*, Aug. 3, 1932.

113 *NHS*, Feb. 10, March 16, 1932.

winter sports team carried away four awards in the skiing competition held at Tolman Pond, Nelson, . . . although none of the entrants . . . had ever tried the events in which they had made such an admirable showing."[114] This winter carnival was enough of a success for Harrisville to organize at least one more in the thirties, and undoubtedly these efforts indicate that there was also a growing popularity for winter sports on a less organized basis.[115]

At the same time, there were some efforts to interest the town's citizenry in the serious consideration of the international scene and the problem of world peace. Lectures were given on the situation in Europe and the League of Nations, and "a temporary committee was formed."[116] Obviously, Harrisville was not converted to internationalism, but *any* organized interest in the question was quite a new note in the town's activities.

This revitalization should not be overstated. Many of these new sprouts soon withered and died. The boy scout troop failed after a year or so from lack of funds and lack of an effective troop committee.[117] The efforts to build a town beach came to naught in the thirties because of failures of leadership and the physical destruction wrought by the hurricane of 1938.[118] And, though the international situation steadily worsened after 1933, there is no further evidence of organized concern in town. It all came down to the paucity of people and resources to sustain the hopeful beginnings. A 1936 study of New Hampshire towns included Harrisville among those described as "Rural, Semi-Recreational Type—Declining."[119]

But "declining" cannot be the last word. The same 1936 study pointed out that one of the main features in the evolution of the towns of New Hampshire had been the varying degrees to which the original settlers or their descendants succeeded in their attempts to use the land and that another feature was the degree to which, in case of a first failure, an adjustment to some other means of livelihood had been accomplished. Harrisville, in rather more than a century and a half, had turned from farming, to manufacturing, to summer residents. At the end of this period its economy still rested on all three, a fact which undoubtedly helped the town weather the Great Depression as well as it did. And Harrisville's potential for economic growth, its growing social and political maturity, and its richer cultural life were hopeful signs for its future.

114 *NHS*, Jan. 18, March 8, 15, 1933.

115 *NHS*, Feb. 5, 12, 1936.

116 *NHS*, April 19, Oct. 18, 1933.

117 Interview with Guy Thayer, Aug. 30, 1966.

118 *Ibid.*

119 Howarth, *op. cit.*, pp. 21–22.

EPILOGUE
HARRISVILLE, 1940–1969

In the last quarter century Harrisville has continued to experience change, but it has been change within a familiar framework. A central role in the town's economic life is still played by the Cheshire Mills. The reason for the company's continued existence is readily apparent. In the textile industry, because of the need for close supervision of the manufacturing processes, management is unusually important. The Cheshire Mills had just three presidents from the time of founding in 1851 until 1955, when John J. Colony, Sr., died. Since then, the principal officer has been his son, John J. Colony, Jr. He, like his father, began working in the mill during his student days. Following his graduation from Harvard, he returned to the company and soon occupied the position of manager and assistant treasurer. Unlike his father and grandfather, however, when he married he brought his bride to live in Harrisville. It was historically fitting that the young Colonys should make their home in the old brick house built by Cyrus Harris and occupied until 1870 by Henry Colony, the first president of the Cheshire Mills. Living in the mill town inevitably meant more interest in, and attention to, the business and town affairs than any president had had since the 1860s. Other members of the Colony family have always held important posts in the company. John Colony, Jr.'s brother Charles also entered the mill in 1937, worked through the departments, and has held the position of treasurer since 1957. The close supervision of knowledgeable and dedicated owners and superintendents has obviously been a vital factor in the company's ability to survive in a period when so many other woolen companies were failing.

In these years there has been relatively little change in the company's plant. In 1947, the famous rope drive was stopped, and thereafter the mill's entire power needs were supplied by purchased electricity. The pure waters of Goose Brook, though no longer providing power, continued to be used for dyeing and finishing operations. Inside the buildings much new machinery has been installed, and the great stainless steel vats and ovens stand in sharp contrast to their antique surroundings. Within its limited resources the company has sought to maintain the structural soundness and appearance of the mill buildings, most of which have been in use for well over a century.

The Second World War ended the business depression for Harrisville as for the nation. Between 1939 and 1941, orders for civilian goods increased, and then with Pearl Harbor there came a flood of both civilian and military orders. Multiple shifts were begun and then retained on a permanent basis. About one third of the mill's wartime production was composed of military items, principally coat lining flannel, sleeping bag material, and a melton for Russian officers' top coats. Since the late 1930s, civilian production has consisted largely of women's wear. Although there was difficulty in getting new machinery during the war, a careful allocation of materials assured the mill of all necessary raw materials.

Business in the immediate postwar years remained good. The woolen industry began to feel a competitive squeeze in 1948, but before it became serious the Korean War intervened. There was little or no war work for the Cheshire Mills, but it kept busy. One effect of the war was that, by saving a number of marginal mills from going under, it ensured that competition would be severe when the war ended.

The worst business conditions for the Cheshire Mills in recent times came in the decade following 1952. Although many mills folded, the increased capacity of those that survived still caused overproduction, and foreign competition made matters worse. By 1957, the situation in Harrisville was serious. Commission work for other mills kept the plant in operation but brought in little profit.

Nonetheless, in a century of making woolen cloth the Cheshire Mills had survived a half dozen periods of serious business recession, and it also survived this latest one. The wholesale demise of woolen mills throughout the country, including the huge American Woolen Company, finally reached the point where it began to help the surviving mills. Since 1965, orders have increased so that commission work is no longer necessary; the shortage now is labor, and the mill runs multiple shifts to meet orders for its highly styled goods for women's wear. Sales figures for earlier years are missing from the records, but in more recent years they have been as follows:

1936	$253,284.29
1940	566,915.26
1945	1,955,761.98

1950	2,322,659.86
1952	2,551,860.81
1955	2,318,481.49

In labor matters the years since 1940 have brought some changes. The decade of the forties saw the payroll lengthen to approximately 200, probably the largest number the mill has ever employed. Then, in seeking economies during the slump of the fifties, it was found that a number of jobs could be eliminated. Improved machinery also allowed a saving in labor. At present, with good business conditions, the payroll is about 135. With business and labor available, full three-shift operation would probably again produce a payroll of slightly over 200.

The composition of the labor force has also changed: it is older, more largely native-born, and more largely nonresident than it used to be. There is almost no influx of young immigrant labor any more. During the slump of the fifties, the young people in Harrisville went to nearby Keene for work, and they have been slow to return to the now-busy Cheshire Mills. At the same time, the new mobility of the population has allowed the company to draw workers from surrounding towns and, in fact, from Vermont. A few years ago not much more than a third of the work force lived in Harrisville. Very recently, some of the younger townspeople have been going into the mill, but the significance of this trend remains in question.

The relations of this work force with its employer appear to be as good as they have usually been and as satisfactory on both sides as they are likely to be in any factory, given the limitations of circumstances and human nature. The president's small, cluttered office could be taken for the shop foreman's. The president and the treasurer, both in shirtsleeves, are readily found there or else striding through the labyrinthine plant. The friendly greetings they exchange with workers, on a perfectly natural first-name basis, probably indicate as well as anything the state of labor relations.

With some qualifications, the outlook for the Cheshire Mills is promising. Foreign competition is still a problem, but it is one now shared with other industries, including those that are, in the terminology of the economist, "capital intensive," like electronics or chemicals, rather than "labor intensive." The principal problem lies in this last-named characteristic. The textile manufacture is one with a high labor cost. Nonetheless, even this has been helped by the development of much more efficient machinery. The mass production of staple goods, such as flannels, is a thing of the past, but there does seem to be a place in the economy for the small woolen mill that can do highly styled goods under very precise conditions, that is favorably located to give quick service, and that can make rapid changes to suit market conditions. These are all attributes of the Cheshire Mills. (Plates XXXI–XXXV)

In other areas of economic life recent years have seen the continued decline of both commerce and agriculture. The brick general store in the village survives as the only long-established business in town. Recently it has been joined by a small pottery in the old school building and a few repair shops in people's homes. As for farming, the town contains not a single taxable farm animal. The only resident who supports himself by agriculture is Ralph Bemis with his chicken farm in Chesham.

Following a long-term trend in the other direction, the industry based upon the summer residents has greatly increased. The year 1940 was a turning point. Then, for the first time, the nonresidents paid more taxes than the residents. The proportion of nonresident valuation and taxes continued to increase until recently. Between 1940 and 1960, the number of camps, cottages, and summer residences rose from 117 to 211, or 80 percent. Most of these are located on water-front property. There are two "lake associations" in town. They each hold summer meetings, at which town officials make statements and answer questions. A rather plaintive concession by a town officer recently that "you've got to keep them happy," indicates as clearly as the statistics the economic importance of the summer residents.

It is interesting to note, however, that within the last few years there has been a reversal of the trend just described. The proportion of resident property owners is now on the increase, and nonresidents are using their property for a greater part of the year. All the same, in 1967 nonresidents still paid 63 percent of the taxes.

Of town affairs a few trends should be mentioned. Remarkably, the physical changes in Harrisville's appearance have thus far not been great. As in New Hampshire generally, the forest area has increased with the disappearance of farms. In addition to the summer cottages, some new dwellings have been built on the fringes of the village, and some summer residences have been adapted to year-round use. Prosperity has allowed the renovation of some structures, and there is little outright dilapidation. Much to be regretted, however, is the loss of many stately elms due to disease, particularly since there has been no systematic replacement with other trees.

At present there is little real estate available in or around the village. Some of the shoreline of the town's ponds remains undeveloped, and there are still large tracts of unoccupied land in town. A zoning ordinance, in effect for several years, has undoubtedly helped the town to maintain its attractive appearance. Now, at this point in its history, the town is considering the establishment of a historic district to protect its unique factory village. Harrisville is one of the first towns in the state to take such action. That a historic district is even under consideration says much about the community. (Plates XXXVI–XL)

Harrisville's roads are no longer the subject of great controversy, and as the town's principal expense they have given way to the schools. The highway expenditure in 1960 constituted a smaller proportion of the town budget than it had in any decennial year since 1890. A subject of some debate, in Harrisville as in neighboring towns, is the black-topping of dirt roads. At present, twenty-six of Harrisville's forty-four miles of road are paved. As state aid is available for paving and such roads require little maintenance, there is an economic argument for paving. For those who are there to escape from urban ugliness and noise, however, paving spells more and faster traffic, with its attendant loss of beauty and quiet.

Periodic summaries of property valuations, the town inventory, and the town budget, for these years as for the earlier ones, are all to be found in the appendices. During the years when the Cheshire Mills was in a precarious position, the town was especially careful to incur no long-term debt. The town is now, as it was in 1940, operating in the black. The median valuation of resident property owners in 1941 fell in the category of $1,000 to $1,999. In 1960, the median was between $2,000 and $2,999. In terms of constant dollars it would appear that the total valuation of the town in 1960 was little more than it was in 1941 and perhaps slightly less than it was in 1900. Any analysis of valuations is complicated by the question of their relation to real market value. In theory they should be the same; in actuality they are not. In recent years Harrisville's valuations have been placed at 37 percent of real market value by the state tax commission. But since it is very difficult to determine what the ratio was in earlier years, no conclusions can be drawn about long-range trends in property values.

The three largest expenditures in the town budget continue to be schools, highways, and payments on the town's borrowings in anticipation of taxes. Reversing an earlier trend in favor of highways, recent budgets show that in Harrisville, as in so many communities, the largest item is schooling. In the last few years this expense has accounted for slightly more than half of the town's total expenditures. To put this another way, it appears that in terms of constant dollars the total budget increased about 57 percent from 1940 to 1965, while in the same period the school budget increased over 300 percent. Again, in terms of constant dollars the total expenditure for all items except schools was actually slightly less in 1965 than it was in 1940.

As to political affairs it can be said that leadership in local government still rests with the old-stock families and those of French–Canadian descent. Thus, Ralph Bemis, whose grandfather was chief selectman for twenty years and whose father was selectman for fifty years, has himself been town moderator for thirty-four years. And Warren Thayer, office manager at the Cheshire Mills, has completed twelve years of active and effective service as selectman. On the other hand, the Irish have not been numerous in recent years, and their influence has waned with the gradual disappearance of the Winn family. The immigrant Finns did not take to political activity as the other national groups did, and with few exceptions they still do not. Despite this pattern, it is still true that, in this century anyway, anyone wanting to hold public office can do so.

Three other characteristics in local politics are apparent. Nationality is still more important than political party (although Finns still tend to vote Republican and French Canadians Democratic), but since the Second World War national antagonisms have tended to lessen. Secondly, since fewer voters work for the Cheshire Mills than formerly, they tend in town meeting to be more independent, outspoken, and critical of the company and its role in town affairs. Finally, in recent years only about half of the registered voters in town attend and cast their votes in town meetings.

In national politics, also, there have been some new trends. Perhaps because there are now fewer immigrants in town and the Winns are no longer a political force, the town goes Democratic less frequently in presidential elections than it did before the Second World War. Until the 1890s, the town usually went Democratic. From that time until World War II, it was both remarkably independent and remarkably true to national sentiment: in the twelve presidential elections from 1896 to 1940, it was "wrong" only once. Harrisville is still much more independent-minded in national politics than either Nelson or Dublin, but it failed to go the way the nation did twice in the last seven elections.

Getting out the vote for national elections has never been a problem in these hill towns of Cheshire County. Sometimes the percentage of registered voters who cast their ballots exceeds 90 percent. In Harrisville the percentage of voters who cast their ballots for president has been slightly lower since World War II than it was before, but even in these last twenty years it has averaged 84 percent. The town's voters are not strongly partisan but, in ways somewhat milder than those of the last century, they still show a lively interest in politics.

More difficult to document than economic and political trends are those that are social or cultural. Nonetheless, what was observable of the town in 1940 can still be maintained: that through the process of time and tribulation, Harrisville has come to be more of a community than ever before in its history.

Not much information about Harrisville's population can be found in the federal censuses of recent years, for the schedules themselves are not open to inspection, and the published reports say little about places with fewer than one thousand people. The long decline in population that began with the century had leveled off by 1940. The prosperity of the forties was reflected in a slight increase in the 1950 population figure. By the same token the slack years for the Cheshire Mills in the fifties corresponded with the town's loss of 11 percent of its population in that decade. Because of current business activity, among other factors, the 1970 Census may show some increase for Harrisville. It is interesting to note that both its older neighbors, Nelson and Dublin, and particularly the latter, have increased in population since their low tide in the period 1920 to 1930.

Following a pattern of a century and a half, it is still the young people who are leaving these towns. Harrisville sent 73 young men and women into service in World War II. For a town of 500 people, that was a respectable delegation. Four of the men were killed in action. Of the returning 69, the list of residents in 1949 showed only 28, or 40 percent, as still living in Harrisville.

The trend in the summer population is quite different. Concrete statistics are lacking, but it seems safe to assume that the number of summer residents must have equaled the number of permanent residents in 1940 and must now be double their number. It is also observable that whereas before approximately 1960 most of these summer people were from Massachusetts, in more recent years they are coming in increasing numbers from New York, New Jersey, and the other Middle Atlantic states. Since a number of the "summer" people return to their cottages for winter sports, while some of the older residents of Harrisville flee to

Florida to escape what the visitors seek, it would seem that the differences between the two categories of population are breaking down.

Apart from the quantitative dimensions, the vital statistics of the town indicate that its residents are older now than the national average and older than was the town's population earlier in its history. The local birthrate in the twentieth century has been generally lower than the national average. Likewise, the crude death rate of the nation has steadily declined, but in Harrisville, since 1940, it has risen to a new high. The explanation is that while young natives of the town have continued to leave, skilled workers at the mill have remained, and some of the summer people have retired to Harrisville.

Harrisville's schools had begun to show improvement in the ten or fifteen years before 1940. So much did this trend accelerate after the war that the improvement of the schools could be said to be the greatest change and most encouraging development in the town's recent history. Most notable in this area have been the vastly increased financial support for education, the construction of a fine new school, and, recently, a significant increase in the number of students.

The town's greatly increased financial support for education since World War II is an experience shared with countless American communities, but it has been unique in terms of Harrisville's own history. The real expenditure per student (in terms of constant dollars) rose about 250 percent in the twenty years after 1940, which was more than it had increased in the fifty years before that date.

At the end of the Second World War, the two school buildings in town were both ancient. The two-story frame school high up the southern slope of the village had been built in 1857, following the bitter public debate over the quality of education in the mill village. At that time it undoubtedly represented an improvement over the scandalous condition of the school held in the vestry, but by 1945 probably even Milan Harris would have agreed that it was obsolete. The other school still operating was the one in Chesham. It housed eight grades in one room and had no running water. It had been remodeled so often that it would be difficult to say just when it was built, but it probably was in no better shape than the Harrisville school. For toilet facilities both schools had outdoor privies.

The improvement that came after the war reflected not only the greater affluence of the town but also a new interest in education. First of all, the two upper grades at Chesham were transported to the larger Harrisville school. Then in 1945 the town decided to pay the transportation costs of students attending high school. This decision represented a great advance for the town's children, the poorest of whom had had in the past no way of continuing their education beyond the eighth-grade level of the local schools.

At this juncture, Mr. Wellington Wells, a long-time summer resident who made his home in Chesham, offered to contribute to the construction of a new school. His eventual insistence that the school be built in Chesham was a hard pill for the Harrisville people to swallow, for they had already picked a good site near the mill village, but there was no refusing the money offered by Mr. Wells. As it turned out, he contributed slightly more than half of the total cost. This, plus contributions by the Cheshire Mills and several individuals, plus a bond issue voted by the town, allowed it to build a spacious one-story brick school complete with auditorium, cafeteria, and library, described when it opened early in 1950 as one of the most modern schools in New Hampshire.

The Wells Memorial School has been the pride and joy of the town ever since. It has been well maintained, and vandalism has been minimal. As visible evidence of the importance the towns-people assign to the matter of education, it undoubtedly has inspired civic pride in Harrisville and especially in Chesham. Despite the presently expanding curriculum, the building is still amply adequate. The building, and particularly the auditorium, serve other groups in town. Soon after its completion, town meeting was moved to the school from aging Eagle Hall, which thus helped the town to avoid once more the question of building a town hall.

The new school did not end the educational problems, but it allowed the town to confront them with some possibility of a satisfactory solution. The school population, after an irregular decline dating from the twenties, began to increase. By 1955, the growth was significant, and by 1960 it was a problem. Four teachers, each with two grades, had a total of ninety-two students, a number that exceeded the planned capacity of the building.

With no relief in sight from the increasing numbers, the superintendent in 1959 warned the town that it must and could meet the problem. He pointed out that of the 230 school districts in New Hampshire, Harrisville was twenty-fourth in taxable wealth per pupil, having $31,000 of taxable property per pupil compared to a state median of $17,000.

In what fashion the town was to meet the problem was

something else. In the early sixties the town voted strongly against joining a consolidated school. Instead, it has reduced the numbers in the Wells School by sending the seventh- and eighth-grade students to school in Keene, just as it had long been sending its high school students there. The elementary students remaining at the Wells School still receive a superior educational program with special instruction in a variety of subjects. The salaries of its teachers were 80 percent of the national average in 1965, the highest they had been for fifty years.

All of these details add up to a transformation of education, one that signifies a new civic interest and pride in the town's school. There is good communication between parents and teachers. Undoubtedly, this community spirit played a part in the decision against joining the consolidated school. And the same spirit is to be seen in the revived activity of the school board. Capable, concerned townspeople, particularly several members of the Bemis family and Mrs. John J. Colony, Jr., have worked to bring about the improvements described. Change is so unremitting in modern America that there is no place for complacency, but the fashion in which they have met the challenge of education in the atomic era certainly entitles the people of Harrisville to a degree of satisfaction.

Since 1959, there has been one other school in town, located on what was formerly the hilltop estate of Arthur Childs. The Thomas More School is a private boys' secondary boarding school conducted by lay Catholics. It now has a faculty of twelve and a student body of sixty-five. There are a number of links between "town and gown," such as the availability of the school's new gymnasium for the town's children for sports activities.

Whether it has been because so much of what might be called the "cultural energy" of the townspeople has gone into the improvement of the schools, because of the greater ease of going to places like Keene or Peterborough, or simply because of the chronic paucity of people and resources, there has been little in the way of organized social or cultural activity. A men's club, formed and active through the fifties, was a casualty of the Cheshire Mills' slump in the fifties. Nonetheless, focal points for social activity are provided by Old Home Day, the town beach, the annual children's Christmas party at Thomas More School, the Fire Department, and the numerous other civic causes.

One of these, the town library, has continued its history of fat years and lean according to the varying concerns of the townspeople. Following the "renaissance" of the thirties, there was a long period of decline. The library report for 1950 showed a sharp drop in circulation from the high of 1940. By 1955, the library's circulation was as small as it had been during the low ebb of cultural life at the beginning of the century. Thereafter, the town simply ceased publishing the circulation figures in its annual reports.

In the last few years, however, there has once again been a change in the library's fortunes. In 1965, the library became affiliated with the state-wide library development program, which provides book funds and consulting services. The collection has been culled and catalogued, the library hours increased, and, as public interest has grown, the circulation figures have risen sharply. In 1968, they reached what must be a town record of 5,570 volumes. Very recently, the town has acquired the brick vestry as a new home for the library and a historical center. Given the funds to renovate the structure, the town should soon have a handsome home for its library as well as an appropriate new use for its vestry. Whether the new interest in the library represents a belated benefit of prolonged prosperity (common enough in university life), a consequence of improved education, or simply a turning away from child-rearing and television still lies in the realm of speculation.

The churches have perhaps better maintained the modest improvement they showed in the thirties. St. Denis's Roman Catholic Church, and its resident pastor, the Reverend Maurice E. Trottier, continue to serve parishioners in Harrisville and Dublin, as well as summer visitors. The church in Chesham, a member of the American Baptist Convention, and the church in the mill village, formerly Congregational and now of the United Church of Christ, plus the church in Nelson, have all shared the services of the same minister, the Reverend Mary B. Upton, of Chesham, for the past eleven years. This sharing of a common minister both indicates and fosters the spirit of cooperation now prevailing between Harrisville's two Protestant churches. Each has its complement of auxiliary organizations. There is little factionalism, and financial support is no longer precarious. Both churches were for a time in disrepair, the Congregational Church so badly so that the vestry was used for services for about seven years. Both churches have since then been restored and are now in good condition. For the Church on Earth there is always ample room for improvement, and the Harrisville churches are certainly not immune to the present and developing stresses in organized religion. Nonetheless, it can be said that these churches have more popular support than they had for long periods in the past, and also that there is a better spirit of cooperation and understanding among them.

Finally, what can be said of Harrisville's prospect and its historical significance? The town was born of change when the industrial revolution first swept over New England, and change confronts it still. A superhighway will soon slice through the outskirts of the village, following the line of the late-lamented railroad. The press of population and taxation may bring a revaluation of property with important social and economic effects. A new immigration could be the answer to the labor needs of the Cheshire Mills and would certainly have no less impact on the town than previous immigrations have had.

Obviously, much of the prospect for the town hangs on the prospect for the nation and mankind. If not overwhelmed by external developments, Harrisville may be expected to continue its pattern of moderate change within a familiar framework. The economic balance between two established industries, the increased sense of community, the responsiveness to the demands of modern life, the recognition of its resources and heritage and the determination to use them wisely, all these elements can serve Harrisville well as it enters upon the second century of its corporate life and the third century of its existence.

Central to both its prospect and its import is the woolen mill, standing symbolically where other New England towns have a common. The survival of the Cheshire Mills is of more than economic significance. Should the mill go under, the town would remain, but it would be at best another New England summer resort. It was the woolen mill of "B. Harris and Company" that fathered this community, and it is the same manufacture that gives the town its identity today.

The significance of this small community lies in its history of adapting to changing circumstances without destroying its past. It is a living reminder of industrial America in its sanguine youth. Most of all, Harrisville is significant as a symbol of the industrial revolution in its mellowness. This mill town with its patina affirms that under certain conditions the machine is man's servant, not his master, and that its end product can be, not the millennium, but comfort, order, community, and beauty.

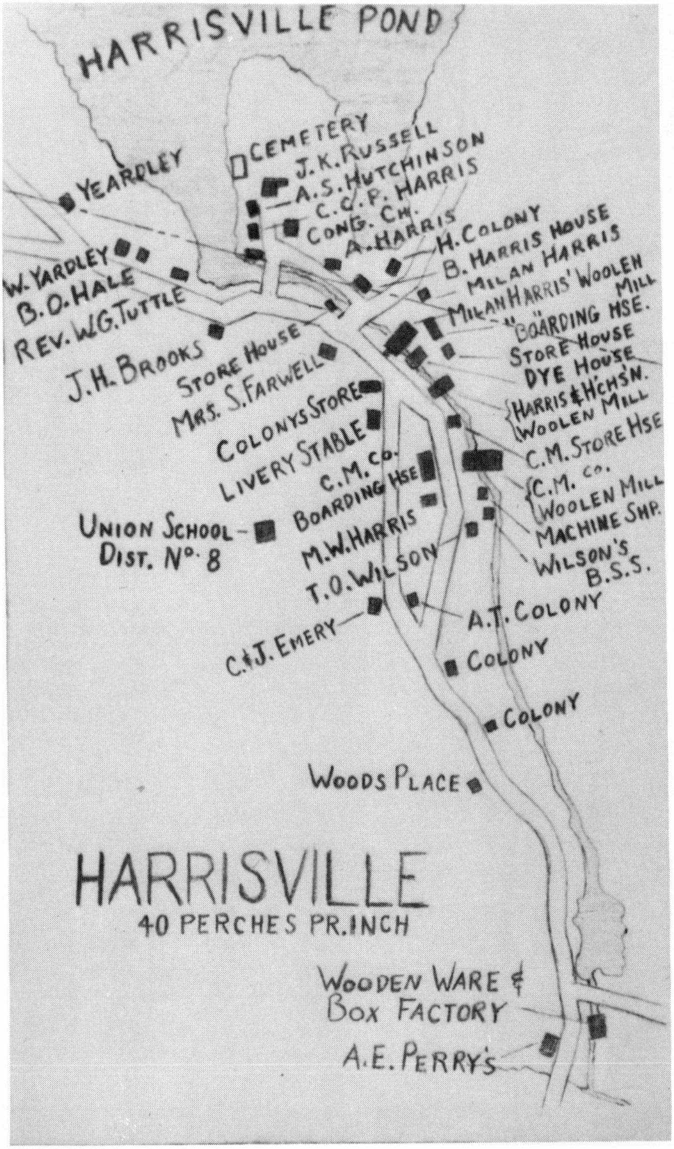
HARRISVILLE POND
YEARDLEY
CEMETERY
J.K. RUSSELL
A.S. HUTCHINSON
C.C.P. HARRIS
CONG. CH.
A. HARRIS
H. COLONY
B. HARRIS HOUSE
MILAN HARRIS
MILAN HARRIS' WOOLEN MILL
BOARDING HSE.
STORE HOUSE
DYE HOUSE
HARRIS & HCHS'N. WOOLEN MILL
C.M. STORE HSE
C.M. CO. WOOLEN MILL
MACHINE SHP.
WILSON'S B.S.S.
W. YARDLEY
B.O. HALE
REV. W.G. TUTTLE
J.H. BROOKS
STORE HOUSE
MRS. S. FARWELL
COLONYS STORE
LIVERY STABLE
C.M. CO. BOARDING HSE
UNION SCHOOL DIST. No. 8
M.W. HARRIS
T.O. WILSON
C. & J. EMERY
A.T. COLONY
COLONY
COLONY
WOODS PLACE
HARRISVILLE
40 PERCHES PR. INCH
WOODEN WARE & BOX FACTORY
A.E. PERRY'S

1 Harrisville Village, 1858. From *Map of Cheshire County, New Hampshire* (Philadelphia, 1858)

2 Harrisville Village, 1877. Reprinted from C. H. Rockwood, *Atlas of Cheshire County, New Hampshire* (New York, 1877), p. 41
Courtesy of New Hampshire State Library

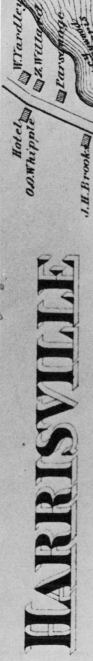

HARRISVILLE

Scale: 150 feet to an Inch.

W.Yardley
Z.Willard
Parsonage
O.O.Whipple
Hotel
J.H.Brooks
A.Hutchinson
J.Farwell
Cong.Ch.
S.Randall
A.R.Harris
Cem.
W.Hall
&.Beals
Harris Mills
W.Harris
M.Harris
Harris B.Shop
J.Hall's Boarding Ho
E.Stratton
Eth.Willis
Store House
W.Inez
Store
Mrs.Farwell
C.Mills
C.Mills Boarding Ho
Atwood
Chesleys Mills
School
C. Mills
R.Winn
C.Mills
C.Mills
Mrs.Wood
J.Blodgett
P.Rousells
W.&A.
Winn
Saloon
W.J.Armstrong
T.Trudell
Winn Fish
Stowell
W.&A.
C.Hester'n
S.M.Powers
Wm.Davis
Willarvie
Atwood

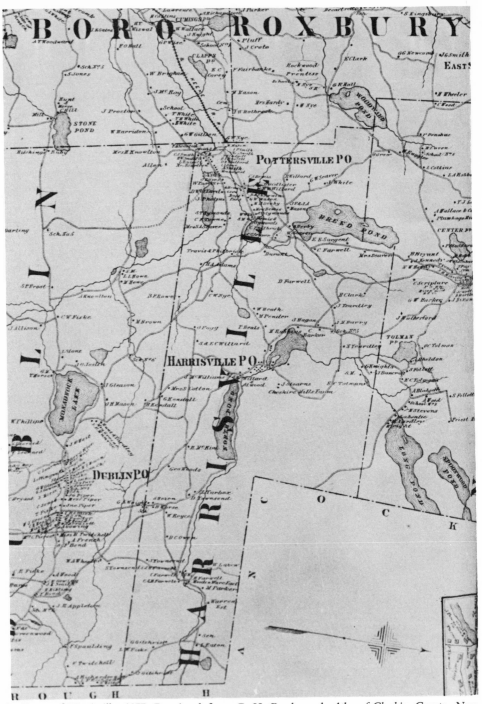

3 Town of Harrisville, 1877. Reprinted from C. H. Rockwood, *Atlas of Cheshire County, New Hampshire* (New York, 1877), p. 41. Courtesy of New Hampshire State Library

4 Map of Southwestern New Hampshire. Reprinted from United States Department of Agriculture, Bureau of Public Roads. New Hampshire transportation map, sheet 2, 1938. Courtesy of New Hampshire State Library

5 Plan of Cheshire Mills, 1892. Reprinted from
Plan of Associated Mutual Insurance Co's.
Surveyed June 6, 1892. Index No. 1122, Serial
No. 3075
Courtesy of John J. Colony, Jr.

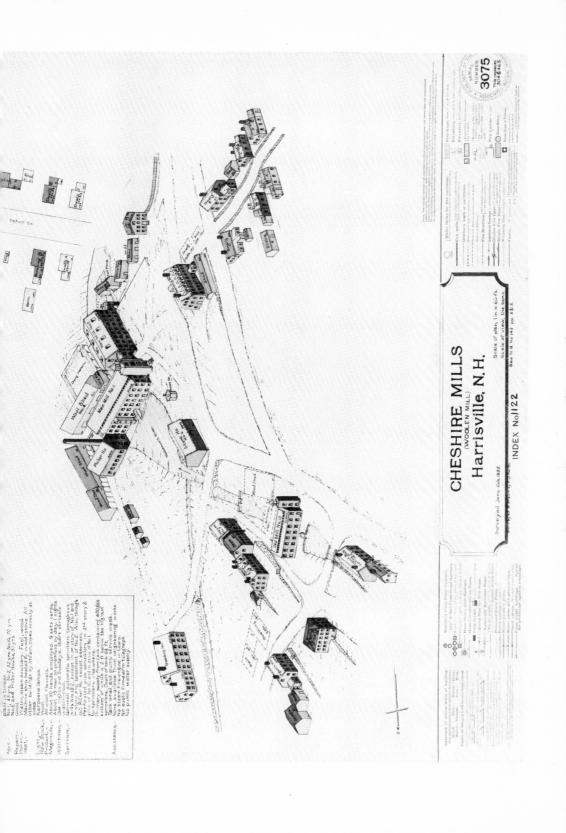

School St.

CHESHIRE MILLS
(WOOLEN MILL)
Harrisville, N.H.

Surveyed June 6th 1892

INDEX No.1122

Scale of plan, 1 in.= 80 ft.
Scale of view, the same.

See N.B. No.149 pp. 4&5.

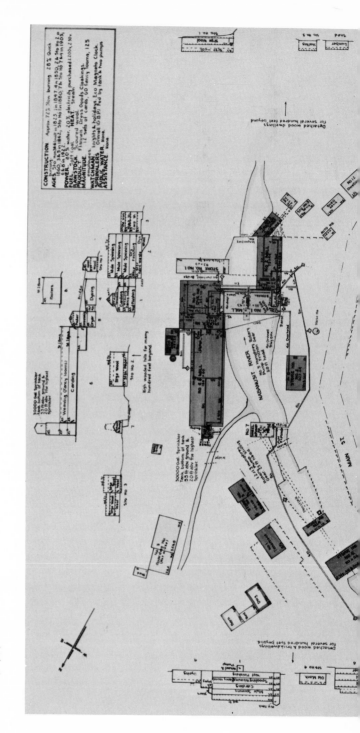

6 Plan of Cheshire Mills, 1924.
Reprinted from Plan of Associated
Mutual Insurance Co's. Surveyed
September 30, 1924. Index No.
7132, Serial No. 16779
Courtesy of John J. Colony, Jr.

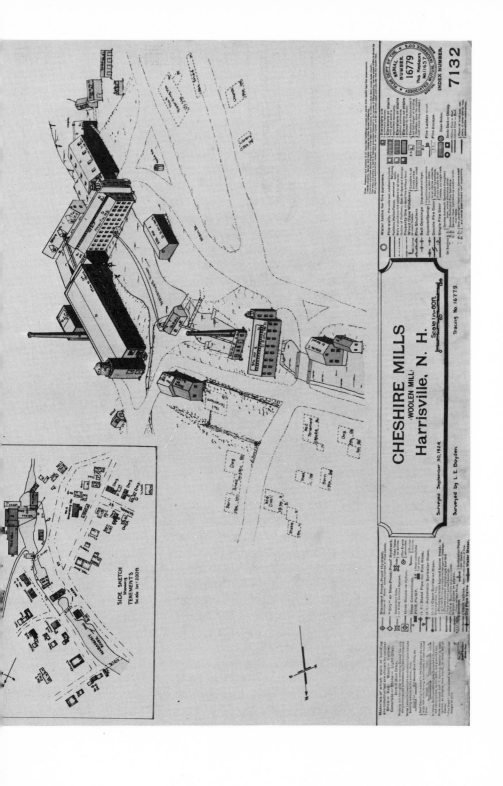

CHESHIRE MILLS
(WOOLEN MILL)
Harrisville, N. H.

Surveyed September 30, 1924

Scale 1in=60ft.

Surveyed by I. E. Boyden

Tracing No 16779

SIDE SKETCH
Showing
TENEMENTS
Scale 1in-200ft

INDEX NUMBER.

7132

APPENDIX 1

Eighteenth-Century Settlers Within Area of Harrisville, N.H.

Name	Year	Rev. Vet.	From	Settled	Left	Source*
Adams, John	by 1773	×	—	S.E. Nelson L.3	×	E, p. 32 F, v. 28, p. 44
Adams, Moses	1763		Sherborn, Mass.	Dublin L.16, R.9		A, p. 179 D, p. 309
Bancroft, James	by 1774	×	Dunstable or Tyngsboro, Mass.	S.E. Nelson L.8	×	E, p. 37 F, v. 28, p. 44
Bancroft, Timothy	1788	×	Dunstable, Mass.	S.E. Nelson		E, p. 37
Beal, Aaron	by 1772	×	Natick, Mass.	S.E. Nelson L.2		E, p. 49 F, v. 28, p. 44
Beal, William	by 1772		—	S.E. Nelson		E, p. 50
Bemis, James	1793	×	Weston, Mass.	Dublin L.17, R.9		D, p. 316
Billhash, Philip	by 1774		—	S.E. Nelson L.2		E, p. 12 F, v. 28, p. 44
Breed, Dr. Nathaniel	1767	×	Sudbury, Mass.	S.E. Nelson		E, p. 54
Brigham, Jonas	1789	×	Sudbury, Mass.	S.E. Nelson	×	E, p. 56
Bryant, James	ca. 1785	×	Reading, Mass.	S.E. Nelson		E, p. 58
Clark, Jonas	1797		Townsend, Mass.	Dublin L.13, R.10	×	D, pp. 322, 419
Cobb, Ebenezer	1778		Temple, N.H.	Dublin L.8, R.9	×	D, pp. 324, 419
Elliot, David	1778	×	Mason, N.H.	Dublin L.10, R.8		D, p. 328
Farwell, John	by 1771	×	Marblehead, Mass. (b. England)	S.E. Nelson L.1		E, pp. 9, 66 F, v. 28, p. 44
Harris, Bethuel	1786		Medway, Mass.	S.E. Nelson		B, p. 213 C, p. 15
Harris, Erastus	ca. 1781	×	Medway, Mass.	S.E. Nelson		C, p. 14 E, p. 91
Harris, Jason	1778		Framingham, Mass.	Dublin L.13, R.10	×	D, pp. 144, 351, 420
Johnson, Simeon	by 1771		—	Dublin L.7, R.8	×	D, p. 355
Kendall, Joel	by 1790		Burlington, Mass.	Dublin L.19, R.10		D, p. 356
Little, Fortune	by 1786		Shirley, Mass. (b. Africa)	Dublin L.3, R.10		D, pp. 289, 361
Marshall, Aaron	1770 or 1778		Holliston, Mass. or Temple, N.H.	Dublin L.8, R.9		D, pp. 144, 361

Name	Year	Rev. Vet.	From	Settled	Left	Source*
Marshall, David	by 1773	×	—	S.E. Nelson L.4		E, p. 108 F, v. 28, p. 44
Mason, Benjamin	ca. 1765	×	Sherborn, Mass.	Dublin L.14, R.9		D, p. 364 F, v. 27, p. 179
Mason, Joseph	by 1774	×	Sherborn, Mass.	S.E. Nelson L.1		E, p. 108 F, v. 28, p. 44
Morse, Daniel	by 1765		Holliston, Mass.	Dublin L.11, R.8	×	D, p. 375
Muzzy, John	by 1769		—	Dublin L.16, R.8		D, p. 376
Newhall, John S.	by 1773		—	S.E. Nelson L.1		E, pp. 13, 114 F, v. 28, p. 44
Nutting, David	1779		Temple, N.H.	Dublin L.6, R.8	×	D, pp. 377, 421
Perry, Ebenezer	by 1777	×	Wilton, N.H.	S.E. Nelson		E, pp. 12, 119
Pratt, Oliver	1790		Shirley, Mass.	Dublin L.4, R.10		D, p. 385
Smith, Abner	1791		Needham, Mass.	Dublin L.22, R.9		A, p. 179 D, p. 392
Spinny, John	by 1773		—	S.E. Nelson L.1		E, pp. 13, 133 F, v. 28, p. 44
Stanford, Joseph	by 1774	×	—	S.E. Nelson L.7		E, p. 134 F, v. 28, p. 44
Stanford, Phineas	1775		Sudbury, Mass.	Dublin L.14, R.10		D, p. 396
Thurston, David	1794		—	Dublin	×	A, p. 177 D, pp. 400, 423
Twitchell, Abel	1774	×	Sherborn, Mass.	Dublin L.13, R.10		D, p. 405
Twitchell, Eleazer	by 1772	×	Sherborn, Mass.	S.E. Nelson L.7	×	D, p. 404, E, p. 147 F, v. 28, p. 44
Warren, Daniel	1782	×	Westborough, Mass.	Dublin L.2, R.9		D, p. 410
Weston, Nathan	by 1782	×	Hollis, N.H.	S.E. Nelson		E, p. 149
Wight, John	1771	×	Medfield, Mass.	Dublin L.17, R.8		D, pp. 142, 150, 412
Yeardley, William	1776		England	Dublin L.14, R.10		A, p. 179 D, p. 415

*A Hamilton Child, *Gazetteer of Cheshire Co., N.H.*
B D. H. Hurd, *History of Cheshire and Sullivan Counties, N.H.*
C Albert Hutchinson, *A Genealogy . . . of Bethuel Harris . . . and His Descendents.*
D Leonard, 1855.
E *Nelson Picnic Assn. Celebration . . . of 150th. Anniversary.*
F N.H. State Papers, Vols. XXVII, XXVIII.

APPENDIX 2

Harris and Colony Genealogies*

Immediate Descendents of Bethuel Harris (1769–1851); m. Deborah Twitchell (1776–1855). Ten children:

I   Cyrus (1797–1848) m. 1st Lydia Wright; 2nd Lucy Cary
       Three children

II   Milan (1799–1884) m. 1st Lois Wright; 2nd Harriet Russell
      1 Milan Walter (1823–1873) m. Lydia Heald
          Three children
      2 Charlotte Elizabeth (1825–?) m. 1st Silas Atwood; 2nd George Burnap
          Four children
      3 Lucretia Jane (1828–1875) m. Joseph K. Russell
          Five children
      4 Alfred Romanzo (1830–1910)
          Four children

III Almon (1800–1876) m. Phebe Sheldon
       Three children

IV Lovell (1802–1888) m. 1st Betsey Felt; 2nd Caroline Burns
       Three children

V   Calmer (1805–1880) m. 1st Lucretia Perry; 2nd Harriet E. Harris; 3rd Mrs. Elizabeth Drake
       Seven children

VI Charles Cotesworth Pinckney (1807–1888) m. 1st Fanny Wilson; 2nd Matilda Stratton
       One child

VII Mary (1809–1895) m. Abner Stiles Hutchinson
       Five children

VIII Sarah (1811–1886) m. Calvin Hayward
       Three children

IX Lydia (1815–1841)

X   Lois (1817–1899) m. Edmund Prouty

* Sources: Harris Family—Leonard, 1920, pp. 788–789.
          Colony Family—State of New Hampshire, Superior Court, Cheshire SS, September term, 1951, No. 8010, *John K. Colony et al., Appellants, vs. John J. Colony and Horatio Colony, Trustees of the Estate of John E. Colony.*

Immediate Descendents of Josiah Colony (1791–1867); m 1st Hannah
Taylor; 2nd Mrs. Jane Briggs Buell. Eight children

I   Timothy (1818–1882) m. Eunice J. Hooper
           1 Josiah T.
           2 George H.
                 Three children
II   George D. (1821–1898) m. Harriet Stevens
           Seven children
III   Henry (1823–1884) m. Mary L. Hayward
           Six children
IV   Mary Ann (1825–1859) m. Joseph Griggs
           No children
V   Alfred T. (1828–1876) m. Fanny Hawkins
           Three children
VI   John E. (1831–1883) m. Julia A. Hills
           One child
VII   Horatio (1835–1917) m. Emeline E. Joslin
           1 John J. (1864–1955) m. Charlotte Whitcomb
                 John J. Jr. (1915——)
                    m. Marjorie P. Page
                    Seven children
                 Charles (1918——)
                    m. Jacqueline Smith
                    Two children
                 Emmeline J. m. John P. Wright
                    Seven children
           2 Charles T. (1867–1944) m. Ellen Warren
                 Horatio (1900——) m. Mary Curtis
           3 Kate (1871–1927) m. Major General James A. Frye
VIII   Josiah D. (1855–1918) m. 1st Rebecca C. Albro; 2nd Katharine
       Simms

APPENDIX 3

The Mills of Harrisville, 1850–1870*

	1850		1860		1870	
Mill	Harris & Hutchinson	Milan Harris & Co.	Cheshire Mills	Milan Harris & Co.	Cheshire Mills	Milan Harris Woolen Co.
Capital Invested	$10,000	$18,000	$65,000	$100,000	$100,000	$160,000
Amt. Wool Used—Tons	15	24	100	55	150	100
Value	$12,000	$19,200	90,000	$5,500 [sic]	$120,000	$100,000
Machinery						
Sets	1	2	—	—	—	—
Cards	—	—	—	—	8	7
Spindles	280	600	1,800	1,200	3,000	2,810
Looms	6	12	30	20	44 Broad	45 Narrow
Hands Employed, Avg.						
Male	10	15	40	30	40	50
Female	7	9	15	30	34	40
Children Under 15	—	—	—	—	6	8
Cost—Avg. Total						
Male	$260	$416	$1,000	$550	}$25,000	$42,000
Female	$100	$150	$270	$500		
Production						
Fabrics	Woolen Cloth	Woolen Cloth	Flannel	Doeskins	Flannel	Fall Cloths, Tricuts, and Beavers
Amt.—Yards	26,000	40,000	425,000	90,000	575,000	150,000
Value	$20,800	$32,000	$140,000	$85,000	$200,000	$225,000

* Source: Census, 1850, Vol. 14, Dublin, Schedule 5; 1860, Nelson, p. 6, Schedule 5; 1870, Vol. 16, Dublin, Schedule 4.

APPENDIX 4

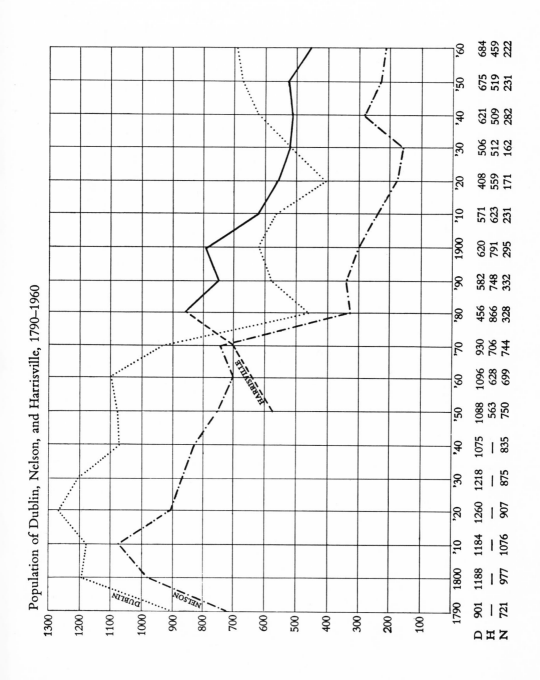

Population of Dublin, Nelson, and Harrisville, 1790–1960

	1790	1800	'10	'20	'30	'40	'50	'60	'70	'80	'90	1900	'10	'20	'30	'40	'50	'60
D	901	1188	1184	1260	1218	1075	1088	1096	930	456	582	620	571	408	506	621	675	684
H	—	—	—	—	—	—	563	628	706	866	748	791	623	559	512	509	519	459
N	721	977	1076	907	875	835	750	699	744	328	332	295	231	171	162	282	231	222

APPENDIX 5

Death Rate in Dublin, N.H., 1820–1852*

Ages	Deaths Jan. 1820– Jan. 1852	Average Death Per Year	Number Living in 1840	Estimated Deaths/1,000 Population in 1840
–1	63	1.97	28	70.36
1–5	75	2.34	112	20.89
5–10	20	.63	119	5.30
10–15	26	.81	115	7.04
15–20	6	.19	127	1.50
20–25	42	1.31	72	18.19
25–30	14	.44	80	5.50
30–35	31	.97	73	13.37
35–40	10	.31	58	5.34
40–45	30	.92	74	12.43
45–50	9	.28	45	6.22
50–55	18	.56	52	10.77
55–60	14	.44	35	12.57
60–65	28	.88	28	35.00
65–70	13	.41	19	21.58
70–75	40	1.25	26	48.08
75–80	23	.72	17	42.35
80–85	34	1.06	11	96.36
85–90	14	.44	5	88.00
90–100	8	.25	1	250.00

* Source: Leonard, 1855, p. 226. *NHS*, March 18, 1840.

APPENDIX 6

Population of Harrisville, 1850–1880*
(By birthplace)

	1850 Total	%	1860 Total	%	1870 Total	%	1880 Total	%
Native-born								
Mass.	54	9.7	99	15.7	41	5.8	81	9.4
N.H.	466	82.5	425	67.7	484	68.6	478	55.2
Vt.	20	3.6	32	5.1	20	2.8	53	6.1
Other	10	1.8	26	4.1	25	3.5	33	3.8
Total	550	97.6	582	92.6	570	80.7	645	74.5
Foreign-born								
England	6	1.1	15	2.3	14	2.0	15	1.7
Ireland	2	.4	24	3.8	52	7.4	58	6.7
Canada	1	.2	2	.4	67	9.7	138	15.9
Other			3	.5	3	.4	10	1.2
Total	9	1.7	44	7.0	136	19.3	221	25.5
Undesignated	4	.7	2	.4				
TOTAL	563	100.0	628	100.0	706	100.0	866	100.0
Foreign Parentage					44	6.2	153	17.6

* Source: Census, 1850, Vol. XIV, Dublin, Vol. XV, Nelson; Census, 1860, N.H., Cheshire Co., Dublin, Nelson; Census, 1870, Vol. XV, Nelson, Vol. XVI, Dublin; Census, 1880, N.H., Cheshire Co., Harrisville.

APPENDIX 7

Cheshire Mills Monthly Payroll[1]*

		1862	1867	1872
SPINNING ROOM	NUMBER EMPLOYED	M: 18	M: 15	M: 18
	DAYS WORKED TYPICAL	$24\frac{1}{2}$[3]	25[3]	25[3]
	AVERAGE	—		
	WAGES — FOREMAN	1.50/D 36.75	2.50/D 62.50[6]	2.50/D 62.50
	2nd HAND			
	SPINNERS			
	WEIGHERS			
	MAXIMUM[2]			
	MINIMUM[2]			
	AVERAGE[2]	25.44	33.15	36.90
CARD ROOM	NUMBER EMPLOYED	M: 13 F: 1	M: 6 F: 8	M: 11 F: 4
	DAYS WORKED TYPICAL	26 (8)[4]	25 (6)	25 (11)
	AVERAGE	20	19	22
	WAGES — FOREMAN	2.50/D 65.00	3.75/D 93.75	3.75/D 93.75
	2nd HAND			
	STRIPPERS			
	STOCKMEN			
	MAXIMUM	M: 15.00[7]	M: 1.25/D F: .66/D	M: 1.92/D
	MINIMUM	M: 8.00[7]	M: .33/D F: .33/D	M: .75/D
	AVERAGE	M: 7.52 F: 1.75/Wk.	M: 20.03 F: 7.95	F: .75/D M: 20.85 F: 18.75
FINISHING ROOM / DYE HOUSE	NUMBER EMPLOYED	M: 15	M: 9	M: 8
	DAYS WORKED TYPICAL	28 (max)	30 (max)	25 (4)
	AVERAGE	21	24	$23\frac{1}{2}$
	WAGES — FOREMAN	75.00	75.00	75.00
	MAXIMUM	22.50[7]	1.73/D	1.58/D
	MINIMUM	12.00[7]	.69/D	1.27/D
	AVERAGE	13.00	27.92	34.16
WEAVE ROOM	NUMBER EMPLOYED	M: 8 F: 29	M: 5 F: 21	M: 5 F: 28
	DAYS WORKED TYPICAL	26 (max)	25 (max)	25 (max)
	AVERAGE			
	FOREMAN	2.25/D 40.50	3.50/D 87.50[6]	3.50/D 87.50
	FIXERS			
	HELPERS			
	WEAVERS			
	WAGES — SPOOLERS			
	MENDERS			
	MAXIMUM	M: 1.25/D	M: 1.75/D	M: 1.75/D
	MINIMUM	M: .62/D	M: 1.42/D	M: 1.25/D
	AVERAGE	M: 25.42 F: 13.66	M: 36.37 F: 20.15	M: 30.25 F: 20.97
SORTING ROOM	NUMBER EMPLOYED	M: 4	M: 4	M: 4
	DAYS WORKED TYPICAL	$25\frac{1}{2}$[3]	23[3]	20 (max)
	AVERAGE			
	MAXIMUM			2.50/D
	WAGES — MINIMUM			2.00/D
	AVERAGE	1.73/D[5] 43.95	2.25/D[5] 45.02	33.98

* SOURCE: CMR, PAYROLL.

(continued)

1877	1882	1887	1892	1897
M: 12 27 (max) 2.25/D 60.75 1.65/D 41.25	M: 13 26 (max) 2.50/D 65.00[6] 1.65/D 30.52	M: 15 F: 1 26 (9) 19½ 2.50/D 65.00[6] 1.15/D[6] 29.90 34.30 .50/D 6.50	M: 14 26 (6) 19 2.50/D 65.00 1.15/D 29.90 26.65 .50/D 13.00	M: 19 30 (max) 17 2.00/D 60.50[6] 1.50/D 45.37 22.85 .50/D 12.50
35.50	34.14			
M: 10 F: 3 25 (7) 23 2.50/D 65.00	M: 9 F: 7 26 (5) 22 3.00/D 78.00	M: 16 F: 1 26 (9) 21½ 3.00/D 78.00[6] 1.50/D 39.00 .85–1.15/D 19.22 .75/D 19.37	M: 15 26 (11) 24 3.00/D 78.00[6] 1.50/D 39.00 .85–1.15/D 24.54 .75/D 16.49	M: 19 25 (8) 20 2.75/D 75.62[6] 1.50/D 38.25 .80–1.25/D 23.26 .75/D 11.52

M: 1.25/D
M: .75/D
F: .75/D   M: 20.59
    F: 17.43
	M: 1.20/D F: .80/D
	M: .75/D F: .75/D
	    M: 22.43
	    F: 14.90			

M: 11
31 (max)
22
1.75/D
.75/D
    29.02
	M: 16
	26 (7)
	22
	2.25/D
	.90/D
	    32.88
		M: 22
		26 (8)
		23
		2.00/D
		.62/D
		1.28/D 31.48
			M: 23
			26 (8)
			23½
			3.00/D
			.75/D
			1.50/D 36.93
				M: 20 F: 6
				25 (10)
				23
				M: 2.50/D
				M: .80/D
				M: 1.38/D F: .96/D

M: 5 F: 28 25 (max) 2.50/D 64.37[6]	M: 11 F: 27 26½ (max) 2.75/D 60.50[6]	M: 11 F: 33 26 (19) 22 2.75/D 71.50[6] 1.85/D 47.30 1.57/D 42.80 29.10 19.56 5.45 (4D)	M: 10 F: 29 26 (8) 18 2.50/D 65.00[6] 1.85/D 32.38 1.67/D 43.33 19.36 12.43	M: 31 F: 17 25 (19) 21 2.50/D 65.00[6] 1.85/D 46.40 1.50/D 34.77 31.45 13.22
M: 1.50/D M: 1.25/D M: 33.14 F: 19.59	M: 1.75/D M: 1.25/D M: 38.15 F: 25.01			

M: 5
25 (max)
16
2.00/D
1.75/D
    28.82
	M: 4
	26 (max)
	21½
	1.75/D
	1.50/D
	    36.15
		M: 6
		26 (3)
		
		    48.55
			M: 3 F: 2
			M: 24 (3)
			F: 69 hrs.
			
			M: 1.75/D M: 41.05
			    F: 6.08
				M: 5
				25 (3)
				
				2.00/D
				.50/D
				1.45/D 35.53

(continued)

		1862	1867	1872
OTHERS	WATCHMAN'S WAGES SUPERINTENDENT COLONYS	26.00	45.00	1.50/D 43.50(2)
MISC.	TOTAL EMPLOYMENT	89	69	80
	NAMES IRISH	6	7	10
	FRENCH	1	8	22
	RENT TYPICAL		5.00	5.00
	AVERAGE	4.80 (6)	5.44 (8)	4.61(14)
	BOARD TYPICAL	M: 7.50 F: 6.44	M: 15.00 F: 7.50	M: 15.00 F: 9.64
	AVERAGE	M: 6.68 F: 5.84 (34)	M: 13.06 F: 6.13 (29)	M: 11.70 F: 8.71 (24)
	GROSS PAYROLL	1749.43	1955.90	2399.76

		1902	1907 (Oct.)	1912
		(MONTH OF SEPTEMBER, 1902–1917; FIRST FULL WEEK OF SEPTEMBER, 1922–1937[1])		
SPINNING R.M.	NO. EMPLOYED M.	23	12	16
	DAYS OR HRS. MED.	19	15	23
	WORKED AVG.	16	15	16
	FOREMAN	2.75/D 76.81[6]	2.75/D 45.92[6]	3.00/D 60.00[6]
	WAGES 2nd HAND	1.60/D 44.32	1.75/D 49.52	2.25/D 52.20
	SPINNERS	31.78	25.72	29.85
CARD R.M.	NO. EMPLOYED M. & F.	M: 21	M: 12	M: 17 F: 3
	DAYS OR HRS. MED.	23	$17\frac{1}{2}$	$8\frac{1}{2}$
	WORKED AVG.	20	$17\frac{1}{2}$	11
	FOREMAN	3.00/D 98.70[6]	4.00/D 107.60[6]	4.00/D 96.00[6]
	WAGES 2nd HAND	1.60/D 17.76	1.75/D 47.07	2.15/D 53.96
	OTHERS RANGE	.90–1.25/D	.80–1.50/D	1.00–1.50/D
	AVG.	21.28	19.66	12.80
FINISHING R.M.	NO. EMPLOYED M. & F.	M: 30 F: 8	M: 16 F: 4	M: 19 F: 9
	DAYS OR HRS. MED.	$25\frac{1}{2}$	$14\frac{1}{2}$	22
	WORKED AVG.	24	15	17
	FOREMAN	3.00/D[8]	3.00/D 71.00[6]	3.50/D
	WAGES OTHERS M.	1.37/D	1.44/D 22.00	1.76/D 38.87
	AVG. F.	.82/D	.96/D 10.45	.98/D 10.33
WEAVE R.M.	NO. EMPLOYED M. & F.	M: 37 F: 29	M: 13 F: 18	M: 46 F: 15
	DAYS OF HRS. MED.	22	$14\frac{1}{2}$	21
	WORKED AVG.	19	14	16
	FOREMAN	2.50/D 70.25[6]	2.75/D 62.97[6]	3.00/D 73.20
	FIXERS	2.25/D 55.46	2.25/D 42.22	2.29/D 52.27
	WAGES HELPERS	1.50/D 38.17	1.50/D	2.00/D
	DRESSERS	1.90/D 48.26	1.50/D	1.78/D
	WEAVERS (PC. RT.)	18.25	16.63	27.39
	SPOOLERS	11.71	20.19	14.02
SRTG. R.M.	NO. EMPLOYED M.	3	2	
	DAYS WORKED AVG.	24	26	
	WAGES AVG.	1.83/D 44.37	1.75/D 44.64	
OTHERS	SUPERINTENDENT	125.00	125.00[6]	
	COLONYS	(C.T. & JJ.) 50.00	(H.W. & C.T.) 66.66 (J.J.) 83.33	(C.T.) 66.66 (H.W. & 83.33 J.J.)

(continued)

1877	1882	1887	1892	1897
		83.33 (C.T. & 41.67 J.J.)	100.00 (C.T. & 50.00 J.J.)	125.00 (C.T. & 50.00 J.J.)
74	87	108	99	120
5	14	11	12	17
19	41	33	29	27
5.00	5.00	5.00	5.00	5.00
4.43 (14)	4.91 (19)	4.77 (18)	4.83 (23)	4.63 (21)
M: 15.00 F: 9.64	M: 15.00	M: 15.00	M: 15.00	M: 15.00 F: 10.80
M: 11.08 F: 6.13 (12)	M: 13.75 F: 5.77 (9)	M: 13.30 (5)	M: 11.22 F: 8.23 (10)	M: 9.25 (20) F: 7.80 (4)
1925.91	2591.82	3267.92	2791.76	3406.58

1917	1922	1927	1932	1937
13	12	12	11	6
23				
19	38 hrs.	19 hrs.	54 hrs.	21 hrs.
3.50/D 81.37	44.00	50.00	50.00	50.00
2.73/D 62.24				
58.06	24.20	13.62	24.28	13.66
M: 15	M: 12	M: 10	M: 7	M: 7
23				
16	38 hrs.	29 hrs.	52 hrs.	14 hrs.
4.40/D 114.40	50.00	55.00	55.00	55.00
3.26/D 84.76				
1.57–2.40/D				
28.87	17.20	12.93	15.20	6.54
M: 28 F: 3	M: 21 F: 3	M: 24 F: 6	M: 21 F: 3	M: 20 F: 8
22				
17	49 hrs.	34 hrs.	54 hrs.	32 hrs.
4.50/D	50.00	55.00		
43.91	24.54	16.11	17.89	16.04
27.07				
M: 41 F: 17	M: 33 F: 21	M: 38 F: 15	M: 17 F: 16	M: 25 F: 12
22				
18	46 hrs.	26 hrs.	51 hrs.	21 hrs.
4.50/D 105.75	50.00	55.00	55.00	55.00
3.15/D 72.84				
2.69/D 60.29	25.86	17.28	17.27	13.08
52.95				
44.58	30.35	13.31	18.35	14.56
35.75	12.30	11.50	11.86	

(C.T.,
H.W., & 200.00
J.J.)

(continued)

		1902	1907	1912
	Total Employed	154	81	128
	Irish	23	3	18
Natl. Names	Fr.–Cdn.	14	20	24
	Finnish	40	14	21
Rent Med.		5.00	5.00	5.00
Avg.[4]		5.02 (28)	5.05 (17)	5.53 (17)
Board Med.[4]		15.00		16.00
Avg.[4]		M: 10.05 (30) F: 7.92 (2)	16.42[9]	15.13 (46)
Gross Payroll		4,205.88	2,288.26	3,567.81

(Misc. is printed vertically at the left of the Rent/Board rows.)

In this table all wage rates shown are per month, unless otherwise indicated (e.g., per day = /D; per week = /Wk.). M and F stand for male and female operatives. Also the following qualifications apply, where indicated by the appropriate superior numbers on the table:

1. Unless otherwise noted, the sampling of wages was taken from the September payroll in each selected year, that month being a good average month in the woolen industry.
2. These averages are exclusive of the foreman's wages.
3. This figure is the only number given in month's payroll.

1917	1922	1927	1932	1937
120	102	105	75	78
9	5	4	1	3
20	21	22	12	20
8	14	19	18	19
5.00				
5.88 (16)	1.40 (20)	1.45 (22)	1.45 (15)	1.35 (16)
		7.21 (17)	7.00 (3)	
18.20 (39)[9]	7.00 (23)			4.35 (5)
5,794.77	2,735.86	1,716.00	1,501.25	1,241.38

4. All numbers in parentheses indicate the number of workers to whom the accompanying figure applies.
5. Operative was on piece rate.
6. New foreman hired since preceding entry.
7. Amount indicates wage rate based on a twenty-six-day month.
8. Operative was either an overseer or a dyer.
9. Average of board was figured only for men who worked most or all of month.

APPENDIX 7a

Cheshire Mills Payroll, Summary

	1862★	1867★	1872★	1877★	1882★	1887★	1892★	1897★
Payroll	89	69	80	74	88	108	99	120
Man–Days Worked	1,851	1,486	1,829	1,770	1,912	2,386	2,066	2,509
Days Worked (avg.)	20.8	21.5	22.9	23.9	21.7	22.1	20.9	20.9
Daily Wage (avg.)	$.95	$1.32	$1.31	$1.09	$1.36	$1.37	$1.35	$1.36
Wage (avg.)	19.66	28.35	30.00	26.03	29.45	30.26	28.20	28.39
Board Paid (avg.)	6.09	9.71	10.58	9.43	10.85	13.30	10.92	9.09

	1902★	Oct. 1907	1912★	1917★	1922†	1927†	1932†	1937†
Payroll	154	81	128	120	102	105	75	78
Man–Days Worked	3,098	1,280	1,994	2,115	571.5	364.5	492.1	237.1
Days Worked (avg.)	20.1	15.8	15.6	17.6	5.6	3.5	6.5	3.0
Daily Wage (avg.)	$1.36	$1.79	$1.79	$2.74	$4.79	$4.71	$3.05	$5.24
Wage (avg.)	27.31	28.25	27.09	48.29	26.82	16.34	20.02	15.92
Board Paid (avg.)	10.05	16.42	15.08	18.20	7.00	7.21	7.00	4.35

★ Based on month of September, 1862–1917.
† Based on first week of September, 1922–1937.

APPENDIX 8

Letter to Henry Colony, Esq., Keene, N.H.★

Harrisville June 20th 1869
H Colony Esq

Sir
I

thought I would write you once more and the last time in
relation to that affair I informed you of last April—in
the first place you did not finde out anything well I am
not much surprised at that for you and John Ed did not act
hardly shrewd enough in the matter, I will tell you why.
in the first place John Ed went up to the upper spinning
one evening as Rutherford told one of his spinners the
next morning, in an awful hurry and wanted to Know what
Kinde of a spinner such a one was and such a one &c he remarked at the time I wonder what he ment by it, an other
evening Clark and one or two others see you and John Ed go
to his clock and Caskins in the evening and hold your lan-
tren up to them Clark remarked at the time I wonder what
they are up to, the next morning John did not have time to
get his coat off till he heard all a bout it and I can as-
sure you it caused quiet a stir a mongst some of your
Spinners, Mc Williams in a day or two after had more to
say then any of the rest he grumbled a bout paying so much
board and wanted Rutherford to go to you and have it cut
down he said this was a d—m poor place now he could do
better anywhere else then he could here, according to what
I can learn himselfe and a few others in the filling room
has had a good thing of it for a long time and has feathered
there nests pretty well at your expense—now Mr. Coloney
you may not believe this but sir Just as true as you run
that mill Just so true has that work been going on—some
of your spinners say for men that has carried on buisness
so long as you have, they think you aught to be a little
shrewder—I will ask you how your filling spinners pay for
the month of april compared with what they had been making
before I give you the information and even up to the present time. an other thing Just go in to the Card Room look
at the roping racks ask purdy how it is he cannot keep up
to the Spinners as he formly did will I am sure he cannot
tell but you can guess—Dave Donohue when he was leaving
said this was a d—m poor place now well I should think
it was a pretty good place for him while he stayd, the
month before he left he drawd fifty two dollars and ten

★ Source: CMR

cents and believe me he did not make $42 honestly—one
more thing and I have don I suppose you Remember Ruther-
fords old Boar Killing one of Clarks colts, do you believe you had to pay for that coalt, perhaps you do not
but I do, clark was on the watch at the time a sertain
man in the village asked him if he ever had a chance
to make any over time now he said yes I worked three days
over time last month and got ten dollars that was considerable over three dollars pr day. dont that look as
though you had to pay for it, I would advise you to keep
your eyes open there is fellows in that mill Watching evry
movement of yours they set on the grass evenings front of
the boarding house and can tell when alf goes through the
filling room and report to John in the morning—here a
short time since for an excuse one or the other of them
would leave a window open or forget there coat for an excuse to go in to the mill evenings to see if you were
still on there track, now I would advise you to keep the
mill door locked and not allow one of them in after bell
hours, also order them to put there curtains down on the
front windows evenings so as they cant see your movements
for be assured they are watching evry movement of yours,
I will now close by informing you since I give you the
information you have a pretty sick lot of customers

Resptly Yours

Spy—

wont you be kinde enough to keep this perfectly private

APPENDIX 9

Insurance Appraisal of Cheshire Mills, 1883*

Floor	Description of Item	Amount
	Buildings 35,844 sq. ft. flooring at 70¢	$25,090.00
1st.	Machinery Stone Mill	
	1–13 plate Steam Press	1,400.00
	1–19 ” ” ”	2,000.00
	1 Cochineal Grinder	100.00
	1 Indigo Mill	125.00
	1–36″ Hy. Extractor	300.00
	1–” ” ” old pattern	200.00
	2 Washing Machines $200, 2 Wool Dryers $90	290.00
	1 Cleaveland Cloth Dryer	1,500.00
	1 Double Screw Press	250.00
2nd.	12 Davis & Furber Flannel Looms @ $125	1,500.00
	8 Pearson Flannel Looms @ $135	1,080.00
	1 D & F Dresser $325	
	1 ” ” 280	605.00
	4–D & F Spoolers @ $55	220.00
	4 new looms @ $175	700.00
3rd.	4 Setts Cards (1000+300)	1,300.00
	5 ” ” (1050+300)	1,350.00
	9 Bramwell Feeds	2,300.00
	1 Duster	40.00
	1 Goddard Burr Picker	600.00
Attic	4 Jacks 200 Spindles ea. @ 450	1,800.00
	2 ” ” ” ” @ 500	1,000.00
	Brick Mill	
1st.	1 Parks & Woolson Double Brush	275.00
2nd.	24 D & F Broad Flannel Looms with drop box @ 215	5,160.00
3rd.	4 Mules, 400 spindles @ 900	3,600.00
Attic	3 Jacks @ 400	1,200.00
	1 ”	450.00
	Boiler House	
	1 Boiler, 4 ft. × 18 ft. C & W	1,200.00
	5 Dye Tubs	270.00
	Shafting, Belting & Piping	8,961.00

* Source: Records of Arkwright–Boston Manufacturers Mutual Insurance Company, Factory Mutual Engineering Corporation, Norwood, Massachusetts.

Automatic Sprinklers	1,100.00
Furniture, Apparatus, etc.	4,480.00
Stock	27,000.00
	—————
	$106,746.00
Repair Shop	2,500.00

APPENDIX 10

Vital Statistics of Harrisville, 1890–1960*
(At five-year intervals)

	1890	1895	1900	1905	1910
Population	748		791		623
Births: Live, Still	L S	L S	L S	L S	L S
Native Parents	1	6 1	9	9	6 1
Foreign, Mixed Parents	10 1	12 1	8	8	2 2
Unknown		2			
Live Births/1,000 Population	14.7		21.5		12.8
Marriages					
Native	4	4	4	3	3
Foreign	1	1		4	2
Mixed		2	1		4
Deaths					
Native	12	11	6	11	12
Foreign	1		3	3	2
Crude Death Rate/1,000 Population	17.3		11.4		22.5

* Source: *HAR*, 1891, 1896, 1901, 1906, 1911, 1916, 1921, 1926, 1931, 1936, 1941, 1951, 1961.

1915	1920	1925	1930	1935	1940	1950	1960
			Nat. 422		425		
	559		F.B. 90		84	519	459
			512		509		
L S	L S	L S	L S	L S	L S	L S	
5 1	7	6	8	6	8 1	11	
6	3	2	2 1	1	2		
							(total) 7
	17.9		19.5		19.7	21.5	15.3
	5	6	2	2	7	9	
	1 1						(total) 11
	1 1			2	1		
	6 6	6	4	4	7	11	(total) 12
	2 1	1		2	2	3	
	12.5		7.8		17.7	27.1	26.1

APPENDIX 11

Town Invoice Summary, 1873–1960*

	1873	1881	1885	1890	1895	1900	1905
Real Estate—Total		$165,005	$168,370	$174,490	$198,500	$209,585	$412,020†
Resident							$293,875
Nonresident							$118,145
Mills & Machinery		$74,500	$52,300	$49,100	$48,100	$48,850	$57,200
Stock in Trade		$30,550	$36,230	$41,900	$31,795	$32,225	$36,050
Electric Plants, Dam							
Oxen		65	60	53	28	14	12
Horses		126	114	132	150	182	203
Cows		265	224	320	228	229	203
Neat Stock		171	96	114	—	86	69
Sheep		552	434	231	393	210	114
Fowls				/250	/250	$153/308	$20/
Wood & Lumber							
Bank Stock		$1,800	$8,200	$1,100	$1,100	$1,789	$262
Interest Money		$23,323	$9,418	$1,500	$6,034	$8,199	$11,125
Total Valuation	$356,935	$341,408	$315,390	$311,416	$322,537	$344,197	$412,020
Taxes Committed to Collector	$8,998	$10,012	$7,967	ca. $4,700	$4,177	$5,336	$6,386
Polls		237	190	193	197	229	237
Tax Rate			$1.50	$1.50	$1.30	$1.55	$1.55
			$.25H'ways				
Town Liabilites	$6,619	$17,771	$22,176	$15,196	$15,000	$15,150	$17,600
Town Assets	$4,714	$1,393	$2,037	$426	$4,422	$1,659	$1,696

* Source: *HAR*, 1874, 1882, 1886, 1891, 1896, 1901, 1906, 1911, 1916, 1921, 1926, 1931, 1936, 1941, 1951, 1961. The list of items in valuation is not complete for all years. Minor items, (e.g., gas tanks, boats) that appeared in valuations for only part of period are not shown. Total valuation figures are, of course, as given.

1910	1915	1920	1925	1930	1935	1940	1950	1960
$320,110	$451,870	$597,185	$667,978	$731,035	$698,550	$696,490	$774,810	$1,085,694
$190,050					$390,975	$338,650		
$130,060					$307,575	$357,840		
$53,600	$60,450	$61,800	$111,000	$110,400	$104,400	$104,000	$129,300	$134,350
					$90,000 Exempted			
$333,625	$31,550	$60,000	$51,200	$53,000	$45,500	$42,800	$106,400	$56,925
					$55,550	$60,000	$110,331	$111,500
2	2	2	—	—	2	2	—	—
159	140	106	91	45	32	17	8	19
139	94	77	93	67	68	61	15	22
38	8	19	4	9	ca. 4	6	—	16
73	11	4	—	8	12	—	—	
$160/230	/125	/130	$315/	$1,460/	$1,092/	$1,450/1,450	$3,840/3,072	$1,875/3,760
	$12,975	$9,400	$49,200	$1,780	$1,100	$4,400		
$1,200	$400	$200						
$12,615	$11,535	$1,975						
$464,885	$600,215	$758,385	$897,488	$909,015	$916,208	$917,510	$1,131,001	$1,406,939
$8,600	$10,323	$19,558	$22,553	$22,566	$22,562	$22,754	$38,326	$81,920
191	210	197	338	284	287	287	232	175
$1.85	$1.65	$2.45	$2.40	$2.42	$2.40	$2.48	$3.46	$5.94
$14,000	$12,023	$17,262	$10,157	$4,124	$3,605	$1,813	$12,859	$38,257
$868	$609	$4,201	$10,724	$7,861	$5,692	$3,214	$5,534	$45,251

† These figures are either erroneous or at least figured differently from other years. The resident and nonresident real estate amounts to the total valuation of the town, making no allowance for other items. Perhaps the other items were somehow included in the real estate. A more likely figure for the total real estate would be the difference between the other items and the total valuation (which is correct) or $258,060.

APPENDIX 12

Valuations of Resident Property Owners, 1906, 1941, 1960*

Valuation	1906 Property Owners		1941 Property Owners		1960 Property Owners	
	Number	% of Total	Number	% of Total	Number	% of Total
$ –999	84	59	37	30	26	18
1,000–1,999	35	25	45	36	31	22
2,000–2,999	8	6	19	15	37	26
3,000–3,999	4	3	8	6	23	16
4,000–4,999	3	2	4	3	9	6
5,000–5,999	1	0.5	3	2	3	2
6,000–6,999	0	0	1	1	3	2
7,000–7,999	0	0	1	1	1	0
8,000–8,999	1	0.5	2	2	4	2
9,000–9,999	2	1	1	1	0	0
10,000 & over	4	3	4	3	5	3
Total	142		125		142	
Ratable Polls	150		214		175	

* Source: Invoice and Taxes of the Town of Harrisville, 1906; HAR, 1942, 1961.

APPENDIX 13

Summary of Town Budgets, 1880–1960*
(Items in current dollars and as approximate percentages of totals)

	1880			1890		1900		1910	
Receipts									
Cash on hand	317	2%		2,137	20%	91	1%	421	4%
Licenses, Fines, Fees	—			—		129	2%	133	1%
Taxes & Interest	5,424	41%		6,456	60%	5,340	79%	8,640	75%
Interest on Funds & Bonds	27			36		23		25	
Miscellaneous	43			28		21		5	
State:									
Taxes, Miscellaneous	807	6%		1,468	14%	383	5%	561	5%
Highways	—			—		—		388	3%
School Fund	—			—		—		446	4%
County:									
Poor	—			—		—		72	
Borrowed	6,523	50%		—		800	12%	849	7%
Liquor Agent	—			552	5%	—		—	
TOTAL	13,141			10,677		6,787		11,540	
Disbursements			††						
State & County Taxes	1,399	18%	10%	1,515	14%	1,279	19%	1,886	16%
Notes & Interest	2,038	25%	15%	4,495	42%	1,159	17%	3,340	29%
Schools	923	12%	7%	1,219	11%	1,625	24%	2,614	23%
Highways & Bridges	1,028	13%	8%	1,643	15%	1,303	19%	2,215	20%
Support of Poor & Social Security	278	3%	2%	63	1%	67	1%	370	3%
Town Officers & Government	430	5%	3%	475	4%	723	11%	512	4%
Library	352	4%	3%	131	1%	72	1%	78	1%
Old Home Day	—			—		—		—	
Recreation	—			—		—		—	
Miscellaneous	5,528†	4%	42%	254	2%	533	7%	119	1%
Liquor Agent	—			456	4%	—		—	
Cash	1,165	15%	9%	426	4%	26		406	3%
TOTAL									
Current Dollars	13,141			10,677		6,787		11,540	
Adjusted to Index A, 1913 = 100	18,431			15,751		10,025		12,395	
Adjusted to Index B, 1947–9 = 100	—			38,685		24,952		31,359	

* Source: *HAR*, 1881, 1891, 1901, 1911, 1921, 1931, 1941, 1951, 1961.
† Includes $5,215 for settlement of lawsuit (Jennie Case).
†† Includes cost of lawsuit.

1920		1930		1940		1950		1960	
819	3%	7,166	15%	2,586	7%	15,226	16.1%	30,512	18.1%
683	2%	970	2%	906	2%	3,683	3.8%	5,260	3.1%
18,952	64%	22,250	47%	23,431	65%	39,736	41.6%	85,594	50.5%
220		—		—		—		—	
13		88		202		669	.7%	4,830	2.8%
496	2%	3,620	8%	2,545	6%	10,472	11.0%	12,142	7.2%
1,185	4%	4,047	8%	—		—		—	
—		—		—		—		—	
253	1%	—		—		—		—	
7,000	23%	9,000	19%	7,500	20%	25,500	26.8%	31,000	18.3%
—		—		—		—		—	
29,621		47,140		37,170		95,286		169,338	
4,456	15%	4,272	9%	2,609	7%	4,443	4.6%	6,506	3.8%
10,251	34%	11,220	24%	7,635	21%	26,345	27.6%	38,765	22.9%
6,079	20%	6,310	13%	7,544	20%	24,462	25.7%	50,845	30.0%
5,833	20%	15,546	33%	13,036	35%	30,535	32.0%	30,290	17.9%
281	1%	239		623	1%	1,063	1.1%	2,060	1.2%
868	3%	1,935	4%	2,949	8%	5,650	5.9%	9,166	5.4%
90		125		225	1%	275	.3%	300	
200	1%	150		100		328	.3%	136	
—		—		411	1%	213	.2%	45	
111		340	1%	1,111	3%	246	.2%	1,786	1.0%
—		—		—		—		—	
1,452	5%	7,003	15%	927	2%	1,726	1.8%	29,439	17.4%
29,621		47,140		37,170		95,286		169,338	
14,541		—		—		—		—	
35,432		75,545		77,761		—		—	

APPENDIX 14

Votes of Harrisville, Nelson, and Dublin in Presidential Elections, 1872–1968*

	1872			1876			1880			1884		
	H	N	D	H	N	D	H	N	D	H	N	D
Republican	95	78	97	93	63	109	82	61	102	68	55	84
Democratic	66	19	15	101	45	29	89	48	24	73	39	17
Prohibition										4	2	
Socialist												
Other												
Total	161	97	112	194	108	138	171	109	126	145	96	101
Names on Checklist												
Population							866	328	456			

	1908			1912			1916			1920		
	H	N	D	H	N	D	H	N	D	H	N	D
Republican	93	51	82	61	31	68	59	31	60	119	66	111
Democratic	73	11	11	79	14	20	83	12	39	104	15	33
Prohibition	4	2	3	2		3						
Socialist		3	2	1	7	2	1	8	1			
Other				8	11	28 (Prog.—Roosevelt)						
Total	170	67	98	151	63	121	143	51	100	223	81	144
Names on Checklist	185		127									
Population										559	171	408

(First vote under Woman Suffrage Amendment)

	1944			1948			1952			1956		
	H	N	D	H	N	D	H	N	D	H	N	D
Republican	97	103	177	114	83	198	144	106	287	162	114	262
Democratic	133	37	82	107	29	70	122	21	79	96	15	80
Prohibition												
Socialist		1							2			
Other				4	1	2 (Wallace)						
Total	232	142	259	228	116	267	270	129	369	260	129	343
Names on Checklist	283	158	323	287	151	361	311	163	428	303	170	394
Population												

* Source: State of New Hampshire, Manuals of the General Court, 1873–1969.

1888			1892			1896			1900			1904		
H	N	D	H	N	D	H	N	D	H	N	D	H	N	D
64	65	97	84	61	88	128	51	87	113	44	72	101	49	78
74	27	22	79	24	18	25	11	7	66	20	17	68	15	15
2		1							5			2		6
												1	2	
						8	2 (Natl. Dm.—Palmer)							
140	92	120	163	85	106	161	64	94	184	66	89	172	66	99
154	104	129							206		118	195		130
									791	295	620			

1924			1928			1932			1936			1940		
H	N	D	H	N	D	H	N	D	H	N	D	H	N	D
97	73	113	128	84	119	113	69	143	114	79	176	115	90	229
125	13	40	122	15	60	129	19	84	124	17	78	138	18	97
						5	5	9						
5	5	3 (Prog.—LaFollette)							1		4 (Union—Lemke)			
227	91	156	250	99	179	247	93	236	239	96	158	253	108	326
284	121	208	285	121	232	271	143	277	196	135	316	293	148	360
												509	282	621

1960			1964			1968		
H	N	D	H	N	D	H	N	D
140	108	266	68	70	132	131	110	239
109	34	112	183	85	226	120	49	179
			(New Party) 1		1			
						5	7	9 (Wallace)
252	142	383	254	158	371	295	168	430
298	172	451	317	187	461	340	209	486
459	222	684						

APPENDIX 15

Harrisville Schools, 1870–1960*

	1870	1875	1881	1885	1890	1895	1900	1905
Students in Harrisville Schools[1]								
No. 1, Village Primary		50	54	41	46	51	?	50
Village	58							
Grammar		36	31	35	30	38	?	28
No. 2, Chesham	38	49	26	29	28	26	?	33
No. 3, (Nelson No. 7)	16	18	18	14	Closed Perm.			
No. 4, (Dublin No. 9)	11	10	7	7	8	Closed Perm.		
No. 5, East Harrisville (Dublin No. 10)	13	17	14	14	12	12	?	15
Total Students	136	180	150	140	124	127	?	126
Students Attending High School	?	?	?	?	?	?	?	?
School Year in Weeks[2]	19	21	19.9	23	19	22	32	33
Budget								
Total Expenditures for Schools[3] of which, Aid from State and other Sources than Property Tax	$1,066	$869	$932	$1,273	$1,173	$1,548	$1,683	
	146	117	133	147	130	75	73	
Expenditure Per Student								
In current Dollars		5.92	5.79	6.66	10.27	9.24	?	13.36
Adj. to Index A, 1913=100[4]		7.29	7.85	10.30	15.15	14.39	?	17.58
Adj. to Index B, 1947–9=100[5]		—	—	—	37.24	35.13	?	44.09
Teachers								
Total Expenditures for Wages[6]		914 (Incl. Board)	806	864	498 277 (Board)	560 272 (Board)	1,097	1,260
Avg. Weekly Wage[7]								
In Current Dollars		7.14	6.75	6.30	8.16	9.45	8.56	9.55
Adj. to Index A[4]		8.79	9.15	9.75	12.03	14.72	12.64	12.57

* Source: *HAR*, 1871, 1876, 1882, 1886, 1891, 1896, 1901, 1906, 1911, 1916, 1921, 1926, 1931, 1936, 1941, 1951, 1961.

1910	1915	1920	1925	1930	1935	1940	1950	1960
44	?	?	36	?	21	34		
28	?	?	40	?	26	19		
25	?	?	32	?	23	27	ca. 55	88
12	Closed	?	13	?	Closed Perm.			
116	?	ca. 121	133	ca. 99	ca. 80	100	ca. 79[9]	ca. 111[9]
7	ca. 5	ca. 11	12	ca. 14	ca. 10	20	ca. 24	23
35	36	37	ca. 36	36	36	36	ca. 36	ca. 35
$2,585	$2,636	$8,761	$6,217	$6,401	$5,694	$7,780	$22,340	$54,172
529	357	64	—	—	—	—	656	2,516
22.28	?	ca. 72.40	46.74	ca. 64.66	ca. 71.18	77.80	282.80	488.04
23.93	?	ca. 35.54	—	—	—	—	—	—
60.54	?	ca. 86.60	71.03	ca. 103.62	ca. 143.22	162.76	279.45	417.49
1,440	1,395	2,967	3,415	3,000	2,708	2,911	6,840	18,259
10.29	12.78	20.05	ca. 23.72	27.77	25.07	26.94	51.39	81.40
11.05	12.64	9.84	—	—	—	—	—	—

	1870	1875	1881	1885	1890	1895	1900	1905
Adj. to Index B[5]	—	—	—	29.57	35.93	31.47	31.52
Avg. Annual Earnings In Current Dol-								
lars	152	134	144	155	208	274	315
Adj. to Index A[4]	187	182	223	229	324	405	415
Adj. to Index B[5]	—	—	—	562	791	1,007	1,040
As % of Natl. Avg. for Tea-								
chers[8]	—	—	—	60	72	83	80

[1] These figures are the average number of students for the two or three terms.

[2] These figures are the average number of weeks for the several schools. Occasionally there is a discrepancy of one or two weeks between the stated number of school weeks and the number for which the teachers were paid. In such cases the latter figures were used. The school year for 1925 was not stated, but it must have been thirty-six weeks or very close to it.

[3] There is a discrepancy in the two sets of figures for the expenditures for the schools in these appendices because of the need for internal consistency in each of the two series. The figures in the "Summary of the Town Budget, 1880–1940" are taken from the Town Treasurer's Report on his payments or disbursements. The figure for this appendix on the schools are taken from the report of the Treasurer of the School District, listing his actual total expenditures. In the years before 1890, some Harrisville students attended school in neighboring towns, and these towns sent some students to Harrisville schools, both on a paying basis. To keep the relationship between total expenditures and students, the money Harrisville paid other towns for its students attending their schools has been deducted from the totals and does not here appear.

[4] Cf. Historical Statistics of the United States, Series E158. Cost-of-Living Indices (Burgess, 1913=100), p. 127.

1910	1915	1920	1925	1930	1935	1940	1950	1960
27.96	31.95	23.98	36.05	44.50	50.44	56.36	50.78	69.63
360	465	742	854	1,000	903	970	1,850	2,849
387	460	364	—	—	—	—	—	—
978	1,163	888	1,298	1,603	1,817	2,029	1,828	2,437
73	80	79	68	70	ca. 72	67	61	55

[5] *Cf.* Historical Statistics of the United States, Series E115 (Food), Consumer Price Indices (BLS, 1947–9 = 100), p. 125. It should be noted that both this and the preceding index were used only to suggest long-term trends in monetary values in Harrisville and should not be taken as reflecting exact prices.

[6] Before 1890, wages and earnings are calculated figures. Town reports gave number of weeks of school in each district and the wage per month.

[7] These average wages and earnings are based on positions in the schools rather than on the individual teachers who filled those positions in the course of a year. These figures before 1900 include board. In the 1890s, board was at the rate of three dollars a week, which the town paid.

[8] *Cf.* Historical Statistics of the United States, Series D603–617. Average Annual Earnings in All Industries and in Selected Industries and Occupations: 1890–1926, (616) Public School Teachers, pp. 91–92; Series D728–734. Earnings in Selected Professional Occupations: 1929–1954, (728) Public School Teachers, p. 97.

[9] Enrollment figures based on school census taken in September of new school year.

APPENDIX 16

Harrisville Public Library, 1880–1960*
(At five-year intervals)

	1880	1885	1890	1895	1900	1905	1910	1915
Volumes								
At Beginning of Yr.	704							
Volumes Added	119				33	25		90
Volumes—Total	823	1239	1370	1563	1802	1989	2244	2495
Volumes Issued	3176	2961	2450	1725	1870	798	1595	1987
Per Capita	3.67		3.27		2.36		2.56	
Total Expenditures for Library	$352	$139	$131	$123	$72	$55	$78	$81

* Source: *HAR* 1881–1961. (The Library Committee's annual report was made in February. Thus the library's year began and ended with that month.)

1920	1925	1930	1935	1940	1950	1960
				2965		
38	68	147	61	169	101	
──	──	──		(1339 discarded)	──	──
2622	2868	3294	1906	3134		
				2112 adult fiction	209 adult fiction	
				1615 juvenile	836 juvenile	
				1031 nonfiction	896 nonfiction	
				──	──	
2229	1200	2000	ca. 3380	4758	1941	
3.99		3.91		9.35	3.7	
$90	$102	$125	$190	$225	$290	$370†

† No report on circulation in 1960; only a treasurer's report.

BIBLIOGRAPHY

Research for the history of Harrisville had its difficulties. There is no earlier history of the town, nor had any preliminary work been done on one. Mention of Harrisville is rare in secondary works, the records of the Harris mills are scant, and the town records are in poor shape. The few old residents with any great fund of knowledge about the town all seem to have died just before this writer commenced his work. There exists very little in the way of letters, diaries, or other personal accounts of life in Harrisville. Of course, these deficiencies reflect the nature of the town in the past—a busy mill town without a great deal of time or inclination for such refinements.

The bulk of the material for writing this account came from mill records, public records, newspapers, and personal interviews. The following were particularly valuable: the extensive records of the Cheshire Mills, to which the writer had complete freedom of access; the town records and town reports of Dublin, Harrisville, and Nelson; the federal census schedules for 1850, 1860, 1870, and 1880; and the nearly complete microfilmed file of the weekly *New Hampshire Sentinel*, published in Keene, the county seat. All issues of this newspaper were examined from its establishment in 1799 through the year 1910. Because the news items concerning Harrisville tapered off after the turn of the century, only selected years were examined for the years after 1910.

In addition, miscellaneous historical materials on Dublin and Nelson were helpful for the years before 1870, when Harrisville was formed from these towns. That the village of Harrisville has changed little in the last century helped a great deal in the reconstruction of its history.

Personal interviews and letters to the writer are not included in the bibliography, as they are documented completely in the notes. Interviews took place in Harrisville unless otherwise noted. All other references are given in full in the bibliography. In the notes, references are given in shortened form, and the following, because of frequency or awkwardness of citation, have been further abbreviated:

Bemis, Charles A. MSS Collection on History of Nelson, N.H., given as Bemis MSS.

Company records of the Cheshire Mills, in Harrisville, including a few miscellaneous records in their possession, given as CMR.

U.S. Bureau of the Census. MS volumes of census schedules for 1850, 1860, 1870, and 1880, for New Hampshire, Cheshire County, given as Census, 1850, Census, 1860, etc., with appropriate volume number, town, and, where applicable, pages.

Harrisville, N.H. Annual Reports of the Town Officers, given as *HAR*, 1871, *HAR*, 1874, etc.

[Leonard, Levi W., *et al.*] (eds.). The History of Dublin, New Hampshire. Boston: Printed by John Wilson and Son, 1855, given as Leonard, 1855.

Leonard, Levi W. (Rev.) Continued and Additional Chapters to 1917 by Rev. Josiah L. Seward, D.D. The History of Dublin, New Hampshire. Dublin, 1920, given as Leonard, 1920.

New Hampshire Sentinel (Keene, N.H.), given as *NHS*. When an item cited was signed or initialed, that identification is included in the note.

Keene Evening Sentinel (Keene, N.H.), given as *KES*.

Unpublished Material

Belknap, Jeremy. Description of Packersfield, N.H. In Massachusetts Historical Society.

Bemis, Charles A. MSS Collection on History of Nelson, N.H. In Nelson, N.H., library.

Book of Records for the Baptist Church of Dublin N.H., 1785–1857. In New Hampshire Historical Society.

Book of Records for the Baptist Society in Dublin [N.H., 1815–1903].

Calvert, Monte A. "The Technology of Woolen Cloth Finishing. . . ." Merrimack Valley Textile Museum, 1963.

Catalogue of Material in C. A. Bemis Collection, History of Nelson, N.H. (six pages, typescript, 1949). In Nelson, N.H., library.

Cheshire County Records. In Cheshire County Courthouse, Keene, N.H.

Cheshire Mills Records:

Boardinghouse Ledger, February 1855 to July 1860. Inscribed "Peter Gilson, Harrisville, N.H."

Cash Book "A," 1850–1858.

Cash Book "B," 1852–?

Cash Book "C," 1882–1889.

Consignments to Faulkner, Kimball, & Co., 1861–1871.

Copybook "A," 1855–1859 (sixty-two pages of letters, signed by Timothy Colony, Treas.).

Copybook "B," 1887–1902 (ca. 1,000 pages of letters, signed by John J. Colony, R. Faulkner, and Horatio Colony).

Craven and Willard, Copybook "C," 1880 (ca. 500 pages of letters signed by Henry Emery, Michael Craven).

Craven and Willard, Copybook "D," 1880–1881 (ca. 1,000 pages of letters signed by Henry Emery, Michael Craven).

General Journal, 1858–1866.

General Journal, April 1872–February 1882.

General Journal, 1883–1893.

Letter from "Spy" to Henry Colony, Esq., June 20, 1869.

Letter Book, 1850–1852.

Letters, 1860 (packets, marked by month).

Milan Harris & Company, Drafts, 1866–1870.

Minutes of Meetings of the Harrisville Manufacturing Company, Nov. 9, 1849, Dec. 13, 1849.

"New Mill Dedication," Jan. 10, 1867. Dance Program.

Payroll Books, 1862–1940.

"Records of the Cheshire Mills." Minutes of Annual Meetings, 1850–present.

Schedule of Wage Rates, June 28, 1933. Six-page typescript and covering letter, for C. M. Faulkner & Colony, Troy Blanket, Homestead Mills, and Keene Silk Fiber Co.

Store Ledger, 1858–1859. Inscribed "Ebenr. Jones, Esq., Harrisville, N.H., July 15, 1858" (173 pages).

Colony, Horatio. Autobiography. Manuscript in possession of its author.

Dublin Town Records:

Town Meetings and Tax Invoices, 1771–1806. In New Hampshire Historical Society.

Town Meetings and Tax Invoices, 1807–1827. In Dublin Town Hall.

Town Meetings and Tax Invoices, 1828–1846. Vol. IV. In Dublin Town Hall.

Town Meetings, 1847–1882. Vol. V. In Dublin Town Hall.

Tax Invoices, 1847–1862. In Dublin Town Hall.

Tax Invoices, 1863–1885. In Dublin Town Hall.

Hall, Michael G. Nelson, New Hampshire, 1780–1870. Unpublished Bachelor's thesis, Department of History, Princeton University, 1948 (typescript). In New Hampshire Historical Society.

Harrisville Town Records:

Invoice and Taxes, 1874. In Harrisville Town Office.

Melville, Henry to John Farmer, Concord, N.H., letter describing Nelson, N.H. (ca. 1823). In New Hampshire Historical Society.

Milan Harris & Company. Payroll Book, Jan. 1861 to June 1867. Formerly in possession of Mrs. Arthur Wright, of Harrisville.

Nelson Literary Society. Records, 1824–1835. (Book also contains records of Philosophical Library, 1808–1824.) In New Hampshire Historical Society.

Nelson Sunday School Association. Records, 1825–1839. In New Hampshire Historical Society.

Nelson Town Records:

Town Records of Monadnock #6 and Packersfield, 1751–1801. A Duplicate of the Records of the Proprietors, the Grantees of the township Monadnock #6. In Nelson Library.

Town Meetings and Tax Invoices, 1792–1809. In Nelson Town Hall.

Town Meetings and Tax Invoices, 1809–1825. Vol. III. In Nelson Town Hall.

Tax Invoices, 1843–1866. In Nelson Town Hall.

Mortgage Deeds, 1842–1892. Vol. XII. In Nelson Town Hall.

Town Meetings, 1841–1852. Vol. XIV. In Nelson Town Hall.

Town Meetings, 1852–1876. Vol. XIII. In Nelson Town Hall.

Tax Invoices, 1867–1889. In Nelson Town Hall.

State of Ohio, Probate Court. Stark County, Application for Letter of Administration. (Doc. G, p. 10—#2319.) Aug. 18, 1884.

U.S. Bureau of the Census. Census of 1850. New Hampshire, Cheshire County, Vol. XIV, Dublin, Houses #549–583, 703–747; Vol. XV, Nelson, Houses #748–770, 832–837, 895. In New Hampshire State Library.

U.S. Bureau of the Census. Census of 1860. New Hampshire, Cheshire County, Dublin, pp. 65–91, Houses #574–605, 608–609, 718, 746–795; Nelson, pp. 49–64, Houses #512–514, 536–565. In New Hampshire State Library.

U.S. Bureau of the Census. Census of 1870. New Hampshire, Cheshire County, Vol. XV, Nelson, pp. 1–24, Houses #85–127, 186–209; Vol. XVI, Dublin, pp. 1–20, Houses #1–44, 80–85, 94, 166–185. In New Hampshire State Library.

U.S. Bureau of the Census. Census of 1880. New Hampshire, Cheshire County, Harrisville (village, pp. 11–21, Houses #105–175, remainder of town, pp. 22–28, Houses #176–252). In National Archives, microfilm copy in possession of author.

Published Material

Abbott, Edith. Women in Industry. New York: D. Appleton & Co., 1910.

Adams, Charles Francis. History of Braintree, Massachusetts. Cambridge: Riverside Press, 1891.

Amerikan Albumi: Kuvia Amerikan Suomalaisten Asuinpaikoilta. [The American Album: Pictures from the living sites of American Finns.] Brooklyn, N.Y.: Suomalainen Kansalliskirjakauppa, [The Finnish National Bookstore] [1904].

Annett, Albert, and Lehtinen, Alice. History of Jaffrey, New Hampshire. Published by the Town, 1937.

Bailyn, Bernard. Education in the Forming of American Society: Needs and Opportunities for Study. Institute of Early American History and Culture. Chapel Hill: University of North Carolina Press, 1960.

Becco Sales Corporation. Becco Echo. Buffalo, New York, Vol. 4, No. 2, June, 1953.

Belknap, Jeremy. The History of New Hampshire. 2nd. ed. 3 vols. Boston: Bradford and Read, 1813.

Benton, C., and Barry, S. F. A Statistical View. . . . Cambridge: Folsom, Wells, & Thurston, 1837.

Billings, Katherine F. The Geology of the Monadnock Quadrangle, New Hampshire. Published by the N.H. Planning and Development Commission, 1949.

Bush, George Garvey. History of Education in New Hampshire. U.S. Bureau of Education. Circular of Information No. 3, 1898. Washington: Government Printing Office, 1898.

Carrigan, Philip. Map of New Hampshire. Concord, [N.H.], 1816.

Catalogue of the Public Library of Harrisville, New Hampshire. Peterboro [N.H.]: Farnum and Scott, 1878.

Catalogue of the Harrisville, N.H., Town Library, 1896. Keene [N.H.]: Sentinel Printing Office, 1896.

Channing, Edward. A History of the United States. Vol. V. New York: The Macmillan Company, 1921.

Charlton, Edwin A. New Hampshire As It Is . . . A Historical Sketch, A Gazetteer, A General View. Claremont, N.H.: Tracy and Sanford, 1855.

Cheshire Republican, Special Trade Edition. Keene, N.H., Aug. 18, 1899.

Child, Hamilton. Gazetteer of Cheshire County, New Hampshire, 1763–1885. Part First. Syracuse, N.Y., 1885.

Christian Science Monitor, Magazine Section. July 27, 1946.

Clark, Victor S. History of Manufactures in the United States. 3 vols., N.Y.: McGraw Hill Book Co., 1929 ed.

Cole, Arthur Harrison. The American Wool Manufacture. 2 vols. Cambridge: Harvard University Press, 1926.

Conant, J. E. & Co. (compiler). Phelps Catalogue. Lowell [Mass.], 1917.

Confession of Faith and Form of Covenant of the Evangelical Congregational Church in Harrisville, N.H., n.p., n.d. In New Hampshire Historical Society.

Cutter, Daniel B. History of Jaffrey, N.H., 1749–1880. Concord, N.H., 1881.

Douglas, Paul H. Real Wages in the United States, 1890–1926. Boston: Houghton Mifflin, 1930.

Dublin School Committee. Centennial History of Education in Dublin. Peterboro [N.H.]: Farnum and Scott, 1876.

Fisk, Thomas, and Wadsworth, Samuel. Map of the Town of Dublin, N.H., including the part set off to the town of Harrisville. Index. From survey of Thomas Fisk, 1853, with additions and corrections by Samuel Wadsworth, 1906. Copyright 1907 by the Town of Dublin.

Fogg, Alonzo J. The Statistics and Gazetteer of N.H. Concord, N.H.: D. L. Guernsey, 1874.

French, James Kip. The Story of Engineering. N.Y.: Doubleday & Co., 1960.

Goldthwait, James Walter. "A Town That Has Gone Downhill," The Geographical Review, Vol. XVII, No. 4 (October 1927), pp. 527–552. (Lyme, N.H.)

Granite Monthly, A N.H. Magazine Devoted to Literature, History, and State Progress (Nov.–Dec. 1917). XLIX.

[Griffin, Simon Goodell, et al.]. Celebration by the Town of Nelson, New Hampshire, of the One Hundred and Fiftieth Anniversary of Its First Settlement, 1767–1917. Nelson Picnic Association, 1917.

Hammond, O. G. Check List of New Hampshire Local History. Concord, N.H., 1925.

Harrisville, N.H. Annual Reports of the Town Officers. 1871, 1874, 1876, 1881, 1882, 1886, 1891, 1896, 1901, 1906, 1911, 1916, 1921, 1926, 1931, 1936, 1941, 1942, 1945, 1949, 1950, 1955, 1958, 1960, 1965.

Harrisville, N.H. Invoice and Taxes of the Town of Harrisville, 1906.

Historical Records Survey. Guide to Depositories of Manuscript Collections in the U.S.—New Hampshire. Manchester [N.H.], 1940.

Howarth, Margery D. New Hampshire, A Study of Its Cities and Towns in Relation to Their Physical Background. Concord, N.H., 1936.

Humphrey, John. Water Power of Nubanisit Lake and River, as used by Cheshire Mills, Harrisville, N.H. [Keene, N.H.], 1903.

Hurd, D. Hamilton (ed.). History of Cheshire and Sullivan Counties, New Hampshire, Philadelphia: J. W. Lewis & Co., 1886.

Hutchinson, Albert. A Genealogy and Ancestral Line of Bethuel Harris of Harrisville, N.H., and His Descendents. Keene, N.H., March 1907.

Huxtable, Ada Louise. "Progressive Architecture in America: New England Mill Village, Harrisville, New Hampshire," *Progressive Architecture*, Vol. 38 (July 1957), pp. 139–140. (Illustrations by William B. Pierson, Jr.)

Journal of the Honorable Senate and House of Representatives of the State of New Hampshire. June Session, 1870. Nashua: Orren C. Moore, State Printers, 1870.

Keene Evening Sentinel. Keene, N.H.

Kolehmainen, John I. Sow the Golden Seed: A History of the Fitchburg (Massachusetts) Finnish–American Newspaper Raivaaja, 1905–1955. Fitchburg, Massachusetts: The Raivaaja Publishing Co., 1955.

Lawrence, Robert F. The New Hampshire Churches. [Claremont, N.H.]: Claremont Manufacturing Co., 1856.

[Leonard, Levi W., et al.]. The History of Dublin, New Hampshire. Boston: Printed by John Wilson and Son, 1855.

Leonard, Levi W. (Rev.). Continued and Additional Chapters to 1917 by Rev. Josiah L. Seward, D. D. The History of Dublin, New Hampshire. Dublin, N.H., 1920.

Manual of the General Court of New Hampshire, 1889–1969.

Map of Cheshire County, New Hampshire, Philadelphia: Smith and Morley, Publishers, 1858.

Marti-Ibanez, Felix, M.D. (ed.). History of American Medicine, A Symposium. New York: M D Publications, Inc., 1959.

Merrimack Valley Textile Museum, Wool Technology and the Industrial Revolution: An Exhibit. [North Andover, Mass., 1965.]

Muzzey, David S. The United States of America. Vol. II. Boston: Ginn and Company, 1924.

Nelson Clarion. March 1870; May 1871.

Nelson Picnic Association. Names and Services of those born or sometime resident in Nelson, New Hampshire, who, as Volunteers, answered the call to arms for the preservation of the Union . . . 1861–1865 (1915).

New Hampshire Laws:
  1797. 11:68. Incorporation of Social Library of Packersfield.
  1836. 32:33. Incorporation of Harrisville Manufacturing Co.
  1839. p. 450. Incorporation of Harrisville Engine Co.
  1847. 39:339. Incorporation of Harrisville Manufacturing Co.
  1850. 42:80. Incorporation of Cheshire Mills.
  1861. 53:193. Incorporation of National Mills.
  1870. XLIV. Incorporation of Town of Harrisville.
New Hampshire Sentinel. Keene, N.H., 1799–1910, 1914–1915, 1918–
  1919, 1924–1925, 1932–1933, 1936–1937, 1940–1950.
New Hampshire State Papers. Vols. VIII, IX, XXVII, XXVIII. Concord,
  N.H., 1874–1896.
New Hampshire Sunday News. Feb. 8, 1953.
Peterborough Transcript. (Begun as Contoocook Transcript.) Peter-
  borough, N.H.: Miller and Scott, June 25, 1851.
Pillsbury, Hobart. New Hampshire . . . A History. 4 vols. and 4 vols.
  biographical. New York: The Lewis Historical Publishing Co., Inc.,
  1927–1929.
Piper, H. H. "A Sketch of Dublin," The Granite Monthly, Vol. XXI,
  No. 2 (Aug. 1896), pp. 79–97.
Podea, Iris. "Quebec to 'Little Canada': The Coming of the French
  Canadians to New England in the Nineteenth Century," New England
  Quarterly, Vol. XXIII, No. 3 (Sept. 1950), pp. 365–380.
Potter, Col. C. E. Military History of New Hampshire . . . from 1623
  to 1861. Report of the Adjutant General of the State of New Hamp-
  shire for . . . 1866. Vol. II. Concord, N.H.: George E. Jenks, 1866.
Rees, Albert. Real Wages in Manufacturing, 1890–1914. Princeton:
  Princeton University Press, 1961.
Robinson, Maurice H. A History of Taxation in New Hampshire.
  Publication of the American Economic Association. Third Series,
  Vol. III, No. 3 (Aug. 1902). New York: Macmillan Co., 1902.
Rockwood, C. H. Atlas of Cheshire County, New Hampshire. New
  York: Comstock and Cline, 1877.
Shryock, Richard H. The Development of Modern Medicine. New
  York: Alfred A. Knopf, 1947.
Smith, Harry Edwin. The United States Federal Internal Tax History
  from 1861 to 1871. Boston: Houghton Mifflin, 1914.
Smith, Page. As a City Upon a Hill, The Town in American History.
  New York: Alfred A. Knopf, 1966.
Squires, James Duane. The Granite State of the United States, A History
  of New Hampshire from 1623 to the Present. 4 vols. New York: The
  American Historical Co., Inc., 1956.
Stackpole, Everett S. History of New Hampshire. 5 vols. New
  York:
  The American Historical Society [1916].
State of New Hampshire, Superior Court, Cheshire SS, Sept. 1951.
  John K. Colony et al., Appellants, vs. John J. Colony and Horatio
  Colony, Trustees of the Estate of John E. Colony. No. 8010.

State Association, Old Home Week in New Hampshire. Annual Reports. Concord, N.H., 1899–1941.

Thompson, Warren S. Population Problems. 4th ed. New York: McGraw-Hill Book Co., Inc., 1953.

Tolman, Newton F. North of Monadnock. Boston: Atlantic Monthly Press Book, Little, Brown and Co., 1957.

U.S. Bureau of the Census. Historical Statistics of the United States. Colonial Times to 1957. Washington, D.C.: U.S. Government Printing Office, 1960.

U.S. Bureau of the Census. Twelfth Census of the United States: 1900. Special Reports. Employees and Wages. By Davis R. Dewey, Ph. D. Washington, D.C.: U.S. Census Office, 1903.

U.S. Bureau of the Census. Fourteenth Census of the United States, 1920, Vol. III, Population. . . .Washington, D.C.: U.S. Government Printing Office, 1922.

U.S. Bureau of the Census. Fifteenth Census of the United States, 1930, Population, Vol. III, pt. 2, p. 170, Table 21. Washington, D.C.: U.S. Government Printing Office, 1932.

U.S. Bureau of the Census. Sixteenth Census of the United States, 1940, Population, Vol. II, pt. 4, p. 796, Table 28. Washington D.C.: U.S. Government Printing Office, 1943.

U.S. Bureau of the Census. Seventeenth Census of the United States, 1950, Population, Vol. II, pt. 29, p. 5, Table 6. Washington D.C.: U.S. Government Printing Office, 1952.

Unwin, William Cawthorne, "Hydraulics," Encyclopaedia Britannica (11th ed.), XIV, 35–110.

Van Cleef, Eugene. "The Finn in America," The Geographical Review, Vol. VI (1918), pp. 185–214.

————. Finland, The Republic Farthest North. Columbus: Ohio State University Press, 1929.

Waring, Janet. Early American Stencils on Walls and Furniture. New York: William R. Scott, Publisher, 1937.

Wilbur, Clifford C. The Old Timer. A Monthly Publication. Keene, N.H., Nos. 1–116 (July 1939–Oct. 1949).

Wilson, Harold Fisher. The Hill Country of Northern New England, Its Social and Economic History, 1790–1930. New York: Columbia University Press, 1936.

INDEX